CAMBRIDGE TRACTS IN MATHEMATICS

General Editors

B. BOLLOBAS, W. FULTON, A. KATOK, F. KIRWAN, P. SARNAK

143   Analysis on Fractals

Jun Kigami

Kyoto University

Analysis on Fractals

CAMBRIDGE
UNIVERSITY PRESS

CAMBRIDGE UNIVERSITY PRESS
Cambridge, New York, Melbourne, Madrid, Cape Town, Singapore, São Paulo

Cambridge University Press
The Edinburgh Building, Cambridge CB2 8RU, UK

Published in the United States of America by Cambridge University Press, New York

www.cambridge.org
Information on this title: www.cambridge.org/9780521793216

First published 2001
This digitally printed version 2008

A catalogue record for this publication is available from the British Library

ISBN 978-0-521-79321-6 hardback
ISBN 978-0-521-05711-0 paperback

To my parents,
Yasuko and Masao

Contents

Introduction

What is "Analysis on Fractals"? Why is it interesting?

To answer those questions, we need to go back to the history of fractals.

Many examples of fractals, like the Sierpinski gasket, the Koch curve and the Cantor set, were already known to mathematicians early in the twentieth century. Those sets were originally pathological (or exceptional) counterexamples. For instance, the Koch curve (see Figure 0.1) is an example of a compact curve with infinite length and the Cantor set is an example of an uncountable perfect set with zero Lebesgue measure. Consequently, they were thought of as purely mathematical objects. In fact, they attracted much interest in harmonic analysis in connection with Fourier transform, and in geometric measure theory. There were extensive works started in the early twentieth century by Wiener, Winter, Erdös, Hausdorff, Besicovich and so on. See [181], [32] and [124]. These sets, however, had never been associated with any objects in nature.

This situation had not changed until Mandelbrot proposed the notion of fractals in the 1970s. In [122, 123] he claimed that many objects in nature are not collections of smooth components. As evidence, using the experiments by Richardson, he showed that some coast lines were not smooth curves but curves which have infinite length like the Koch curve. Choosing

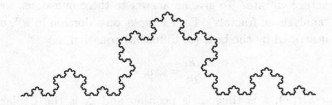

Fig. 0.1. Koch curve

1

Fig. 0.2. Sierpinski gasket

words more carefully and accurately, we need to say that some coast lines should be modeled by curves with infinite length rather than (compositions of) smooth curves.

Mandelbrot coined this revolutionary idea and introduced the notion of fractals as a new class of mathematical objects which represent nature. The importance of his proposal was soon recognized in many areas of science, for example, physics, chemistry and biology. In mathematics, a new area called fractal geometry developed quickly on the foundation of geometric measure theory, harmonic analysis, dynamical systems and ergodic theory. Fractal geometry treats the properties of (fractal) sets and measures on them, like the Hausdorff dimension and the Hausdorff measure. From the viewpoint of applications, it concerns the static aspects of the objects in nature.

How about the dynamical aspects? There occur (physical) phenomena on those objects modeled by fractals. How can we describe them? More precisely, how does heat diffuse on fractals and how does a material with a fractal structure vibrate? To give an answer to these questions, we need a theory of "analysis on fractals". For example, on a domain in $\mathbb{R}^n$, diffusion of heat is described by the heat (or diffusion) equation,

$$\frac{\partial u}{\partial t} = \Delta u,$$

where $u = u(t, x)$, t is time, x is position and Δ is the Laplacian defined by $\sum_{i=1}^{n} \frac{\partial^2}{\partial x_i^2}$. If our domain is a fractal, we need to know what the

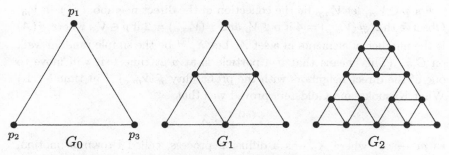

Fig. 0.3. Approximation of the Sierpinski gasket by graphs G_m

"Laplacian" on it is. This problem contains somewhat contradictory factors. Since fractals like the Sierpinski gasket and the Koch curve do not have any smooth structures, to define differential operators like the Laplacian is not possible from the classical viewpoint of analysis. To overcome such a difficulty is a new challenge in mathematics and this is what analysis on fractals is about.

During the 1970s and 1980s, physicists tried to describe phenomena on fractals. They succeeded in calculating some of the physical characteristics of fractals, for example, the spectral exponent, which should describe the distribution of the eigenstates. (See, for example, [118] and [75] for reviews of studies in physics.) However they did not know how to define "Laplacians" on fractals. See Note and References of Chapter 4 for details.

Motivated by studies in physics, Kusuoka [106] and Goldstein [51] independently took the first step in the mathematical development. They constructed "Brownian motion" on the Sierpinski gasket. Their method of construction is now called the probabilistic approach. First they considered a sequence of random walks on the graphs which approximate the Sierpinski gasket and showed that by taking a certain scaling factor, those random walks converged to a diffusion process on the Sierpinski gasket. To be more precise, let us define the Sierpinski gasket. Let $\{p_1, p_2, p_3\}$ be a set of vertices of an equilateral triangle in $\mathbb{C}$. Define $f_i(z) = (z - p_i)/2 + p_i$ for $i = 1, 2, 3$. Then The Sierpinski gasket K is the unique non-empty compact subset K of $\mathbb{R}$ that satisfies

$$K = f_1(K) \cup f_2(K) \cup f_3(K).$$

See Figure 0.2. Let $V_0 = \{p_1, p_2, p_3\}$. Define a sequence of finite sets $\{V_m\}_{m \geq 0}$ inductively by $V_{m+1} = f_1(V_m) \cup f_2(V_m) \cup f_3(V_m)$. Then we have the natural graph G_m whose set of vertices is V_m. (See Figure 0.3.)

For $p \in V_m$, let $V_{m,p}$ be the collection of the direct neighbors of p in V_m. Observe that $\#(V_{m,p}) = 4$ if $p \notin V_0$ and $\#(V_{m,p}) = 2$ if $p \in V_0$, where $\#(A)$ is the number of elements in a set A. Let $X_t^{(m)}$ be the simple random walk on G_m. (This means that if a particle is at p at time t, it will move to one of the direct neighbors with the probability $\#(V_{m,p})^{-1}$ at time $t+1$.) What Kusuoka and Goldstein proved was that

$$X_{5^{-m}t}^{(m)} \to X_t$$

as $m \to \infty$, where X_t was a diffusion process, called Brownian motion, on the Sierpinski gasket. In this probabilistic approach, a Laplacian is the infinitesimal generator of the semigroup which is associated with the diffusion process.

Barlow and Perkins [20] followed the probabilistic approach and obtained an Aronson-type estimate of the heat kernel associated with Brownian motion on the Sierpinski gasket. See Notes and References of Chapter 5. Then, in [116], Lindstrøm extended this construction of Brownian motion to nested fractals, which is a class of finitely ramified self-similar sets with strong symmetry. See 3.8 for the definition of nested fractals. (Roughly speaking, finitely ramified self-similar sets are the self-similar sets which become disconnected if one removes a finite number of points. See 1.3 for details.)

On the other hand, in [82], a direct definition of the Laplacian on the Sierpinski gasket was proposed. Under this direct definition, one could describe the structures of harmonic functions, Green's function and solutions of Poisson's equations. This alternative approach is called the analytical approach. Instead of the sequence of random walks, one considered a sequence of discrete Laplacians on the graphs and then proved that by choosing a proper scaling, those discrete Laplacians would converge to a good operator, called the Laplacian on the Sierpinski gasket. More precisely, let $\ell(V_m) = \{f : f \text{ maps } V_m \text{ to } \mathbb{R}\}$. Then define a linear operator $L_m : \ell(V_m) \to \ell(V_m)$ by

$$(L_m u)(p) = \sum_{q \in V_{m,p}} (u(q) - u(p))$$

for any $u \in \ell(V_m)$ and any $p \in V_m$. This operator L_m is the natural discrete Laplacian on the graph G_m. Then the Laplacian on the Sierpinski gasket, denoted by Δ, is defined by

$$5^m (L_m u)(p) \to (\Delta u)(p)$$

as $m \to \infty$. This Δ is now called the standard Laplacian on the Sierpinski gasket. (Of course, it needs to be shown that Δ is a meaningful operator in some sense with a non-trivial domain, as we will show in the course of this book. Also we will explain why 5^m is the proper scaling. See 3.7, in particular, Example 3.7.3.) This analytical approach was followed by Kusuoka [107] and Kigami [83], where they extended the construction of the Laplacians to more general class of finitely ramified fractals.

Since those early studies, many people have studied analysis on fractals and obtained numerous results using both approaches. Naturally the two approaches are complementary to each other and share the same goal. In this book, we will basically follow the analytical approach and study Dirichlet forms, Laplacians, eigenvalues of Laplacians and heat kernels on post critically finite self-similar sets. (Post critically finite self-similar sets are the mathematical formulation of finite ramified self-similar sets. See 1.3.) The advantage of the analytical approach is that one can get concrete and direct description of harmonic functions, Green's functions, Dirichlet forms and Laplacians. On the other hand, however, studying the detailed structure of the heat kernels, like the Aronson-type estimates, we need to employ the probabilistic approach. (Barlow's lecture note [6] is a self-contained and well-organized exposition in this direction. See also Kumagai [104] for a review of recent results.) Moreover, the probabilistic approach can be applied to infinitely ramified self-similar sets, for example, the Sierpinski carpet (Figure 0.4) as well. In the series of papers, [7, 8, 9, 10, 11, 12], Barlow and Bass constructed Brownian motions on the (higher dimensional) Sierpinski carpets and obtained the Aronson-type estimate of the associated heat kernels by using the probabilistic approach. Except for Kusuoka and Zhou [109], so far, the analytical approach has not succeeded in studying analysis on infinitely ramified fractals.

One may ask "why do you only study self-similar sets?". Indeed, self-similar sets are a special class of fractals and there are no objects in nature which have the exact structures of self-similar sets. The reason is that self-similar sets are perhaps the simplest and the most basic structures in the theory of fractals. They should give us much information on what would happen in the general case of fractals. Although there have been many studies on analysis on fractals, we are still near the beginning in the exploration of this field. We hope that this volume will contribute to fruitful developments in the future.

The organization of this book is as follows. In the first chapter, we will explain the basics of the geometry of self-similar sets. We will give the definition of self-similar sets, study topological structures of self-similar sets and introduce self-similar measures on them. The key notion is the self-

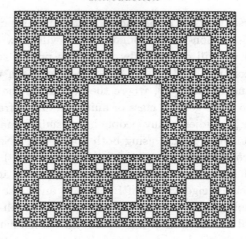

Fig. 0.4. Sierpinski carpet

similar structure which is a purely topological description of a self-similar set. See 1.3. Also, we will define post critically finite self-similar structure in 1.3, which will be our main stage of analysis on fractals.

In Chapter 2, we will study analysis on finite sets, namely, Dirichlet forms and Laplacians. The important fact is that those notions are closely related to electrical networks and that the effective resistance associated with them gives a distance on the finite set. Getting much help from this analogy with electrical networks, we will study the convergence of Dirichlet forms on a sequence of finite sets. This convergence theory will play an essential role in constructing Dirichlet forms and Laplacians on post critically finite self-similar sets in the next chapter.

Chapter 3 is the heart of this book, where we will explain how to construct Dirichlet forms, harmonic functions, Green's functions and Laplacians on post critically finite self-similar sets. The key notion here is the "harmonic structure" introduced in 3.1. In this chapter, we will spend many pages to argue how to deal with the case when a harmonic structure is not regular and also when $K \backslash V_0$ is not connected, where K is the self-similar set and V_0 corresponds to the boundary of K. These cases are of interest and sometimes really make a difference. However one would still get most of the essence of the theory by assuming that the harmonic structure is regular and that $K \backslash V_0$ is connected. So the reader may do so to avoid too many proofs.

In Chapter 4, we will study eigenvalues and eigenfunctions of Laplacians on post critically finite self-similar sets. We will obtain a Weyl-type estimate of the distribution of eigenvalues in 4.1 and show the existence of localized eigenfunctions in 4.4.

In the final chapter, we will study (Dirichlet or Neumann) heat kernels associated with Laplacians (or Dirichlet forms). In 5.2, it will be shown that the parabolic maximum principle holds for solutions of the heat equations. In 5.3, we will get on-diagonal estimates of heat kernels as time goes to zero.

This book is based on my graduate course at Cornell University in the fall semester, 1997. I would like to thank the Department of Mathematics, Cornell University for their hospitality. In particular, I would like to express my sincere gratitude to Professor R. S. Strichartz, who suggested that I wrote these lecture notes, and gave me many fruitful comments on the manuscript. I also thank Dr. C. Blum and Dr. A. Teplyaev who attended my lecture and gave me many useful suggestions. I am also grateful to the Isaac Newton Institute of Mathematical Science, University of Cambridge, where a considerable part of the manuscript was written during my stay. I would express my special thanks to Professors M. T. Barlow and R. F. Bass who carefully read the whole manuscript and helped me to improve my written English. I would also like to thank all the people who gave me valuable comments on the material; among them are Professors M. L. Lapidus, B. M. Hambly, V. Metz, T. Kumagai, Mr. T. Shimono and Mr. K. Kuwada. Finally I would like thank the late Professor Masaya Yamaguti, who was my thesis adviser and introduced me to the study of analysis on fractals.

1
Geometry of Self-Similar Sets

In this chapter, we will review some basics on the geometry of self-similar sets which will be needed later. Specifically, we will explain what a self-similar set is (in 1.1), how to understand the structure of a self-similar set (in 1.2 and 1.3) and how to calculate the Hausdorff dimension of a self-similar set (in 1.5).

The key notion is that of a "self-similar structure" introduced in 1.3, which is a description of a self-similar set from a purely topological viewpoint. As we will explain in 1.3, the topological structure of a self-similar set is essential in constructing analytical structures like Laplacians and Dirichlet forms. More precisely, if two self-similar sets are topologically the same (i.e., homeomorphic), then analytical structure on one self-similar set can be transferred to another self-similar set through the homeomorphism.

In particular, we will introduce the notion of post critically finite self-similar structures, on which we will construct the analytical structures like Laplacians and Dirichlet forms in Chapter 3.

1.1 Construction of self-similar sets

In this section, we will define self-similar sets on a metric space and show an existence and uniqueness theorem for self-similar sets. First we will introduce the notion of contractions on a metric space.

Notation. Let (X, d) be a metric space. For $x \in X$ and $r > 0$,

$$B_r(x) = \{y : y \in X, d(x, y) \leq r\}$$

Definition 1.1.1. Let (X, d_X) and (Y, d_Y) be metric spaces. A map f :

$X \to Y$ is said to be (uniformly) Lipschitz continuous on X with respect to d_X, d_Y if

$$L = \sup_{x,y \in X, x \neq y} \frac{d_Y(f(x), f(y))}{d_X(x, y)} < \infty.$$

The above constant L is called the Lipschitz constant of f and is denoted by $L = \mathrm{Lip}(f)$.

Obviously, by the above definition, a Lipschitz continuous map is continuous.

Definition 1.1.2 (Contraction). Let (X, d) be a metric space. If $f : X \to X$ is Lipschitz continuous on X with respect to d and $\mathrm{Lip}(f) < 1$, then f is called a contraction with respect to the metric d with contraction ratio $\mathrm{Lip}(f)$. In particular, a contraction f with contraction ratio r is called a similitude if $d(f(x), f(y)) = rd(x, y)$ for all $x, y \in X$.

Remark. If f is a similitude on $(\mathbb{R}^n, d_E)$, where d_E is the Euclidean distance on $\mathbb{R}^n$, then there exist $a \in \mathbb{R}^n$, $U \in O(n)$ and $r > 0$ such that $f(x) = rUx + a$ for all $x \in \mathbb{R}^n$. (Exercise 1.1)

The following theorem is called the "contraction principle".

Theorem 1.1.3 (Contraction principle). *Let (X, d) be a complete metric space and let $f : X \to X$ be a contraction with respect to the metric d. Then there exists a unique fixed point of f, in other words, there exists a unique solution to the equation $f(x) = x$. Moreover if x_* is the fixed point of f, then $\{f^n(a)\}_{n \geq 0}$ converges to x_* for all $a \in X$ where f^n is the n-th iteration of f.*

Proof. If r is the ratio of contraction of f, then for $m > n$,

$$d(f^n(a), f^m(a)) \leq d(f^n(a), f^{n+1}(a)) + \cdots + d(f^{m-1}(a), f^m(a))$$
$$\leq (r^n + \cdots + r^{m-1})d(a, f(a)) \leq \frac{r^n}{(1-r)}d(a, f(a)).$$

Hence $\{f^n(a)\}_{n \geq 0}$ is a Cauchy sequence. As (X, d) is complete, there exists $x_* \in X$ such that $f^n(a) \to x_*$ as $n \to \infty$. Using the fact that $f^{(n+1)}(a) = f(f^n(a))$, we can easily deduce that $x_* = f(x_*)$.

Now, if $f(x) = x$ and $f(y) = y$, then $d(x, y) = d(f(x), f(y)) \leq rd(x, y)$. Therefore $d(x, y) = 0$ and $x = y$. So we have uniqueness of fixed points. $\square$

Remark. In general, for a mapping f from a set to itself, a solution of $f(x) = x$ is called a fixed point or an equilibrium point of f.

We now state the main theorem of this section, which ensures uniqueness and existence of self-similar sets.

Theorem 1.1.4. *Let (X, d) be a complete metric space. If $f_i : X \to X$ is a contraction with respect to the metric d for $i = 1, 2, \ldots, N$, then there exists a unique non-empty compact subset K of X that satisfies*

$$K = f_1(K) \cup \cdots \cup f_N(K).$$

K is called the self-similar set with respect to $\{f_1, f_2, \ldots, f_N\}$.

Remark. In other literature, the name "self-similar set" is used in a more restricted sense. For example, Hutchinson [76] uses the name "self-similar set" only if all the contractions are similitudes. Also, in the case that all the contractions are affine functions on $\mathbb{R}^n$, the associated set may be called a self-affine set.

The contraction principle is a special case of Theorem 1.1.4 where $N = 1$. In the rest of this section, we will give a proof of Theorem 1.1.4. Define

$$F(A) = \bigcup_{1 \le j \le N} f_j(A)$$

for $A \subseteq X$. The main idea is to show existence of a fixed point of F. In order to do so, first we choose a good domain for F, defined by

$$\mathcal{C}(X) = \{A : A \text{ is a non-empty compact subset of } X\}.$$

Obviously F is a mapping from $\mathcal{C}(X)$ to itself. Next we define a metric δ on $\mathcal{C}(X)$, which is called the Hausdorff metric on $\mathcal{C}(X)$.

Proposition 1.1.5. *For $A, B \in \mathcal{C}(X)$, define*

$$\delta(A, B) = \inf\{r > 0 : U_r(A) \supseteq B \text{ and } U_r(B) \supseteq A\},$$

where $U_r(A) = \{x \in X : d(x, y) \le r \text{ for some } y \in A\} = \cup_{y \in A} B_r(y)$. Then δ is a metric on $\mathcal{C}(X)$. Moreover if (X, d) is complete, then $(\mathcal{C}(X), \delta)$ is also complete.

Before giving a proof of the above proposition, we recall some standard definitions in general topology.

Definition 1.1.6. Let (X, d) be a metric space and let K be a subset of X.
(1) A finite set $A \subset K$ is called an r-net of K for $r > 0$ if and only if $\cup_{x \in A} B_r(x) \supseteq K$.
(2) K is said to be totally bounded if and only if there exists an r-net of K for any $r > 0$.

It is well-known that a metric space is compact if and only if it is complete and totally bounded.

Proof of Proposition 1.1.5. Obviously, we see that $\delta(A,B) = \delta(B,A) \geq 0$ and $\delta(A,A) = 0$.

$\delta(A,B) = 0 \Rightarrow A = B$: For any n, $U_{1/n}(B) \supseteq A$. Therefore for any $x \in A$, we can choose $x_n \in B$ such that $d(x,x_n) \leq 1/n$. As B is closed, $x \in B$. Hence we have $A \subseteq B$. One can obtain $B \subseteq A$ in exactly the same way.

Triangle inequality: If $r > \delta(A,B)$ and $s > \delta(B,C)$, then $U_{r+s}(A) \supseteq C$ and $U_{r+s}(C) \supseteq A$. Hence $r+s \geq \delta(A,C)$. This implies $\delta(A,B)+\delta(B,C) \geq \delta(A,C)$.

Next we prove that $(\mathcal{C}(X),\delta)$ is complete if (X,d) is complete. For a Cauchy sequence $\{A_n\}_{n\geq 1}$ in $(\mathcal{C}(X),\delta)$, define $B_n = \overline{\cup_{k\geq n}A_k}$. First we will show that B_n is compact. As B_n is a monotonically decreasing sequence of closed sets, it is enough to show that B_1 is compact. For any $r > 0$, we can choose m so that $U_{r/2}(A_m) \supseteq A_k$ for all $k \geq m$. As A_m is compact, there exists a $r/2$-net P of A_m. We can immediately verify that $\cup_{x\in P}B_r(x) \supseteq U_{r/2}(A_m) \supseteq \cup_{k\geq m}A_k$. As $\cup_{x\in P}B_r(x)$ is closed, it is easy to see that P is an r-net of B_m. Adding r-nets of $A_1, A_2, \ldots, A_{m-1}$ to P, we can obtain an r-net of B_1. Hence B_1 is totally bounded. Also, B_1 is complete because it is a closed subset of the complete metric space X. Thus it follows that B_n is compact.

Now as $\{B_n\}$ is a monotonically decreasing sequence of non-empty compact sets, $A = \cap_{n\geq 1}B_n$ is compact and non-empty. For any $r > 0$, we can choose m so that $U_r(A_m) \supseteq A_k$ for all $k \geq m$. Then $U_r(A_m) \supseteq B_m \supseteq A$. On the other hand, $U_r(A) \supseteq B_m \supseteq A_m$ for sufficiently large m. Thus we have $\delta(A,A_m) \leq r$ for sufficiently large m. Hence $A_m \to A$ as $m \to \infty$ in the Hausdorff metric. So $(\mathcal{C}(X),\delta)$ is complete. $\square$

Theorem 1.1.4 can be stated in the following way using the Hausdorff metric $(\mathcal{C}(X),\delta)$.

Theorem 1.1.7. *Let (X,d) be a complete metric space and let $f_i : X \to X$ be a contraction for $i = 1,2,\ldots,n$. Define $F : \mathcal{C}(X) \to \mathcal{C}(X)$ by $F(A) = \cup_{1\leq i\leq N}f_i(A)$. Then F has a unique fixed point K. Moreover, for any $A \in \mathcal{C}(X)$, $F^n(A)$ converges to K as $n \to \infty$ with respect to the Hausdorff metric.*

Lemma 1.1.8. *For $A_1, A_2, B_1, B_2 \in \mathcal{C}(X)$,*

$$\delta(A_1 \cup A_2, B_1 \cup B_2) \leq \max\{\delta(A_1,B_1),\delta(A_2,B_2)\}$$

Proof. If $r > \max\{\delta(A_1, B_1), \delta(A_2, B_2)\}$, then $U_r(A_1) \supseteq B_1$ and $U_r(A_2) \supseteq B_2$. Hence $U_r(A_1 \cup A_2) \supseteq B_1 \cup B_2$. A similar argument implies $U_r(B_1 \cup B_2) \supseteq A_1 \cup A_2$. Hence $r \geq \delta(A_1 \cup A_2, B_1 \cup B_2)$. This completes the proof.   □

Lemma 1.1.9. *If f is a contraction with contraction ratio r, then, for any $A, B \in \mathcal{C}(X)$, $\delta(f(A), f(B)) \leq r\delta(A, B)$.*

Proof. If $U_s(A) \supseteq B$ and $U_s(B) \supseteq A$, $U_{sr}(f(A)) \supseteq f(U_s(A)) \supseteq f(B)$. Also the same argument implies $U_{sr}(f(B)) \supseteq f(A)$. Therefore, $\delta(f(A), f(B)) \leq rs$ and this completes the proof.   □

Proof of Theorem 1.1.7. Using Lemma 1.1.8 repeatedly, we obtain

$$\delta(F(A), F(B)) = \delta(\cup_{1 \leq j \leq N} f_j(A), \cup_{1 \leq j \leq N} f_j(B)) \leq \max_{1 \leq j \leq N} \delta(f_j(A), f_j(B)).$$

By Lemma 1.1.9, $\delta(f_i(A), f_i(B)) \leq r_i \delta(A, B)$, where r_i is the contraction ratio of f_i. If $r = \max_{1 \leq i \leq N} r_i$, then $\delta(F(A), F(B)) \leq r\delta(A, B)$. Therefore F turns out to be a contraction with respect to the Hausdorff metric. By Proposition 1.1.5, we see that $(\mathcal{C}(X), \delta)$ is complete. Now the contraction principle (Theorem 1.1.3) implies Theorem 1.1.7 immediately.   □

1.2 Shift space and self-similar sets

In this section, we will introduce the shift space, which is the key to understanding the topological structure of self-similar sets. In fact, Theorem 1.2.3 will show that every self-similar set is a quotient space of a shift space.

Definition 1.2.1. Let N be a natural number.
(1) For $m \geq 1$, we define

$$W_m^N = \{1, 2, \ldots, N\}^m = \{w_1 w_2 \ldots w_m : w_i \in \{1, 2, \ldots, N\}\}.$$

$w \in W_m^N$ is called a word of length m with symbols $\{1, 2, \ldots, N\}$. Also, for $m = 0$, we define $W_0^N = \{\emptyset\}$ and call $\emptyset$ the empty word. Moreover, set $W_*^N = \cup_{m \geq 0} W_m^N$ and denote the length of $w \in W_*^N$ by $|w|$.
(2) The collection of one-sided infinite sequences of symbols $\{1, 2, \ldots, N\}$ is denoted by Σ^N, which is called the shift space with N-symbols. More precisely,

$$\Sigma^N = \{1, 2, \ldots, N\}^{\mathbb{N}} = \{\omega_1 \omega_2 \omega_3 \ldots : \omega_i \in \{1, \ldots, N\} \text{ for } i \in \mathbb{N}\}.$$

For $k \in \{1, 2, \ldots, N\}$, define a map $\sigma_k : \Sigma^N \to \Sigma^N$ by $\sigma_k(\omega_1 \omega_2 \omega_3 \ldots) =$

$k\omega_1\omega_2\omega_3\ldots$. Also define $\sigma : \Sigma^N \to \Sigma^N$ by $\sigma(\omega_1\omega_2\omega_3\ldots) = \omega_2\omega_3\omega_4\ldots$. σ is called the shift map.

Remark. The two sided infinite sequence of $\{1, 2, \ldots, N\}$,

$$\{1, 2, \ldots, N\}^{\mathbb{Z}} = \{\ldots \omega_{-2}\omega_{-1}\omega_0\omega_1\omega_2\ldots : \omega_i \in \{1, 2, \ldots, N\} \text{ for } i \in \mathbb{Z}\}$$

may also be called the shift space with N-symbols. If one wants to distinguish the two, the above Σ^N should be called the one-sided shift space with N-symbols. In this book, however, we will not treat the two-sided symbol space.

For ease of notation, we write W_m, W_* and Σ instead of W_m^N, W_*^N and Σ^N.

Obviously, σ_k is a branch of the inverse of σ for any $k \in \{1, 2, \ldots, N\}$. If we choose an appropriate distance, it turns out that σ_k is a contraction and the shift space Σ is the self-similar set with respect to $\{\sigma_1, \sigma_2, \ldots, \sigma_N\}$.

Theorem 1.2.2. *For $\omega, \tau \in \Sigma$ with $\omega \neq \tau$ and $0 < r < 1$, define $\delta_r(\omega, \tau) = r^{s(\omega, \tau)}$, where $s(\omega, \tau) = \min\{m : \omega_m \neq \tau_m\} - 1$. (i.e., $n = s(\omega, \tau)$ if and only if $\omega_i = \tau_i$ for $1 \leq i \leq n$ and $\omega_{n+1} \neq \tau_{n+1}$.) Also define $\delta_r(\omega, \tau) = 0$ if $\omega = \tau$. δ_r is a metric on Σ and (Σ, δ_r) is a compact metric space. Furthermore, σ_k is a similitude with $\mathrm{Lip}(\sigma_k) = r$ and Σ is the self-similar set with respect to $\{\sigma_1, \sigma_2, \ldots, \sigma_N\}$.*

Proof. It is obvious by the definition that $\delta_r(\omega, \tau) \geq 0$ and $\delta_r(\omega, \tau) = 0$ implies $\omega = \tau$. As $\min\{s(\omega, \tau), s(\tau, \kappa)\} \leq s(\omega, \kappa)$ for $\omega, \tau, \kappa \in \Sigma$, we can see that $\delta_r(\omega, \tau) + \delta_r(\tau, \kappa) \geq \delta_r(\omega, \kappa)$.

Now for every $w = w_1w_2\ldots w_m \in W_*$, we define

$$\Sigma_w = \{\omega = \omega_1\omega_2\omega_3\ldots \in \Sigma : \omega_1\omega_2\ldots\omega_m = w_1w_2\ldots w_m\}.$$

Let $\{\omega^n\}_{n \geq 1}$ be a sequence in Σ. Using induction on m, we can choose $\tau \in \Sigma$ so that $\{n \geq 1 : (\omega^n)_j = \tau_j \text{ for } j = 1, 2, \ldots, m\}$ is an infinite set for any $m \geq 1$. So there exists a subsequence of $\{\omega^n\}$ that converges to τ as $n \to \infty$. Hence (Σ, δ_r) is compact.

Finally, it is obvious that σ_k is a similitude with $\mathrm{Lip}(\sigma_k) = r$. Also we can easily see that $\Sigma = \sigma_1(\Sigma) \cup \cdots \cup \sigma_N(\Sigma)$. This implies that Σ is the self-similar set with respect to $\{\sigma_1, \sigma_2, \ldots, \sigma_N\}$. $\square$

Σ is called the (topological) Cantor set with N-symbols. See Example 1.2.6.

For the rest of this section, we assume that (X, d) is a complete metric space, $f_i : X \to X$ is a contraction with respect to (X, d) for every $i \in \{1, 2, \ldots, N\}$ and that K is the self-similar set with respect to

$\{f_1, f_2, \ldots, f_N\}$. Also, for $A \subseteq X$, the diameter of A, $\mathrm{diam}(A)$, is defined by $\mathrm{diam}(A) = \sup_{x,y \in A} d(x,y)$.

The following theorem shows that every self-similar set is a quotient space of a shift space by a certain equivalence relation.

Theorem 1.2.3. *For $w = w_1 w_2 \ldots w_m \in W_*$, set $f_w = f_{w_1} \circ f_{w_2} \circ \ldots \circ f_{w_m}$ and $K_w = f_w(K)$. Then for any $\omega = \omega_1 \omega_2 \omega_3 \ldots \in \Sigma$, $\cap_{m \geq 1} K_{\omega_1 \omega_2 \ldots \omega_m}$ contains only one point. If we define $\pi : \Sigma \to K$ by $\{\pi(\omega)\} = \cap_{m \geq 1} K_{\omega_1 \omega_2 \ldots \omega_m}$, then π is a continuous surjective map. Moreover, for any $i \in \{1, 2, \ldots, N\}$, $\pi \circ \sigma_i = f_i \circ \pi$.*

Proof. Note that

$$K_{\omega_1 \omega_2 \ldots \omega_m \omega_{m+1}} = f_{\omega_1 \omega_2 \ldots \omega_m}(f_{\omega_{m+1}}(K)) \subseteq f_{\omega_1 \omega_2 \ldots \omega_m}(K) = K_{\omega_1 \omega_2 \ldots \omega_m}.$$

As $K_{\omega_1 \omega_2 \ldots \omega_m}$ is compact, $\cap_{m \geq 1} K_{\omega_1 \omega_2 \ldots \omega_m}$ is a non-empty compact set. Set $R = \max_{1 \leq i \leq N} \mathrm{Lip}(f_i)$. Then it follows that $\mathrm{diam}(f_i(A)) \leq R\,\mathrm{diam}(A)$. Hence $\mathrm{diam}(K_{\omega_1 \omega_2 \ldots \omega_m}) \leq R^m \mathrm{diam}(K)$. So $\mathrm{diam}(\cap_{m \geq 1} K_{\omega_1 \omega_2 \ldots \omega_m}) = 0$. Therefore $\cap_{m \geq 1} K_{\omega_1 \omega_2 \ldots \omega_m}$ should contain only one point.

If $\delta_r(\omega, \tau) \leq r^m$, then $\pi(\omega), \pi(\tau) \in K_{\omega_1 \omega_2 \ldots \omega_m} = K_{\tau_1 \tau_2 \ldots \tau_m}$. Therefore $d(\pi(\omega), \pi(\tau)) \leq R^m \mathrm{diam}(K)$. This immediately implies that π is continuous.

Using

$$\{\pi(\sigma_i(\omega))\} = \cap_{m \geq 1} K_{i\omega_1 \omega_2 \ldots \omega_m} = \cap_{m \geq 1} f_i(K_{\omega_1 \omega_2 \ldots \omega_m}) = \{f_i(\pi(\omega))\},$$

we can easily verify that $\pi \circ \sigma_i = f_i \circ \pi$.

Finally we must show that π is surjective. Note that

$$\pi(\Sigma) = \pi(\sigma_1(\Sigma) \cup \cdots \cup \sigma_N(\Sigma))$$
$$= \pi(\sigma_1(\Sigma)) \cup \cdots \cup \pi(\sigma_N(\Sigma)) = f_1(\pi(\Sigma)) \cup \cdots \cup f_N(\pi(\Sigma)).$$

As $\pi(\Sigma)$ is a non-empty compact set, the uniqueness of self-similar sets (Theorem 1.1.4) implies that $\pi(\Sigma) = K$. $\qquad\square$

Proposition 1.2.4. *Define $\dot{w} = www \ldots$ if $w \in W_*$ and $w \neq \emptyset$. Then $\pi(\dot{w})$ is the unique fixed point of f_w.*

Proof. As f_w is a contraction, it has a unique fixed point. By Theorem 1.2.3, $\pi(\dot{w}) = \pi(w\dot{w}) = f_w(\pi(\dot{w}))$. Hence $\pi(\dot{w})$ is the fixed point of f_w. $\qquad\square$

Using the above proposition, we see that $\pi(v_1 v_2 \ldots v_k \dot{w}) = f_v(p_w)$ where $w \in W_*$, $w \neq \emptyset$, $v = v_1 v_2 \ldots v_k \in W_*$ and p_w is the fixed point of f_w.

This relation helps us to understand π in many examples. Moreover, since periodic sequences are dense in Σ, we have

$$K = \overline{\{p_w : w \in W_*, w \neq \emptyset\}}.$$

In fact, π determines a topological structure on self-similar sets.

Proposition 1.2.5. *Suppose f_i is injective for every $i \in \{1, 2, \ldots, N\}$. Then, $\pi(\omega) = \pi(\tau)$ for $\omega \neq \tau \in \Sigma$ if and only if $\pi(\sigma^m \omega) = \pi(\sigma^m \tau)$, where $m = s(\omega, \tau)$.*

Proof. If $w = \omega_1 \omega_2 \ldots \omega_m = \tau_1 \tau_2 \ldots \tau_m$, then $\pi(\omega) = f_w(\pi(\sigma^m \omega))$ and $\pi(\tau) = f_w(\pi(\sigma^m \tau))$. As f_w is injective, we have $\pi(\omega) = \pi(\tau)$ for $\omega, \tau \in \Sigma$ if and only if $\pi(\sigma^m \omega) = \pi(\sigma^m \tau)$. The other direction is obvious. $\qquad \square$

Note that if $\pi(\omega) = \pi(\tau)$, then $\pi(\sigma^m(\omega)) = \pi(\sigma^m(\tau)) \in C_K$, where $m = s(\omega, \tau)$ and $C_K = \cup_{1 \leq i < j \leq N}(K_i \cap K_j)$.

Example 1.2.6 (Cantor set). Let $X = [0, 1]$. Choose positive real numbers a and b so that $a + b < 1$. Define $f_1(x) = ax$ and $f_2(x) = b(x - 1) + 1$. If K is the self-similar set with respect to $\{f_1, f_2\}$, $K_1 \subset [0, a]$ and $K_2 \subset [1 - b, 1]$. Hence $C_K = K_1 \cap K_2 = \emptyset$. Therefore $\pi : \Sigma \to K$ is injective. By Theorem 1.2.3, π is also surjective and hence it is a homeomorphism between Σ and K. In particular, if $a = b = 1/3$, K is called the Cantor ternary set or the Cantor middle third set.

Example 1.2.7 (Koch curve). Let $X = \mathbb{C}$. Suppose that $\alpha \in \{z : z \in \mathbb{C}, |z|^2 + |1 - z|^2 < 1\}$. Set $f_1(z) = \alpha \bar{z}$ and $f_2(z) = (1 - \alpha)(\bar{z} - 1) + 1$. Let D be a triangle domain with vertices $\{0, \alpha, 1\}$, including the boundary. Then it follows that $f_1(D) \cup f_2(D) \subseteq D$ and $f_1(D) \cap f_2(D) = \{\alpha\}$. See Figure 1.1. Hence $K(\alpha) \subseteq D$, where $K(\alpha)$ is the self-similar set with respect to $\{f_1, f_2\}$. Also note that $f_1(0) = 0$, $f_2(1) = 1$ and $f_2(0) = f_1(1) = \alpha$. Denoting $\pi = \pi_\alpha$, we obtain that $\pi_\alpha(\dot{1}) = 0, \pi_\alpha(\dot{2}) = 1$ and $\pi_\alpha(1\dot{2}) = \pi_\alpha(2\dot{1}) = \alpha$. Moreover, $C_K = K_1 \cap K_2 = \{\alpha\}$. Hence we can deduce that $\pi_\alpha(\omega) = \pi_\alpha(\tau)$ and $\omega \neq \tau$ if and only if there exists $w \in W_*$ such that $\{\omega, \tau\} = \{w1\dot{2}, w2\dot{1}\}$. In particular, $K(1/2) = [0, 1]$ and $K(\alpha)$ is called the Koch curve if $\alpha = \frac{1}{2} + \frac{1}{2\sqrt{3}}\sqrt{-1}$. See Figure 0.1. Note that $\pi_\alpha \circ \pi_{1/2}^{-1}$ is a homeomorphism between $[0, 1]$ and $K(\alpha)$.

Example 1.2.8 (Sierpinski gasket). Let $X = \mathbb{C}$ and let $\{p_1, p_2, p_3\}$ be a set of vertices of an equilateral triangle. Define $f_j(z) = (z - p_j)/2 + p_j$ for $j = 1, 2, 3$. The self-similar set with respect to $\{f_1, f_2, f_3\}$ is called the Sierpinski gasket. It is easy to see that $\pi(\dot{j}) = p_j$ for $j = 1, 2, 3$. Let T be

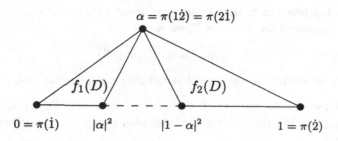

Fig. 1.1. $f_1(D) \cup f_2(D)$ $(\alpha = 0.4 + 0.3\sqrt{-1})$

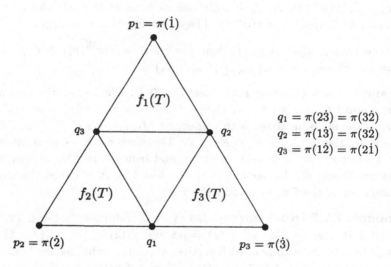

Fig. 1.2. Topological structure of the Sierpinski gasket

the equilateral triangle with vertices $\{p_1, p_2, p_3\}$, including the boundary. Then $f_1(T) \cup f_2(T) \cup f_3(T) \subseteq T$. Hence $K \subset T$. Also $f_1(K) \cap f_2(K) = f_1(T) \cap f_2(T)$ and this set contains only one point, which is denoted by $\{q_3\}$. Then $\pi^{-1}(q_3) = \{2\dot{1}, 1\dot{2}\}$. In the same way, if $f_2(K) \cap f_3(K) = \{q_1\}$ and $f_3(K) \cap f_1(K) = \{q_2\}$ then $\pi^{-1}(q_1) = \{2\dot{3}, 3\dot{2}\}$ and $\pi^{-1}(q_2) = \{3\dot{1}, 1\dot{3}\}$. See Figure 1.2. By those facts, if $\pi(\omega) = \pi(\tau)$ and $\omega \neq \tau$, there exists $w \in W_*$ such that $\{\omega, \tau\} = \{w1\dot{2}, w2\dot{1}\}$ or $\{w2\dot{3}, w3\dot{2}\}$ or $\{w3\dot{1}, w1\dot{3}\}$.

Example 1.2.9 (Hata's tree-like set). Let $X = \mathbb{C}$. Set $f_1(z) = c\bar{z}$, $f_2(z) = (1 - |c|^2)\bar{z} + |c|^2$, where $|c|, |1 - c| \in (0, 1)$. The self-similar set K with respect to $\{f_1, f_2\}$ is called Hata's tree-like set. Let $A = \{t :$

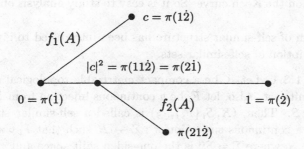

Fig. 1.3. Hata's tree-like set; $f_1(A) \cup f_2(A)$

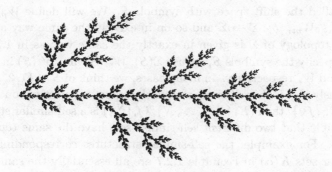

Fig. 1.4. Hata's tree-like set($c = 0.4 + 0.3\sqrt{-1}$)

$0 \leq t \leq 1\} \cup \{tc, 0 \leq t \leq 1\}$. Then it follows that $f_1(A) \cup f_2(A) \supset A$. Hence if $A_m = \cup_{w \in W_m} f_w(A)$, then $\{A_m\}_{m \geq 0}$ is an increasing sequence and $K = \overline{\cup_{m \geq 0} A_m}$. Also we can easily observe that $f_1(K) \cap f_2(K) = \{|c|^2\}$, $f_1(0) = 0, f_2(1) = 1$ and $f_1(f_1(1)) = f_2(0) = |c|^2$. Hence $\pi^{-1}(0) = \{\dot{1}\}$, $\pi^{-1}(1) = \{\dot{2}\}$, $\pi^{-1}(c) = \{1\dot{2}\}$ and $\pi^{-1}(|c|^2) = \{11\dot{2}, 2\dot{1}\}$. See Figure 1.3. Moreover, if $\pi(\omega) = \pi(\tau)$ and $\omega \neq \tau$, there exists $w \in W_*$ such that $\{\omega, \tau\} = \{w11\dot{2}, w2\dot{1}\}$.

1.3 Self-similar structure

From the viewpoint of analysis, only the topological structure of a self-similar set is important. For example, suppose you want to study analysis on the Koch curve. Recalling Example 1.2.7, there exists a natural homeomorphism between $[0, 1]$ and the Koch curve. Through this homeomorphism, any kind of analytical structure on $[0, 1]$ can be translated to its counterpart on the Koch curve. So it is easy to study analysis on the Koch curve.

The notion of self-similar structure has been introduced to give a topological description of self-similar sets.

Definition 1.3.1. Let K be a compact metrizable topological space and let S be a finite set. Also, let F_i be a continuous injection from K to itself for any $i \in S$. Then, $(K, S, \{F_i\}_{i \in S})$ is called a self-similar structure if there exists a continuous surjection $\pi : \Sigma \to K$ such that $F_i \circ \pi = \pi \circ \sigma_i$ for every $i \in S$, where $\Sigma = S^{\mathbb{N}}$ is the one-sided shift space and $\sigma_i : \Sigma \to \Sigma$ is defined by $\sigma_i(w_1 w_2 w_3 \dots) = i w_1 w_2 w_3 \dots$ for each $w_1 w_2 w_3 \dots \in \Sigma$.

Σ is called the shift space with symbols S. We will define $W_m = S^m$, $W_* = \cup_{m \geq 0} W_m$, $\sigma : \Sigma \to \Sigma$ and so on in exactly the same way as in 1.2. Also the topology of Σ is given in exactly the same way as in 1.2. If we need to specify the symbols S, we use $\Sigma(S)$, $W_m(S)$ and $W_*(S)$ in place of Σ, W_m and W_* respectively. In many cases, we think of $S = \{1, 2, \dots, N\}$.

Obviously if K is the self-similar set with respect to injective contractions $\{f_1, f_2, \dots, f_N\}$, then $(K, \{1, 2, \dots, N\}, \{f_i\}_{i=1}^N)$ is a self-similar structure. It is possible that two different self-similar sets have the same topological structure. For example, the self-similar structures corresponding to the self-similar sets $K(\alpha)$ in Example 1.2.7 are all essentially the same. More precisely, they are isomorphic in the following sense.

Definition 1.3.2. Let $\mathcal{L}_j = (K_j, S_j, \{F_i^{(j)}\}_{i \in S_j})$ be self-similar structures for $j = 1, 2$. Also let $\pi_j : \Sigma(S_j) \to K_j$ be the continuous surjection associated with $\mathcal{L}_i$ for $j = 1, 2$. We say that $\mathcal{L}_1$ and $\mathcal{L}_2$ are isomorphic if there exists a bijective map $\rho : S_1 \to S_2$ such that $\pi_2 \circ \iota_\rho \circ \pi_1^{-1}$ is a well-defined homeomorphism between K_2 and K_1, where ι_ρ is the natural bijective map induced by γ, i.e., $\iota(\omega_1 \omega_2 \dots) = \rho(\omega_1) \rho(\omega_2) \dots$.

We say that two self-similar structures are the same if they are isomorphic.

Proposition 1.3.3. *If $(K, S, \{F_i\}_{i \in S})$ is a self-similar structure, then π is unique. In fact,*

$$\{\pi(\omega)\} = \bigcap_{m \geq 0} F_{\omega_1 \omega_2 \dots \omega_m}(K)$$

for any $\omega = \omega_1 \omega_2 \dots \in \Sigma$.

Proof. By the above definition, we have $F_w \circ \pi = \pi \circ \sigma_w$ for any $w \in W_*$. Hence, $\pi(\omega) \in \cap_{m \geq 0} F_{\omega_1 \omega_2 \dots \omega_m}(K)$. For $x \in \cap_{m \geq 0} F_{\omega_1 \omega_2 \dots \omega_m}(K)$, there exists $x_m \in \Sigma_{\omega_1 \omega_2 \ldots \omega_m}$ such that $\pi(x_m) = x$. Note that π is continuous. Since $x_m \to \omega$ as $m \to \infty$, it follows that $x = \pi(x_m) \to \pi(\omega)$ as $m \to \infty$. Hence $x = \pi(\omega)$. $\qquad\square$

Definition 1.3.4. Let $\mathcal{L} = (K, S, \{F_i\}_{i \in S})$ be a self-similar structure. We define $C_{\mathcal{L},K} = \cup_{i,j \in S, i \neq j}(F_i(K) \cap F_j(K))$, $\mathcal{C}_\mathcal{L} = \pi^{-1}(C_{\mathcal{L},K})$ and $\mathcal{P}_\mathcal{L} = \cup_{n \geq 1}\sigma^n(\mathcal{C}_\mathcal{L})$. $\mathcal{C}_\mathcal{L}$ is called the critical set of $\mathcal{L}$ and $\mathcal{P}_\mathcal{L}$ is called the post critical set of $\mathcal{L}$. Also we define $V_0(\mathcal{L}) = \pi(\mathcal{P}_\mathcal{L})$.

For ease of notation, we use $\mathcal{C}$, $\mathcal{P}$ and V_0 instead of $\mathcal{C}_\mathcal{L}$, $\mathcal{P}_\mathcal{L}$ and $V_0(\mathcal{L})$ as long as it can not cause any confusion.

The critical set and the post critical set play an important role in determining the topological structure of a self-similar set. For example, if $\mathcal{C} = \emptyset$, (and hence $\mathcal{P}$, V_0 are all empty sets), then K is homeomorphic to the (topological) Cantor set Σ.

Also V_0 is thought of as a "boundary" of K. For example, define $F_1(x) = \frac{1}{2}x$ and $F_2(x) = \frac{1}{2}x + \frac{1}{2}$ and recall Example 1.2.7. Then we find that $\mathcal{C} = \{1\dot{2}, 2\dot{1}\}$ and $\mathcal{P} = \{\dot{1}, \dot{2}\}$. Hence $V_0 = \{0, 1\}$. See also Exercise 1.3 for another example.

Proposition 1.3.5. *Let* $\mathcal{L} = (K, S, \{F_i\}_{i \in S})$ *be a self-similar structure. Then*
(1) $\pi^{-1}(V_0) = \mathcal{P}$.
(2) *If* $\Sigma_w \cap \Sigma_v = \emptyset$ *for* $w, v \in W_*$, *then* $K_w \cap K_v = F_w(V_0) \cap F_v(V_0)$, *where* $K_w = F_w(K)$.
(3) $\mathcal{C} = \emptyset$ *if and only if* π *is injective*.

Proof. (1) If $\pi(\omega) \in V_0$, then there exist $\tau \in \mathcal{C}$ and $m \geq 1$ such that $\pi(\sigma^m \tau) = \pi(\omega)$. Set $\omega' = \tau_1 \tau_2 \ldots \tau_m \omega$. Then

$$\pi(\omega') = F_{\tau_1 \tau_2 \ldots \tau_m}(\pi(\omega)) = F_{\tau_1 \tau_2 \ldots \tau_m}(\pi(\sigma^m \tau)) = \pi(\tau) \in C_{\mathcal{L},K}.$$

Hence $\omega' \in \mathcal{C}$ and $\omega \in \mathcal{P}$.
(2) It is obvious that $F_w(V_0) \cap F_v(V_0) \subseteq K_w \cap K_v$. For $x \in K_w \cap K_v$, we can choose $\omega, \tau \in \Sigma$ so that $x = \pi(w\omega) = \pi(v\tau)$. As $\Sigma_w \cap \Sigma_v = \emptyset$, there exists $k < \min\{|w|, |v|\}$ such that $w_1 w_2 \ldots w_k = v_1 v_2 \ldots v_k$ and $w_{k+1} \neq v_{k+1}$. As $F_{w_1 w_2 \ldots w_k}$ is injective, it follows that $\pi(\sigma^k(w\omega)) = \pi(\sigma^k(v\tau))$. Hence we can conclude that $\sigma^k(w\omega), \sigma^k(v\tau) \in \mathcal{C}$ and therefore $\omega, \tau \in \mathcal{P}$.
(3) If π is injective, then K is homeomorphic to Σ and hence $\mathcal{C} = \emptyset$. Conversely, if π is not injective, we can use the same argument as in Proposition 1.2.5 to show that $\mathcal{C} \neq \emptyset$. $\qquad\square$

The following proposition describes the local topology of a self-similar structure.

Proposition 1.3.6. *Let $\mathcal{L} = (K, S, \{F_i\}_{i \in S})$ be a self-similar structure. For any $x \in K$ and any $m \geq 0$, define*

$$K_{m,x} = \bigcup_{w \in W_m : x \in K_w} K_w.$$

Then $\{K_{m,x}\}_{m \geq 0}$ is a fundamental system of neighborhoods of x.

Proof. Let d be a metric on K which is compatible with the original topology. First we show that $\max_{w \in W_m} \operatorname{diam}(K_w) \to 0$ as $m \to \infty$. If not, there exists $\{w(m)\}_{m \geq 0}$ such that $w(m) \in W_m$ for any m, and $\inf_{m \geq 0} \operatorname{diam}(K_{w(m)}) > 0$. Choose $\omega(m) \in \Sigma_{w(m)}$ for any $m \geq 0$. Then since Σ is compact, there exists a subsequence $\{\omega(m_i)\}_{i \geq 1}$ which converges to some $\omega \in \Sigma$ as $i \to \infty$. Note that $\operatorname{diam}(K_{w_1 w_2 \ldots w_m}) \leq \operatorname{diam}(K_{w_1 w_2 \ldots w_n})$ if $m \geq n$. It follows that $\liminf_{m \to \infty} \operatorname{diam}(K_{\omega_1 \omega_2 \ldots \omega_m}) > 0$. This contradicts Proposition 1.3.3.

Secondly, we show that $K_{m,x}$ is a neighborhood of x. Let $\{x_m\}_{m \geq 1}$ be a sequence in K which converges to x as $m \to \infty$. Choose $\omega^m \in \pi^{-1}(x_m)$ for any $m \geq 1$. Then there exists a subsequence $\{\omega^{m_i}\}_{i \geq 1}$ that converges to some $\omega \in \Sigma$ as $i \to \infty$. Since π is continuous, $\pi(\omega) = x$. Hence $x_{m_i} \in K_{m,x}$ for sufficiently large i. Therefore $K_{m,x}$ is a neighborhood of x.

Combining the two facts, we conclude that $\{K_{m,x}\}_{m \geq 0}$ is a fundamental system of neighborhoods of x. $\square$

A self-similar structure $(K, S, \{F_i\}_{i \in S})$ may contain an unnecessary symbol. For example, let $K = [0, 1]$ and define $S = \{1, 2, 3\}$, $F_1(x) = \frac{1}{2}x$, $F_2(x) = \frac{1}{2}x + \frac{1}{2}$ and $F_3(x) = \frac{1}{2}x + \frac{1}{4}$. Then obviously $K = F_1(K) \cup F_2(K)$ and we don't need F_3 to describe K. This example may be a little artificial but there are more natural examples. To explain such examples, we need to introduce some notation.

Let $\mathcal{L} = (K, S, \{F_i\}_{i \in S})$ be a self-similar structure. Let W be a finite subset of $W_* \backslash W_0$. Then $\Sigma(W) = W^{\mathbb{N}}$ can be identified as a subset of $\Sigma(S) = S^{\mathbb{N}}$ in a natural manner. Set $K(W) = \pi(\Sigma(W))$. Then $(K(W), W, \{F_w\}_{w \in W})$ is a new self-similar structure. We denote this self-similar structure by $\mathcal{L}(W)$.

Using this notation, we can rephrase the above example as $K(\{1, 2\}) = K(S)$. The following is a more natural example.

Example 1.3.7. Let $K = [0, 1]$ and define $S = \{1, 2\}$, $F_1(x) = \frac{3}{4}x$ and $F_2(x) = \frac{3}{4}x + \frac{1}{4}$. Then $\mathcal{L} = (K, S, \{F_1, F_2\})$ is a self-similar structure. Set $W = \{11, 22\}$. Then $K(W) = K$ because $K = F_{11}(K) \cup F_{22}(K)$. This means that to describe K, we don't need the words $\{12, 21\}$.

You may notice that this kind of unnecessary symbol (or word) occurs when the overlap set $C_{\mathcal{L},K}$ (or equivalently $C_{\mathcal{L}}$) is "large". The following theorem justifies such an intuition.

Theorem 1.3.8. *Let $\mathcal{L} = (K, S, \{F_i\}_{i \in S})$ be a self-similar structure. The following conditions are equivalent. If $\mathcal{L}$ satisfies one of the following conditions, we say that $\mathcal{L}$ is minimal.*

(Mi1) *If $\pi(A) = K$ for a closed set $A \subseteq \Sigma$, then $A = \Sigma$.*

(Mi2) *For any $w \in W_*$, K_w is not contained in $\cup_{v \in W_m \setminus \{w\}} K_v$, where $m = |w|$.*

(Mi3) *If $K(W) = K$ for $W \subseteq W_m$, then $W = W_m$.*

(Mi4) *K_w is not contained in $C_{\mathcal{L},K}$ for any $w \in W_*$.*

(Mi5) *$\mathrm{int}(C_{\mathcal{L}}) = \emptyset$.*

(Mi6) *$\mathrm{int}(\mathcal{P}_{\mathcal{L}}) = \emptyset$.*   (Mi6*) *$\mathcal{P}_{\mathcal{L}} \neq \Sigma$.*

(Mi7) *$\mathrm{int}(V_0) = \emptyset$.*   (Mi7*) *$V_0 \neq K$.*

As we can see from (Mi3), a minimal self-similar structure does not have any unnecessary symbol (or word). It is easy to see that the self-similar structures corresponding to the self-similar sets in 1.2 are all minimal.

Proof.   (Mi1) $\Rightarrow$ (Mi4) Assume that $C \supset K_w$ for some $w \in W_*$. Let $k \in S$ be the first symbol of w. Then for any $x \in K_w$, there exists some $j \neq k$ such that $x \in K_j$. If $m = |w|$ and $A = \cup_{v \in W_m \setminus \{w\}} \Sigma_v$, then A is closed and $\pi(A) = K$.

(Mi4) $\Rightarrow$ (Mi5) Assume that $\mathrm{int}(C) \neq \emptyset$. Then $C \supset \Sigma_w$ for some $w \in W_*$. Hence $C \supset K_w$.

(Mi5) $\Rightarrow$ (Mi6*) Assume that $\mathcal{P} = \Sigma$. Then as $\mathcal{P} = \cup_{m \geq 1} \sigma^m C$, Baire's category argument shows that $\mathrm{int}(\sigma^m C) \neq \emptyset$ for some m. (See, for example, [186] about Baire's category argument.) Hence, $\sigma^m C \supseteq \Sigma_w$ for some $w \in W_*$. Therefore $\sigma^k C = \Sigma$ for $k = m + |w|$. Now $\sigma^k C = \cup_{v \in W_k} \sigma^k(\Sigma_v \cap C)$. Again using Baire's category argument, it follows that $\sigma^k(\Sigma_v \cap C) \supseteq \Sigma_u$ for some $v \in W_k$ and $u \in W_*$. Therefore $C \supseteq \Sigma_{vu}$.

(Mi6*) $\Rightarrow$ (Mi6) Assume that $\mathrm{int}(\mathcal{P}) \neq \emptyset$. Then $\mathcal{P} \supset \Sigma_w$ for some $w \in W_*$. Since $\sigma^m \mathcal{P} \subset \mathcal{P}$ for $m = |w|$, we have $\Sigma = \mathcal{P}$.

(Mi6) $\Rightarrow$ (Mi7) As $\pi^{-1}(V_0) = \mathcal{P}$, we have $\pi^{-1}(\mathrm{int}(V_0)) \subseteq \mathrm{int}(\mathcal{P})$.

(Mi7) $\Rightarrow$ (Mi7*) $\Rightarrow$ (Mi6*) This is obvious by the fact that $\pi^{-1}(V_0) = \mathcal{P}$.

(Mi6*) $\Rightarrow$ (Mi2) Assume that $K_w \subseteq \cup_{v \in W_m \setminus \{w\}} K_v$ for some m and $w \in W_m$. Then for any $\omega \in \Sigma$, there exist $v \in W_m \setminus \{w\}$ and $\tau \in \Sigma$ such that $\pi(w\omega) = \pi(v\tau)$. As $w \neq v$, we can choose $k \leq m$ so that $w_1 w_2 \ldots w_{k-1} = v_1 v_2 \ldots v_{k-1}$ and $w_k \neq v_k$. Since $F_{w_1 w_2 \ldots w_{k-1}}$ is injective, we see that $\sigma^k(w\omega) \in C$. Therefore $\omega \in \mathcal{P}$. So $\mathcal{P} = \Sigma$.

(Mi2) $\Rightarrow$ (Mi1) If there exists a closed subset $A \subset \Sigma$ with $\pi(A) = K$, then A^c is a non-empty open set and so it should contain Σ_w for some $w \in W_*$. Since $A \supset \cup_{v \in W_m \setminus \{w\}} \Sigma_v$, where $m = |w|$, we have $K_w \in \cup_{v \in W_m \setminus \{w\}} K_v$.

(Mi2) $\Rightarrow$ (Mi3) Let W be a proper subset of W_m and assume $K(W) = K$. Then for $w \in W_m \setminus W$, $K_w \subset K = \cup_{v \in W} K_v$. Hence (Mi2) does not hold.

(Mi3) $\Rightarrow$ (Mi2) If $K_w \subset \cup_{v \in W_m \setminus \{w\}} K_v$, where $m = |w|$, then $K = \cup_{v \in W} F_v(K)$, where $W = W_m \setminus \{w\}$. Hence, for any $x \in K$, there exists $\omega \in \Sigma(W)$ such that $\pi(\omega) = x$. Therefore $K(W) = K$. $\square$

Remark. It seems quite possible that the condition $\mathrm{int}(C_{\mathcal{L},K}) = \emptyset$ is also equivalent to those conditions in Theorem 1.3.8. Unfortunately this is not true. In fact, there is an example where $\mathrm{int}(C_{\mathcal{L},K}) = \emptyset$ but $\mathrm{int}(C_{\mathcal{L}}) \neq \emptyset$. See Exercise 1.5.

Definition 1.3.9. Let S be a finite set. We say that a finite subset $\Lambda \subset W_*(S)$ is a partition of $\Sigma(S)$ if $\Sigma_w \cap \Sigma_v = \emptyset$ for any $w \neq v \in \Lambda$ and $\Sigma = \cup_{w \in \Lambda} \Sigma_w$. A partition Λ_1 is said to be a refinement of a partition Λ_2 if and only if either $\Sigma_w \subseteq \Sigma_v$ or $\Sigma_w \cap \Sigma_v = \emptyset$ for any $(w, v) \in \Lambda_1 \times \Lambda_2$.

W_m is a partition for any $m \geq 0$ and W_n is a refinement of W_m if (and only if) $n \geq m$.

Lemma 1.3.10. *Let $\mathcal{L} = (K, S, \{F_i\}_{i \in S})$ be a self-similar structure. Define $V(\Lambda, \mathcal{L}) = \cup_{w \in \Lambda} F_w(V_0)$ if Λ is a partition of Σ. Then $V(\Lambda_1, \mathcal{L}) \supseteq V(\Lambda_2, \mathcal{L})$ if Λ_1 is a refinement of Λ_2.*

Proof. Assume that Λ_1 is a refinement of Λ_2. Set $x = \pi(w\omega)$ for $w \in \Lambda_2$ and $\omega \in \mathcal{P}$. Then there exists $v \in \Lambda_1$ such that $v = w\omega_1 \ldots \omega_k$. As $\omega_{k+1}\omega_{k+2}\ldots \in \mathcal{P}$, we can see that $x = \pi(w\omega) \in V(\Lambda_1, \mathcal{L})$. $\square$

Lemma 1.3.11. *Let $\mathcal{L} = (K, S, \{F_i\}_{i \in S})$ be a self-similar structure. Define $V_m(\mathcal{L}) = V(W_m, \mathcal{L})$. Then $V_m(\mathcal{L}) \subseteq V_{m+1}(\mathcal{L})$ and*

$$V_{m+1}(\mathcal{L}) = \cup_{i \in S} F_i(V_m(\mathcal{L})).$$

Furthermore, set $V_*(\mathcal{L}) = \cup_{m \geq 0} V_m(\mathcal{L})$. If $V_0 \neq \emptyset$, then $V_*(\mathcal{L})$ is dense in K.

Proof. The proof of the first statement is immediate from Lemma 1.3.10. If $x = \pi(\omega) \in K$, then for $\tau \in \mathcal{P}$, $x_n = \pi(\omega_1 \ldots \omega_n \tau)$ converges to x as $n \to \infty$. Hence $V_*(\mathcal{L})$ is dense in K. $\square$

We write $V(\Lambda)$, V_m and V_* instead of $V(\Lambda, \mathcal{L})$, $V_m(\mathcal{L})$ and $V_*(\mathcal{L})$ respectively if no confusion can occur.

Let Λ be a partition of $\Sigma(S)$. If $\mathcal{L} = (K, S, \{F_i\}_{i \in S})$ is a self-similar structure and $\Lambda \neq W_0$, then we can define a self-similar structure $\mathcal{L}(\Lambda) = (K(\Lambda), \Lambda, \{F_w\}_{w \in \Lambda})$ as before. (See the definition of $\mathcal{L}(W)$ between Proposition 1.3.6 and Example 1.3.7.) Immediately, by Definition 1.3.9, it follows that $K(\Lambda) = K$ and $\Sigma(S) = \Sigma(\Lambda)$. (More precisely, $\Sigma(\Lambda)$ can be identified with $\Sigma(S)$ in a natural manner.) Of course, the topological structures of K and $K(\Lambda)$ are expected to be the same since they are virtually the same self-similar structures.

Proposition 1.3.12. *Let $\mathcal{L} = (K, S, \{F_i\}_{i \in S})$ be a self-similar structure and let Λ be a partition of $\Sigma(S)$. Then $\mathcal{P}_\mathcal{L} \supseteq \mathcal{P}_{\mathcal{L}(\Lambda)}$, where we identify $\Sigma(S)$ and $\Sigma(\Lambda)$ through the natural mapping. Furthermore, if $\Lambda = W_m(S)$ for $m \geq 1$, then $\mathcal{P}_\mathcal{L} = \mathcal{P}_{\mathcal{L}(\Lambda)}$.*

Proof. Let $\alpha = \alpha_1 \alpha_2 \ldots \in \mathcal{P}_{\mathcal{L}(\Lambda)}$, where $\alpha_i \in \Lambda$. Then there exists $\beta_1 \beta_2 \ldots \beta_m \in W_*(\Lambda) \backslash W_0(\Lambda)$ and $\gamma = \gamma_1 \gamma_2 \ldots \in \Sigma(\Lambda)$ such that $\pi(\beta) = \pi(\gamma)$ and $\beta_1 \neq \gamma_1$, where $\beta = \beta_1 \beta_2 \ldots \beta_m \alpha \in \Sigma(\Lambda)$. Hence, if $\beta_1 = w_1 w_2 \ldots w_m \in W_m(S)$ and $\gamma_1 = v_1 v_2 \ldots v_n \in W_n(S)$, we can find k so that $w_1 w_2 \ldots w_k = v_1 v_2 \ldots v_k$ and $w_{k+1} \neq v_{k+1}$. Therefore, as elements in $\Sigma(S)$, $\pi(\sigma^k \beta) = \pi(\sigma^k \gamma)$ and hence $\sigma^k \beta \in \mathcal{C}_\mathcal{L}$. This implies that $\alpha \in \mathcal{P}_\mathcal{L}$.

Next let $\Lambda = W_m$ for $m \geq 1$. For $\omega = \omega_1 \omega_2 \ldots \in \mathcal{P}_\mathcal{L}$, there exists $w \in W_*(S) \backslash W_0(S)$ and $\tau \in \Sigma(S)$ such that $\pi(w\omega) = \pi(\tau)$ and $w_1 \neq \tau_1$. Now we can choose $v \in W_*(S)$ so that $vw = \beta_1 \beta_2 \ldots \beta_j$ and $v\tau = \gamma_1 \gamma_2 \ldots$ with $\beta_i, \gamma_i \in \Lambda$ and $\beta_1 \neq \gamma_1$. If $\omega = \alpha_1 \alpha_2 \ldots$, where $\alpha_i \in \Lambda$, then it follows that $\beta_1 \beta_2 \ldots \beta_j \alpha_1 \alpha_2 \ldots \in \mathcal{C}_{\mathcal{L}(\Lambda)}$. Therefore $\omega = \alpha_1 \alpha_2 \ldots \in \mathcal{P}_{\mathcal{L}(\Lambda)}$. $\square$

Even if $\Lambda \neq W_m(S)$, $\mathcal{P}_{\mathcal{L}(\Lambda)}$ often coincides with $\mathcal{P}_\mathcal{L}$. In general, however, this is not true. See Exercise 1.6 and 1.7 for examples.

Finally, we will give the definition of post critically finite (p. c. f. for short) self-similar structure, which is one of the key notions in this book.

Definition 1.3.13. Let $\mathcal{L} = (K, S, \{F_i\}_{i \in S})$ be a self-similar structure. $\mathcal{L}$ is said to be post critically finite or p. c. f. for short if and only if the post critical set $\mathcal{P}_\mathcal{L}$ is a finite set.

If $\mathcal{L} = (K, S, \{F_i\}_{i \in S})$ is post critically finite, V_m is a finite set for all m. In particular, $K_w \cap K_v$ is a finite set for any $w \neq v \in W_m$. Such a self-similar set is often called a finitely ramified self-similar set. Obviously, a p. c. f. self-similar set is finitely ramified. The converse is, however, not true.

Later, in Chapter 3, we will mainly study analysis on post critically finite self-similar sets.

Lemma 1.3.14. *Let $\mathcal{L} = (K, S, \{F_i\}_{i \in S})$ be post critically finite and let $p \in K$. If $F_w(p) = p$ for some $w \in W_*$ and $w \neq \emptyset$, then $\pi^{-1}(p) = \{\dot{w}\}$.*

Proof. Obviously, $\dot{w} \in \pi^{-1}(p)$. First we consider the case when $w = k \in W_1$. Assume that there exists $\tau = \tau_1 \tau_2 \ldots \in \Sigma$ such that $\pi(\tau) = p$ and $\tau \neq \dot{k}$. Without loss of generality, we may suppose that $\tau_1 \neq k$. Let $\tau^n = (\sigma_k)^n \tau$ for any $n \geq 1$. Then $\tau^n \in \pi^{-1}(p)$ and hence $\pi^{-1}(p)$ is an infinite set. On the other hand, since $p \in K_k \cap K_{\tau_1}$, $\pi^{-1}(p)$ is contained in the critical set. As $\pi^{-1}(p)$ is an infinite set, this contradicts the fact that $\mathcal{L}$ is post critically finite.

Now for general case, let $w \in W_m$. Then by Proposition 1.3.12, $\mathcal{L}_m = (K, W_m, \{F_v\}_{v \in W_m})$ is also post critically finite. So applying the above argument to $\mathcal{L}_m$, we see that $\pi^{-1}(p)$ contains only one element. Hence $\pi^{-1}(p) = \{\dot{w}\}$. □

Example 1.3.15 (Sierpinski gasket). Let K be the Sierpinski gasket defined in Example 1.2.8. Then $\mathcal{L} = (K, S, \{f_i\}_{i \in S})$, where $S = \{1, 2, 3\}$ and the f_i are the same maps as in Example 1.2.8, is a post critically finite self-similar structure. In fact, $C_{\mathcal{L},K} = \{q_1, q_2, q_3\}$, $C_{\mathcal{L}} = \{1\dot{2}, 2\dot{1}, 2\dot{3}, 3\dot{2}, 3\dot{1}, 1\dot{3}\}$ and $\mathcal{P}_{\mathcal{L}} = \{\dot{1}, \dot{2}, \dot{3}\}$. Also $V_0 = \{p_1, p_2, p_3\}$ and $V_1 = V_0 \cup \{q_1, q_2, q_3\}$. See Figure 1.2.

Example 1.3.16 (Hata's tree-like set). Let f_1, f_2 and K be the same as in Example 1.2.9. Then $\mathcal{L} = (K, \{1, 2\}, \{f_1, f_2\})$ is a p. c. f. self-similar structure. In fact, $C_{\mathcal{L},K} = \{|c^2|\}$, $C_{\mathcal{L}} = \{11\dot{2}, 2\dot{1}\}$ and $\mathcal{P}_{\mathcal{L}} = \{1\dot{2}, \dot{2}, \dot{1}\}$. See Figure 1.3. Hence $V_0 = \{c, 0, 1\}$. Also $V_1 = \{c, 0, 1, |c|^2, f_2(c)\}$. Note that self-similar structures are isomorphic for all c with $|c|, |1 - c| \in (0, 1)$.

Of course there are numerous examples of non-p. c. f. self-similar structures. One easy example is the unit square. (See Exercise 1.3.) Another famous example is the Sierpinski carpet, which may be thought of as the simplest non-trivial non-p. c. f. self-similar structure.

Example 1.3.17 (Sierpinski carpet). Let $p_1 = 0$, $p_2 = 1/2$, $p_3 = 1$, $p_4 = 1 + \sqrt{-1}/2$, $p_5 = 1 + \sqrt{-1}$, $p_6 = 1/2 + \sqrt{-1}$, $p_7 = \sqrt{-1}$ and $p_8 = \sqrt{-1}/2$. Set $f_i(z) = (z - p_i)/3 + p_i$ for $i = 1, 2, \ldots, 8$. The self-similar set K with respect to $\{f_i\}_{i=1,2,\ldots,8}$ is called the Sierpinski carpet. See Figure 0.4. Let $\mathcal{L}$ be the corresponding self-similar structure. The $\mathcal{L}$ is not post critically finite. In fact, $C_{\mathcal{L},K}$, $C_{\mathcal{L}}$ and $\mathcal{P}_{\mathcal{L}}$ are infinite sets. In particular, V_0 equals the boundary of the unit square $[0, 1] \times [0, 1]$.

1.4 Self-similar measure

In this section, we will introduce an important class of measures on a self-similar structure, that is, self-similar measures. First we will recall some of the fundamental definitions in measure theory.

$(X, \mathcal{M})$ is called a measurable space if X is a set and $\mathcal{M}$ is a σ-algebra whose elements are subsets of X. A measure μ on a measurable space $(X, \mathcal{M})$ is a non-negative σ-additive function defined on $\mathcal{M}$.

Definition 1.4.1. Let (X, d) be a metric space and let μ be a measure on a measurable space $(X, \mathcal{M})$.
(1) The Borel σ-algebra, $\mathcal{B}(X, d)$, is the minimal σ-algebra which contains all open subsets of X. An element of $\mathcal{B}(X, d)$ is called a Borel set. If no confusion can occur, we write $\mathcal{B}(X)$ instead of $\mathcal{B}(X, d)$.
(2) μ is called a Borel measure if $\mathcal{M}$ contains $\mathcal{B}(X)$.
(3) μ is called a Borel regular measure if it is a Borel measure and, for any $A \in \mathcal{M}$, there exists $B \in \mathcal{B}(X)$ such that $\mu(A) = \mu(B)$ and $A \subseteq B$.
(4) We say that μ is complete if any subset of a null set is measurable, i.e., $B \in \mathcal{M}$ if $B \subseteq A \in \mathcal{M}$ and $\mu(A) = 0$.
(5) μ is called a probability measure if and only if $\mu(X) = 1$.

The following proposition is one of the most important facts about Borel regular measures. See, for example, [124] and [158].

Proposition 1.4.2. *Let (X, d) be a metric space and let μ be a Borel regular measure on $(X, \mathcal{M})$. Assume that $\mu(X) < \infty$. Then, for any $A \in \mathcal{M}$,*

$$\mu(A) = \inf\{\mu(U) : U \text{ is a open set that contains } A\}$$
$$= \sup\{\mu(F) : F \text{ is a closed set that is contained in } A\}$$

Proposition 1.4.3 (Bernoulli measure). *Let S be a finite set. If $p = (p_i)_{i \in S}$ satisfies $\sum_{i \in S} p_i = 1$ and $0 < p_i < 1$ for any $i \in S$, then there exists a unique complete Borel regular measure μ^p on $(\Sigma, \mathcal{M}^p)$, where $\Sigma = S^{\mathbb{N}}$, that satisfies $\mu^p(\Sigma_w) = p_{w_1} p_{w_2} \dots p_{w_m}$ for any $w = w_1 w_2 \dots w_m \in W_*$. This measure μ^p is called the Bernoulli measure on Σ with weight p.*

Remark. In this book, all the measures we will encounter are supposed to be complete unless otherwise stated.

Also the Bernoulli measure with weight p is characterized as the unique Borel regular probability measure on Σ that satisfies

$$\mu(A) = \sum_{i \in S} p_i \mu(\sigma_i^{-1}(A))$$

for any Borel set $A \subset \Sigma$.

Proposition 1.4.4 (Self-similar measures). *Let* $\mathcal{L} = (K, S, \{F_i\}_{i \in S})$ *be a self-similar structure and let* π *be the natural map from* Σ *to* K *associated with* $\mathcal{L}$. *If* $p = (p_i)_{i \in S} \in \mathbb{R}^S$ *satisfies* $\sum_{i \in S} p_i = 1$ *and* $0 < p_i < 1$ *for any* $i \in S$, *then we define* ν^p *by* $\nu^p(A) = \mu^p(\pi^{-1}(A))$ *for* $A \in \mathcal{N}^p = \{A : A \subseteq K, \pi^{-1}(A) \in \mathcal{M}^p\}$. *Then,* ν^p *is a Borel regular measure on* $(K, \mathcal{N}^p)$. ν^p *is called the self-similar measure on* K *with weight* p.

It is known that ν^p is the unique Borel regular probability measure on K that satisfies

$$\nu(A) = \sum_{i \in S} p_i \nu(F_i^{-1}(A))$$

for any Borel set $A \subset K$. See [34, Chapter 2] and [76] for the proof of this fact.

By definition,

$$\nu^p(K_w) \geq p_{w_1} p_{w_2} \cdots p_{w_m} \tag{1.4.1}$$

for any $w = w_1 w_2 \ldots w_m \in W_*$. Intuitively, it seems that equality holds in (1.4.1) rather than inequality if the overlapping set $C_{\mathcal{L}}$ is small enough. More precisely, we have the following theorem.

Theorem 1.4.5. *Let* $\mathcal{L} = (K, S, \{F_i\}_{i \in S})$ *be a self-similar structure and let* π *be the natural map from* Σ *to* K *associated with* $\mathcal{L}$. *Also let* $p = (p_i)_{i \in S}$ *satisfy* $\sum_{i \in S} p_i = 1$ *and* $0 < p_i < 1$ *for any* $i \in S$. *Then*

$$\nu^p(K_w) = p_{w_1} p_{w_2} \cdots p_{w_m}$$

for any $w = w_1 w_2 \ldots w_m \in W_*$ *if and only if* $\mu^p(\mathcal{I}_\infty) = 0$, *where*

$$\mathcal{I}_\infty = \{\omega \in \Sigma : \#(\pi^{-1}(\pi(\omega))) = +\infty\}.$$

Remark. We will show that $\mathcal{I}_\infty \in \mathcal{M}^p$.

Lemma 1.4.6. *For any* $A \in \mathcal{M}^p$, *define*

$$A_\circ = \{\omega \in \Sigma : \sigma^m \omega \in A \text{ for infinitely many } m \in \mathbb{N}.\}.$$

Then $A_\circ \in \mathcal{M}^p$ *and* $\mu^p(A_\circ) \geq \mu^p(A)$. *In particular, if* $A \in \mathcal{B}(\Sigma)$ *then* $A_\circ \in \mathcal{B}(\Sigma)$.

Proof. Define $\sigma_w = \sigma_{w_1} \circ \cdots \circ \sigma_{w_m}$ for $w = w_1 w_2 \ldots w_m \in W_*$. Set $A_m = \cup_{w \in W_m} \sigma_w(A)$. Then $A_\circ = \limsup_{m \to \infty} A_m$. Hence $A_\circ \in \mathcal{M}^p$ and by Fatou's lemma, we have $\mu^p(A_\circ) \geq \limsup_{m \to \infty} \mu^p(A_m)$. (Note that μ^p is a finite measure.) On the other hand, $\mu^p(A_m) = \sum_{w \in W_m} \mu^p(\sigma_w(A)) = \sum_{w \in W_m} p_w \mu^p(A) = \mu^p(A)$, where $p_w = p_{w_1} p_{w_2} \cdots p_{w_m}$. Hence it follows that $\mu^p(A_\circ) \geq \mu^p(A)$. $\square$

Lemma 1.4.7. *Define* $\mathcal{I} = \{\omega \in \Sigma : \#(\pi^{-1}(\pi(\omega))) > 1\}$. *Then* $\mathcal{I} \in \mathcal{B}(\Sigma)$, $\mathcal{I}_\infty \in \mathcal{M}^p$, $\mathcal{I}_\circ \subseteq \mathcal{I}_\infty \subseteq \mathcal{I}$ *and* $\mu^p(\mathcal{I}_\circ) = \mu^p(\mathcal{I}_\infty) = \mu^p(\mathcal{I})$.

Proof. Set $I_m = \cup_{w \neq v \in W_m}(K_w \cap K_v)$. Then it follows that I_m is closed and $\mathcal{I} = \cup_{m \geq 1}\pi^{-1}(I_m)$. Hence $\mathcal{I} \in \mathcal{B}(\Sigma)$. Now if $\omega \in \mathcal{I}_\circ$, by using induction, we can choose $\{m_k\}_{k \geq 1}$, $\{n_k\}_{k \geq 1}$ and $\{\omega^{(k)}\}_{k \geq 1}, \{\tau^{(k)}\}_{k \geq 1} \subset \Sigma$ so that

$$1 \leq m_1 < n_1 < m_2 < n_2 < \cdots < m_k < n_k < m_{k+1} < \cdots,$$
$$\sigma^{m_k}\omega \in \mathcal{I}, \sigma^{m_k}\omega \neq \tau^{(k)}, \pi(\sigma^{m_k}\omega) = \pi(\tau^{(k)}),$$
$$\omega^{(k)} = \omega_1\omega_2\ldots\omega_{m_k}\tau^{(k)}, \omega_1\omega_2\ldots\omega_{n_{k-1}} = \omega^{(k)}{}_1\omega^{(k)}{}_2\ldots\omega^{(k)}{}_{n_{k-1}}$$

and $\omega_{n_k} \neq \omega^{(k)}{}_{n_k}$. This implies that $\pi(\omega^{(k)}) = \pi(\omega)$ and hence $\omega \in \mathcal{I}_\infty$. Thus we have shown that $\mathcal{I}_\circ \subseteq \mathcal{I}_\infty \subseteq \mathcal{I}$. By Lemma 1.4.6, $\mu^p(\mathcal{I}) \leq \mu^p(\mathcal{I}_\circ)$, we can see that $\mu^p(\mathcal{I}) = \mu^p(\mathcal{I}_\circ)$. As μ^p is complete, $\mathcal{I}_\infty \in \mathcal{M}^p$ and $\mu^p(\mathcal{I}_\infty) = \mu^p(\mathcal{I})$. $\square$

Proof of Theorem 1.4.5. By the definition of $\mathcal{I}$, we can easily see that $\mu^p(\mathcal{I}) = 0$ if and only if $\mu^p(\Sigma_w) = \nu^p(K_w) = p_w$ for any $w \in W_*$. This along with Lemma 1.4.7 implies the theorem. $\square$

Remark. It is well-known that μ^p is ergodic with respect to the shift map σ. This means that if $A \in \mathcal{M}^p$ and $\sigma^{-1}(A) = A$ then $\mu^p(A) = 0$ or 1. Since $\sigma^{-1}(\mathcal{I}_\circ) = \mathcal{I}_\circ$, $\mu^p(\mathcal{I}_\circ) = \mu^p(\mathcal{I}_\infty) = \mu^p(\mathcal{I}) = 0$ or 1.

Corollary 1.4.8. *If* $\pi^{-1}(x)$ *is a finite set for any* $x \in K$, *then* $\nu^p(K_w) = p_w$ *for all* $w \in W_*$.

Since $\mathcal{I} = \cup_{w \in W_*}\sigma_w(\mathcal{C}_\mathcal{L})$, $\mu^p(\mathcal{C}_\mathcal{L}) > 0$ implies $\mu^p(\mathcal{I}) > 0$. Hence by Theorem 1.3.8, we have the following corollary.

Corollary 1.4.9. *If* $\nu^p(K_w) = p_w$ *for any* $w \in W_*$, *then* $\mathcal{L}$ *is minimal.*

Although the next theorem does not directly relate to self-similar measures, it tells us a useful fact: two Borel regular measures on a self-similar set are comparable if they are comparable on K_w for all $w \in W_*$.

Theorem 1.4.10. *Let* $\mathcal{L} = (K, S, \{F_i\}_{i \in S})$ *be a self-similar structure. Let* μ *and* ν *be Borel regular measures on* $(K, \mathcal{M}(\mu))$ *and* $(K, \mathcal{M}(\nu))$ *respectively. Assume that* $\nu(K) < \infty$ *and* $\nu(\mathcal{I}) = 0$. *If there exists* $c > 0$ *such that* $\mu(K_w) \leq c\nu(K_w)$ *for any* $w \in W_*$, *then* $\mu(A) \leq c\nu(A)$ *for any* $A \in \mathcal{M}(\mu) \cap \mathcal{M}(\nu)$. *In particular,* $\mu(\mathcal{I}) = 0$.

Proof. Let U be an open subset of K. Set $W(U) = \{w \in W_* : K_w \subset U\}$. For $w, v \in W(U)$, we define $w \geq v$ if and only if $\Sigma_w \supseteq \Sigma_v$. Then $\geq$ is a partial order on $W(U)$. If $W^+(U)$ is the collection of maximal elements in

$W(U)$ with respect to this order, then $U = \cup_{w \in W^+(U)} K_w$ and $K_w \cap K_v \subset \mathcal{I}$ for $w \neq v \in W^+(U)$. Therefore

$$\mu(U) \leq \sum_{w \in W^+(U)} \mu(K_w) \leq c \sum_{w \in W^+(U)} \nu(K_w) = c\nu(U).$$

Now, by Proposition 1.4.2, for any $A \in \mathcal{M}(\mu) \cap \mathcal{M}(\nu)$, there exists a decreasing sequence of open sets $\{O_k\}_{k \geq 1}$ such that $A \subseteq O_k$ for any k, $\mu(\cap_{k \geq 1} O_k) = \mu(A)$ and $\nu(\cap_{k \geq 1} O_k) = \nu(A)$. As $\mu(O_k) \leq c\nu(O_k)$, we have $\mu(A) \leq c\nu(A)$. $\qquad\qquad\qquad\qquad\qquad\qquad\qquad\qquad\qquad\qquad\qquad\square$

1.5 Dimension of self-similar sets

In this section, we will introduce the notion of the Hausdorff dimension of metric spaces and show how to calculate the Hausdorff dimension of self-similar sets.

Definition 1.5.1. Let (X, d) be a metric space. For any bounded set $A \subset X$, we define

$$\mathcal{H}_\delta^s(A) = \inf\{\sum_{i \geq 1} \operatorname{diam}(E_i)^s : A \subset \cup_{i \geq 1} E_i, \operatorname{diam}(E_i) \leq \delta\},$$

where $\operatorname{diam}(E)$ is the diameter of E defined in 1.2. Also, we define $\mathcal{H}^s(A) = \limsup_{\delta \downarrow 0} \mathcal{H}_\delta^s(A)$.

It is well-known that $\mathcal{H}^s$ is a complete Borel regular measure for any $s > 0$. See Rogers [157] and Falconer [32] for example. $\mathcal{H}^s$ is called the s-dimensional Hausdorff measure of (X, d).

Lemma 1.5.2. *Let (X, d) be a metric space. For $0 \leq s < t$,*

$$\mathcal{H}_\delta^t(E) \leq \delta^{t-s} \mathcal{H}_\delta^s(E)$$

for any $E \subseteq X$.

Proof. If $E \subseteq \cup_{i \geq 1} E_i$ and $\operatorname{diam}(E_i) \leq \delta$ for any i, then

$$\sum_{i \geq 1} \operatorname{diam}(E_i)^t \leq \sum_{i \geq 1} \operatorname{diam}(E_i)^{t-s} \operatorname{diam}(E_i)^s \leq \delta^{t-s} \sum_{i \geq 1} \operatorname{diam}(E_i)^s.$$

$\qquad\qquad\qquad\qquad\qquad\qquad\qquad\qquad\qquad\qquad\qquad\qquad\qquad\square$

By Lemma 1.5.2, we can obtain the following proposition.

Proposition 1.5.3. *For any $E \subseteq X$,*

$$\sup\{s : \mathcal{H}^s(E) = \infty\} = \inf\{s : \mathcal{H}^s(E) = 0\}. \qquad (1.5.1)$$

Proof. By Lemma 1.5.2, if $s < t$, then $\mathcal{H}^s(E) < \infty$ implies $\mathcal{H}^t(E) = 0$ and also $\mathcal{H}^t(E) > 0$ implies $\mathcal{H}^s(E) = \infty$. Now it is easy to see (1.5.1). □

Definition 1.5.4 (Hausdorff dimension). The quantity given by the equality (1.5.1) is called the Hausdorff dimension of E, which is denoted by $\dim_H E$.

Remark. The Hausdorff measure and the Hausdorff dimension depend on a metric d. In this sense, if we need to specify which metric we are looking at, we will use the notation of $\dim_H(E, d)$ instead of $\dim_H E$.

The following lemma is often useful to calculate the Hausdorff dimension of a metric space. It is often called "Frostman's lemma". See, for example, Mattila [124]. It is also called the "mass distribution principle" in Falconer [33].

Lemma 1.5.5. *Let (K, d) be a compact metric space. If $\mathcal{H}^\alpha(K) < \infty$ and there exist positive constants c, l_0 and a probability measure μ on K such that*

$$\mu(B_l(x)) \leq cl^\alpha$$

for all $x \in K$ and any $l \in (0, l_0)$, then

$$\mu(A) \leq c\mathcal{H}^\alpha(A)$$

for any Borel set $A \subset K$. In particular, $0 < \mathcal{H}^\alpha(K) < \infty$.

Remark. According to the discussion of Moran [134], the converse of the above lemma is true: if $0 < \mathcal{H}^\alpha(K) < \infty$, then there exists a probability measure μ on K such that, for some $c > 0$,

$$\mu(B_l(x)) \leq cl^\alpha$$

for all $x \in K$ and $l > 0$. Moran proved this fact if K was a compact subset of Euclidean space. His argument, however, can be easily extended to this case.

Proof. For $U \subset K$ and $x \in U$, note that $U \subset B_{\mathrm{diam}(U)}(x)$. Hence, if $A \subseteq \cup_i U_i$, then

$$\mu(A) \leq \sum_i \mu(B_{\mathrm{diam}(U_i)}(x_i)) \leq c \sum_i \mathrm{diam}(U_i)^\alpha,$$

where $x_i \in U_i$. Therefore $\mu(A) \leq c\mathcal{H}_l^\alpha(A)$. Letting $l \to 0$, it follows that $\mu(A) \leq c\mathcal{H}^\alpha(A)$. □

Now let $(K, \{1, 2, \ldots, N\}, \{F_i\}_{1 \le i \le N})$ be a self-similar structure and let d be a metric on K which is compatible with the original topology of K. In general, it is not easy to evaluate the Hausdorff dimension $\dim_H(K, d)$. Moran [134] introduced what is now called "the open set condition", which ensures that the intersections $K_i \cap K_j$ for $i \ne j \in \{1, 2, \ldots, N\}$ are "small". Under this condition, he gave a formula for the Hausdorff dimension of K when K was a subset of $\mathbb{R}^k$, d was the Euclidean metric on $\mathbb{R}^k$ and F_i were similitudes with respect to d. See Proposition 1.5.8 and Corollary 1.5.9 for Moran's result. His result is useful in calculating the Hausdorff dimension of many well-known examples of self-similar sets. See Exercise 1.9.

Remark. Moran published his paper [134] long before the notion of "fractal" existed. Of course, he didn't use the terminology "self-similar set" but he had exactly the same notion of self-similar set as we have today. Hutchinson [76] rediscovered Moran's result about 40 years later and introduced the name "open set condition".

Unfortunately we can apply Moran's result only when K is a subset of $\mathbb{R}^k$, d is the Euclidean metric on $\mathbb{R}^k$ and the F_i are similitudes with respect to d. Later in this book, a metric called an effective resistance metric, which satisfies none of those requirements, will become important from the analytical point of view. Here, we will introduce an extended version of Moran's theorem (Theorem 1.5.7) that can be applied in more general situations.

Definition 1.5.6. For $\mathbf{r} = (r_1, r_2, \ldots, r_N)$ where $0 < r_i < 1$ and for $0 < a < 1$,

$$\Lambda(\mathbf{r}, a) = \{w : w = w_1 w_2 \ldots w_m \in W_*, r_{w_1 w_2 \ldots w_{m-1}} > a \ge r_w\},$$

where $r_v = r_{v_1} r_{v_2} \ldots r_{v_k}$ for $v = v_1 v_2 \ldots v_k \in W_k$.

Remark. $\Lambda(\mathbf{r}, a)$ is a partition of Σ.

The following is our main theorem. This theorem was introduced in [88]. The ideas are, however, essentially the same as in Moran [134].

Theorem 1.5.7. *Assume that there exist $\mathbf{r} = (r_1, r_2, \ldots, r_N)$ where $0 < r_i < 1$ and positive constants c_1, c_2, c_* and M such that*

$$\mathrm{diam}(K_w) \le c_1 r_w \tag{1.5.2}$$

for all $w \in W_*$ and

$$\#\{w : w \in \Lambda(\mathbf{r}, a), d(x, K_w) \le c_2 a\} \le M \tag{1.5.3}$$

for any $x \in K$ and any $a \in (0, c_*)$, where $d(x, K_w) = \inf_{y \in K_w} d(x, y)$. Then there exist constants $c_3, c_4 > 0$ such that for all $A \in \mathcal{B}(K, d)$,

$$c_3 \nu(A) \leq \mathcal{H}^\alpha(A) \leq c_4 \nu(A), \tag{1.5.4}$$

where ν is a self-similar measure on K with weight $(r_i{}^\alpha)_{1 \leq i \leq N}$ and α is the unique positive number that satisfies

$$\sum_{i=1}^N r_i{}^\alpha = 1. \tag{1.5.5}$$

In particular, $0 < \mathcal{H}^\alpha(K) < \infty$ and $\dim_{\mathrm{H}}(K, d) = \alpha$.

Remark. Under the assumption (1.5.3), it is easy to see that $\#(\pi^{-1}(x)) \leq M$ for any $x \in K$. Hence, by Corollary 1.4.8,

$$\nu(K_w) = r_w{}^\alpha$$

for any $w \in W_*$. Also $\nu(\mathcal{I}) = 0$.

Proof. We write $\Lambda_a = \Lambda(\mathbf{r}, a)$. First we show that $\mathcal{H}^\alpha(K_w) \leq (c_1)^\alpha \nu(K_w)$ for all $w \in W_*$. For $w = w_1 w_2 \ldots w_m \in W_*$, we define $\Lambda_a(w) = \{v = v_1 v_2 \ldots v_k : wv \in \Lambda_a\}$, where $wv = w_1 w_2 \ldots w_m v_1 v_2 \ldots v_k$. Then we can see that $\Lambda_a(w)$ is a partition for sufficiently small a. Hence

$$r_w{}^\alpha = \sum_{v \in \Lambda_a(w)} r_{wv}{}^\alpha. \tag{1.5.6}$$

By (1.5.2), it follows that $\mathrm{diam}(K_{wv}) \leq c_1 r_{wv} \leq c_1 a$ for $v \in \Lambda_a(w)$. Also note that $K_w = \cup_{v \in \Lambda_a(w)} K_{wv}$. Then

$$\mathcal{H}^\alpha_{c_1 a}(K_w) \leq c_1{}^\alpha \sum_{v \in \Lambda_a(w)} r_{wv}{}^\alpha = (c_1 r_w)^\alpha.$$

Letting $a \to 0$, we obtain

$$\mathcal{H}^\alpha(K_w) \leq (c_1)^\alpha r_w{}^\alpha = (c_1)^\alpha \nu(K_w).$$

Next we show that $\nu(K_w) \leq M c_2{}^{-\alpha} \mathcal{H}^\alpha(K_w)$. Let μ be the Bernoulli measure on Σ with weight $(r_i{}^\alpha)_{1 \leq i \leq N}$. For every $x \in K$,

$$\pi^{-1}(B_{c_2 a}(x)) \subset \bigcup_{w \in \Lambda_{a,x}} \Sigma_w,$$

where $\Lambda_{a,x} = \{w : w \in \Lambda_a, d(x, K_w) \leq c_2 a\}$. Hence it follows that

$$\nu(B_{c_2 a}(x)) \leq \sum_{w \in \Lambda_{a,x}} \mu(\Sigma_w).$$

Since $\mu(\Sigma_w) = r_w^\alpha \leq a^\alpha$ and $\#(\Lambda_{a,x}) \leq M$ by (1.5.3), we have

$$\nu(B_{c_2 a}(x)) \leq M c_2^{-\alpha}(c_2 a)^\alpha.$$

Lemma 1.5.5 implies that

$$\nu(A) \leq M c_2^{-\alpha} \mathcal{H}^\alpha(A).$$

for any $A \in \mathcal{B}(K, d)$. Hence there exist $c_3, c_4 > 0$ such that

$$c_3 \nu(K_w) \leq \mathcal{H}^\alpha(K_w) \leq c_4 \nu(K_w).$$

By Theorem 1.4.10, we can verify (1.5.4). □

In the rest of this section, we show that the open set condition implies (1.5.2) and (1.5.3) of Theorem 1.5.7.

Proposition 1.5.8. *Suppose K is a subset of $\mathbb{R}^k$, d is the Euclidean metric of $\mathbb{R}^k$ and $F_i : \mathbb{R}^k \to \mathbb{R}^k$ is an r_i-similitude for $i = 1, 2, \ldots, N$ with respect to d. If the open set condition holds, i.e., there exists a bounded non-empty open set $O \subset \mathbb{R}^k$ such that*

$$\bigcup_{i=1}^N F_i(O) \subset O \quad and \quad F_i(O) \cap F_j(O) = \emptyset \quad for\ i \neq j,$$

then there exist constants $c_1, c_2, M > 0$ such that

$$\mathrm{diam}(K_w) \leq c_1 r_w$$

for all $w \in W_*$ and

$$\#\{w : w \in \Lambda(\mathbf{r}, a), d(x, K_w) \leq c_2 a\} \leq M$$

for all $0 < a < 1$ and $x \in K$.

Proof. We can see that $K_w \subset \overline{O}_w$ for any $w \in W_*$, where $O_w = F_w(O)$. (By Exercise 1.2, it follows that $\overline{O} \supseteq K$.) Without loss of generality, we may assume that $\mathrm{diam}(O) \leq 1$. Then, for all $w \in W_*$, $\mathrm{diam}(K_w) \leq \mathrm{diam}(\overline{O}_w) \leq r_w$. Let m be the k-dimensional Lebesgue measure and let $\Lambda_{a,x} = \{w : w \in \Lambda(\mathbf{r}, a), d(x, K_w) \leq a\}$. Then $\cup_{w \in \Lambda_{a,x}} O_w \subset B_{2a}(x)$. Since the O_w are mutually disjoint, we have $\sum_{w \in \Lambda_{a,x}} \mathrm{m}(O_w) \leq \mathrm{m}(B_{2a}(x))$. Hence it follows that $\#(\Lambda_{a,x}) r_w^k \mathrm{m}(O) \leq 2^k C a^k$, where $C = \mathrm{m}(\text{unit ball})$. Since $r_w \geq aR$ where $R = \min\{r_1, r_2, \ldots, r_N\}$, we see that $\#(\Lambda_{a,x}) \leq 2^k C R^{-k} \mathrm{m}(O)^{-1}$. □

Corollary 1.5.9 (Moran's theorem). *If K satisfies the open set condition, then $\mathrm{dim}_H(K, d) = \alpha$, where α is given by (1.5.5) with $r_i = \mathrm{Lip}(F_i)$.*

1.6 Connectivity of self-similar sets

Let $(K, S, \{F_i\}_{i \in S})$ be a self-similar structure. In this section we will give a simple condition for connectivity of K and also show that K is connected if and only if it is arcwise connected. As a reminder, the definition of connectivity is as follows.

Definition 1.6.1. Let (X, d) be a metric space.
(1) (X, d) is said to be connected if and only if any closed and open subset of X is X or the empty set. Also a subset A of X is said to be connected if and only if the metric space $(A, d|_A)$ is connected.
(2) A subset A of X is said to be arcwise connected if and only if there exists a path between x and y for any $x, y \in A$: there exists a continuous map $\gamma : [0, 1] \to A$ such that $\gamma(0) = x$ and $\gamma(1) = y$.
(3) (X, d) is said to be locally connected if and only if, for any $x \in X$ and any neighborhood U of x, there exists a connected neighborhood V of x with $V \subseteq U$.

Of course, arcwise connectivity implies connectivity, but the converse is not true in general. Now we come to the main theorem of this section.

Theorem 1.6.2. *The following are equivalent.*
(1) *For any $i, j \in S$, there exists $\{i_k\}_{k=0,1,\dots,n} \subseteq S$ such that $i_0 = i$, $i_n = j$ and $K_{i_k} \cap K_{i_{k+1}} \neq \emptyset$ for any $k = 0, 1, \dots, n - 1$.*
(2) *K is arcwise connected.*
(3) *K is connected.*

Proof. Obviously (2) $\Rightarrow$ (3). So let us show (3) $\Rightarrow$ (1). Choose $i \in S$ and define $A \subseteq S$ by

$$A = \{j \in S : \text{there exists } \{i_k\}_{k=0,1,\dots,n} \subseteq S \text{ such that}$$
$$i_0 = i, \; i_n = j \text{ and } K_{i_k} \cap K_{i_{k+1}} \neq \emptyset \text{ for any } k = 0, 1, \dots, n - 1\}.$$

If $U = \cup_{j \in A} K_j$ and $V = \cup_{j \notin A} K_j$, then $U \cap V = \emptyset$ and $U \cup V = K$. Also both U and V are closed sets because K_i is closed and A is a finite set. Hence U is an open and closed set. Hence $U = K$ or $U = \emptyset$. Obviously $K_i \subseteq U$ and hence $U = K$. Therefore $V = \emptyset$ and hence $A = S$.

To prove (1) $\Rightarrow$ (2), we need the following lemma.

Lemma 1.6.3. *For a map $u : [0, 1] \to K$ and for $t \in [0, 1]$, we define*

$$D(u, t) = \sup\{\limsup_{n \to \infty} d(u(t_n), u(s_n)) : \lim_{n \to \infty} t_n = \lim_{n \to \infty} s_n = t\}.$$

If $f_n : [0, 1] \to K$ is uniformly convergent to $f : [0, 1] \to K$ as $n \to \infty$ and $\lim_{n \to \infty} D(f_n, s) = 0$, then f is continuous at s.

Proof of Lemma 1.6.3. Let d be a metric on K that is compatible with the original topology of K. If $t_n \to s$ and $s_n \to s$ as $n \to \infty$, then

$$d(f(t_n), f(s_n)) \leq d(f(t_n), f_m(t_n)) + d(f_m(t_n), f_m(s_n)) + d(f_m(s_n), f(s_n)).$$

Set $r_m = \sup\{d(f_m(t), f(t)) : 0 \leq t \leq 1\}$. Then the above inequality implies $D(f, s) \leq 2r_m + D(f_m, s)$. Now letting $m \to \infty$, we can see that $D(f, s) = 0$. Hence f is continuous at s. $\qquad\square$

Now we return to the proof of (1) $\Rightarrow$ (2). Define

$$P = \{f : K^2 \times [0,1] \to K : f(p,q,0) = p \text{ and}$$
$$f(p,q,1) = q \text{ for any } (p,q) \in K^2\}.$$

Also for $f, g \in P$, set

$$d_P(f, g) = \sup\{d(f(p,q,t), g(p,q,t)) : (p,q,t) \in K^2 \times [0,1]\}.$$

Then (P, d_p) is a complete metric space. By (1), for any $(p,q) \in K^2$, we can choose $n(p,q)$, $\{i_k(p,q)\}_{0 \leq k \leq n(p,q)-1} \subseteq S$ and $\{x_k(p,q)\}_{0 \leq k \leq n(p,q)} \subseteq K$ so that $x_0(p,q) = p$, $x_{n(p,q)}(p,q) = q$ and $x_k(p,q), x_{k+1}(p,q) \in K_{i_k(p,q)}$ for $k = 0, \ldots, n(p,q) - 1$. For $f \in P$, define $Gf \in P$ by, for $k/n(p,q) \leq t \leq (k+1)n(p,q)$,

$$(Gf)(p,q,t) = F_{i_k(p,q)}(f(y_k(p,q), z_k(p,q), n(p,q)t - k)),$$

where $y_k(p,q) = F_{i_k(p,q)}^{-1}(x_k(p,q))$ and $z_k(p,q) = F_{i_k(p,q)}^{-1}(x_{k+1}(p,q))$. Then it follows that $d_P(G^m f, G^m g) \leq r_m$, where $r_m = \max_{w \in W_m} \operatorname{diam}(K_w)$. Since $r_m \to 0$ as $m \to \infty$, there exists $f_* \in P$ such that $G^m f \to f_*$ as $m \to \infty$ in P. Also set $D(f) = \sup\{D(f_{(p,q)}, t) : (p,q,t) \in K^2 \times [0,1]\}$ for $f \in P$, where $f_{(p,q)}(t) = f(p,q,t)$. Then $D(G^m f) \leq r_m D(f)$. Hence by Lemma 1.6.3, $f_*(p,q,t)$ is continuous with respect to t. As $f_*(p,q,t)$ is a continuous path between p and q, K is arcwise connected. $\qquad\square$

Proposition 1.6.4. *If K is connected, then K is locally connected.*

Proof. By Proposition 1.3.6, $\{K_{m,x}\}_{m \geq 0}$ is a system of fundamental neighborhoods of x. If K is connected then K_w is connected for any $w \in W_*$. Hence $K_{m,x}$ is connected. $\qquad\square$

For p. c. f. self-similar structures, Theorem 1.6.2 can be written as follows.

Corollary 1.6.5. *If $(K, S, \{F_i\}_{i \in S})$ is post critically finite, then K is connected if and only if, for any $p, q \in V_1$, there exist $\{p_i\}_{0 \leq i \leq m} \subset V_1$ and $\{k_i\}_{0 \leq i \leq m-1} \subseteq S$ such that $p_0 = p$, $p_m = q$ and $p_i, p_{i+1} \in F_{k_i}(V_0)$ for $i = 0, \ldots, m - 1$.*

In the rest of this section, we study connectivity of $K \backslash V_0$ for a connected p. c. f. self-similar structure.

Proposition 1.6.6. *Let* $(K, S, \{F_i\}_{i \in S})$ *be a connected post critically finite self-similar structure. Let* C *be a connected component of* $K \backslash V_0$. *Then* C *is arcwise connected. Moreover, for any* $x \in C$ *and any* $p \in \overline{C}$, *there exists a path between* x *and* p *in* $C \cup \{p\}$.

Proof. Let $x \in C$. Note that $K_{m,x}$ is arcwise connected. (See Proposition 1.6.4 and its proof.) Since C is open, $K_{m,x}$ is contained in C for sufficiently large m. Hence if $O = \{y : y \in C, \text{there exists a path between } x \text{ and } y \text{ in } C\}$, then O is open. Therefore, $O = C$. So C is arcwise connected.

Let $p \in \overline{C}$. If $p \in C$, the statement is obvious because C is arcwise connected. So assume that $p \in V_0$. Choose m so that $K_{m,p} \cap V_0 = \{p\}$. Let $y \in K_{m,p} \cap C$. Since $K_{m,p}$ is (arcwise) connected, there exists a path between y and p. This path is included in $C \cup \{p\}$. Also there exists a path between x and y which is included in C. Joining those two paths, we obtain the desired path between x and p. $\square$

Proposition 1.6.7. *Let* $(K, S, \{F_i\}_{i \in S})$ *be a connected post critically finite self-similar structure. Let* p *be the fixed point of* F_i. *If* C *is a connected component of* $K \backslash \{p\}$, *then* $C \cap V_0 \neq \emptyset$. *In particular, the number of connected components of* $K \backslash \{p\}$ *is finite. Moreover, let* $\{C_j\}_{j=1,2,\dots,m}$ *be the collection of all connected components of* $K \backslash \{p\}$. *Then there exists a permutation of* $\{1, 2, \dots, m\}$, ρ, *such that* $F_i(C_k) = C_{\rho(k)} \cap K_i$.

Proof. Suppose that $U_1, \dots, U_l$ are connected components of $K \backslash \{p\}$. Then we may choose n so that U_j is not contained in $F_i^n(K)$ for all $j = 1, 2, \dots, l$. By Proposition 1.3.5-(2), $U_j \cap F_i^n(V_0) \neq \emptyset$ for any $j = 1, 2, \dots, l$. Therefore $l \leq \#(V_0)$. This implies that the number of connected components of $K \backslash \{p\}$ is finite.

Now, let $C_1, \dots, C_m$ be the collection of all connected components of $K \backslash \{p\}$. Note that $F_i(C_j)$ is connected and $\cup_{j=1}^m F_i(C_j) = K_i \backslash \{p\}$. Therefore, there exists $\rho(j)$ such that $F_i(C_j) \subset C_{\rho(j)}$. Lemma 1.3.14 implies that $K_{1,p} = K_i$. Hence $C_k \cap K_i \neq \emptyset$ for any k and there exists j such that $C_j \cap F_i^{-1}(C_k) \neq \emptyset$. This implies that ρ is a permutation of $\{1, 2, \dots, m\}$. As $K_i \backslash \{p\} = \cup_{j=1}^m (K_i \cap C_i)$, we see that $K_i \cap C_{\rho(j)} = F_i(C_j)$ for any j.

Next choose n so that C_j is not contained in $F_i^n(K)$ for any j. Then, it follows that $C_{\rho^n(k)} \cap F_i^n(V_0) \neq \emptyset$ for any k. Since $C_{\rho^n(k)} \cap F_i^n(V_0) = F_i^n(V_0 \cap C_k)$, we obtain $C_k \cap V_0 \neq \emptyset$ for any k. $\square$

Proposition 1.6.8. *Let* $\mathcal{L} = (K, S, \{F_i\}_{i \in S})$ *be a connected post critically* *finite self-similar structure.*

(1) *For any $p \in V_0$, let $J(p)$ be the collection of all connected components of $K\backslash\{p\}$. Then $J(p)$ is a finite set.*

(2) *The number of connected components of $K\backslash V_0$ is finite.*

(3) *For $p \in V_0$, define*

$$J(p, V_0) = \{C : C \text{ is a connected component of } K\backslash V_0, p \in \overline{C}\}.$$

Set $m(p, V_0) = \#(J(p, V_0))$ and $m(p) = \#(J(p))$. If $p \in V_0$ is a fixed point of F_w for some $w \in W_* \backslash W_0$, then $m(p, V_0) = m(p)$.

Proof. (1) Suppose that $J(p)$ is an infinite set. Let $\{C_i\}_{i \geq 1}$ be a collection of connected components of $K\backslash\{p\}$. Assume that $C_j \neq C_j$ if $i \neq j$.

Claim 1. $\#\{j : \mathrm{diam}(C_j) \geq \epsilon\} < \infty$ for any $\epsilon > 0$.

Suppose that $\mathrm{diam}(C_{j_k}) \geq \epsilon$ for any $k \geq 0$. Then for any k, there exists $x_k \in C_{j_k}$ such that $d(p, x_k) \geq \epsilon/2$. Since K is compact, we may choose a subsequence $\{x_{k_n}\}_{n \geq 1}$ so that $x_{k_n} \to x$ as $n \to \infty$ for some $x \in K$. Let C be the connected component of $K\backslash\{p\}$ with $x \in C$. Since C is a neighborhood of x, x_{k_n} belongs to C for sufficiently large n. This contradicts the fact that each x_{k_n} belongs to a different C_j. Hence we have the claim.

Now since $\mathcal{L}$ is p.c.f., we may write $\pi^{-1}(p) = \{\omega(1), \dots, \omega(m)\}$. Then there exists k such that, for any $j = 1, 2, \dots, m$, $\omega(j) = w(j)v(j)$ for some $w(j) \in W_k$ and some $v(j) \in W_*$. Let $p_j = \pi(v(j))$ for $j = 1, 2, \dots, m$. Note that $p_j \in V_0$ and $F_{w(j)}(p_j) = p$ for all j. Also, by Proposition 1.6.7, the number of connected components of $K\backslash\{p_j\}$ is finite. (We may change the self-similar structure to $\mathcal{L}_n = (K, W_n, \{F_w\}_{w \in W_n})$, where $n = |v(j)|$. Then apply Proposition 1.6.7 to p_j, which is a fixed point of $F_{v(j)}$.) Since $K_{k,p} = \cup_{j=1}^{m} F_{w(j)}(K)$, it follows that the number of connected components of $K_{k,p}\backslash\{p\}$ is finite. (See Proposition 1.3.6 for the definition of $K_{k,p}$.)

By the way, since $K_{k,p}$ is a neighborhood of p, Claim 1 implies that $K_{k,p}$ contains infinitely many connected components of $K\backslash\{p\}$. This contradiction shows that the number of connected components of $K\backslash\{p\}$ is finite.

(2) Suppose that the number of connected components of $K\backslash V_0$ is infinite. Then there exists $p \in V_0$ such that $J(p, V_0)$ is an infinite set. By the same discussion as in the proof of (1), it follows that, for any $\epsilon > 0$, $\#\{C | C \in J(p, V_0), \mathrm{diam}(C) < \epsilon\} = \infty$. On the other hand, let $\epsilon = \min_{q \in V_0\backslash\{p\}} d(p, q)/2$ and assume that $C \in J(p, V_0)$ and $\mathrm{diam}(C) < \epsilon$. Then C is a connected component of $K\backslash V_0$. By (1), $\{C | C \in J(p, V_0), \mathrm{diam}(C) < \epsilon\}$ is a finite set.

(3) Let $k = m(p, V_0)$ and let $U = (\cup_{C \in J(p, V_0)} C) \cup \{p\}$. Then U is a

neighborhood of p. Hence, if $w(n) = \underbrace{w \ldots w}_{n \text{ times}}$, then $K_{w(n)} \subset U$ for sufficiently large n. Therefore if k' is the number of connected components of
$K_{w(n)} \backslash \{p\}$, then $k' \geq k$. Since $m(p) = k'$, we see that $m(p) \geq m(p, V_0)$.

On the other hand, a connected component of $K \backslash \{p\}$ contains at least
one $C \in J(p, V_0)$. Hence $m(p) \leq m(p, V_0)$. $\qquad \square$

The next proposition also concerns a connected p. c. f. self-similar structure. It gives a sufficient condition for $K \backslash V_0$ to be connected.

Proposition 1.6.9. *Let* $(K, S, \{F_i\}_{i \in S})$ *be a connected post critically finite self-similar structure. Assume that, for any* $p, q \in V_0$, *there exists a
homeomorphism* $g : K \to K$ *such that* $g(V_0) = V_0$ *and* $g(p) = q$. *Then
$K \backslash V_0$ is connected.*

If a connected p. c. f. self-similar structure satisfies the assumption of the
above proposition, we say that the self-similar structure is weakly symmetric.

To prove the above proposition, we need the following lemmas.

Lemma 1.6.10. *Assume the conditions in Proposition 1.6.9. Let C be a
connected component of $K \backslash V_0$. Then $\#(\overline{C} \cap V_0) \geq 2$.*

Proof. Suppose that there exists a connected component of $K \backslash V_0$ satisfying
$\#(\overline{C} \cap V_0) = 1$. Let p_0 be the unique $p_0 \in V_0$ with $p_0 \in \overline{C}$. For any
$p \in V_0$, there exists a homeomorphism $g : K \to K$ such that $g(p_0) = p$.
Letting $C_p = g(C)$, we see that C_p is a connected component of $K \backslash V_0$ and
$\overline{C}_p \cap V_0 = \{p\}$. Then it follows that C_p is a connected component of $K \backslash \{p\}$.

Now, since $\mathcal{L}$ is post critically finite, there exists $p \in V_0$ such that p is
a fixed point of F_w for some $w \in W_* \backslash W_0$. Let $m = |w|$. By exchanging
the self-similar structure $\mathcal{L}$ with $\mathcal{L}_m = (K, W_m, \{F_v\}_{v \in W_m})$, we can use
Proposition 1.6.7 and obtain that $C_p \cap V_0 \neq \emptyset$. This contradicts the fact
that $\overline{C}_p \cap V_0 = \{p\}$. $\qquad \square$

Lemma 1.6.11. *Assume the conditions in Proposition 1.6.9. Then, for all
$p \in V_0$, $m(p, V_0) = m(p)$.*

Proof. Since $\mathcal{L}$ is weakly symmetric, $m(p)$ and $m(p, V_0)$ are independent
of the choice of $p \in V_0$. Hence, as in the proof of the last lemma, we
may assume that $F_w(p) = p$ for some $w \in W_* \backslash W_0$. Then the statement is
immediate by Proposition 1.6.8-(3). $\qquad \square$

Proof of Proposition 1.6.9. Let $\mathcal{J}$ be the collection of connected compo-
nents of $K \backslash V_0$. Define $V = V_0 \cup \mathcal{J}$ and $E = \{(p, C) : p \in V_0, C \in J(p, V_0)\}$.

Let $G = (V, E)$ be a undirected graph, where V is the set of vertices and E is the set of edges. Note that G is connected.

First we show that this graph G contains no loop. Suppose that there exists a loop in G : there exist $\{p_i\}_{i=1,2,\ldots,n} \subset V_0$ and $\{C_i\}_{i=1,2,\ldots,n} \subset \mathcal{J}$ such that $(p_i, C_i), (p_{i+1}, C_i) \in E$ for $i = 1, 2, \ldots, n$, where $p_{n+1} = p_1$. Then C_i and C_{i+1} are connected components of $K \backslash V_0$ whose closures contain p_i. Let $A = \cup_{j=1}^n (C_j \cup \{p_j\})$. Then for any i, $A \backslash \{p_i\}$ is connected and is contained in one connected component of $K \backslash \{p_i\}$. Hence, C_i and C_{i+1} are contained in the same connected component of $K \backslash \{p_i\}$. This contradicts Lemma 1.6.11.

Since G contain no loop, G is a tree : for any $x, y \in V$, there exists a unique sequence of edges from x to y. Hence G has an end point, i.e., there exists $x \in V$ such that $\#\{y \in V : (x, y) \in E$ or $(y, x) \in E\} = 1$. By Lemma 1.6.10, we see that $x \in V_0$. Hence $m(x, V_0) = 1$. Therefore, $m(p, V_0) = m(x, V_0) = 1$ for any $p \in V_0$. On the other hand, if $\#\mathcal{J} \geq 2$, then there exists $p \in V_0$ such that $m(p, V_0) \geq 2$ because G is connected. Hence $\#\mathcal{J} = 1$. $\square$

Notes and references

The general references to this chapter are Falconer[32, 33] and Yamaguti *et al.*[184].

1.1   The notion of self-similar sets could be traced back to the Cantor set, which is the first mathematical example of what are now called "fractals". In [134], Moran considered a class of sets which are extension of the Cantor set and calculated the Hausdorff dimensions of them. His class includes, for example, the Koch curve and the Sierpinski gasket. After Mandelbrot proposed the notion of fractals in [122, 123], Hutchinson[76] formulated the mathematical definition of self-similar sets, which is more restrictive than what we call "self-similar sets" in this book. See the remark after Theorem 1.1.4. Theorem 1.1.4 was essentially obtained in [76]. See also Hata[64] for an extension of Theorem 1.1.4.

1.2   The shift space Σ and the shift map σ are important concepts in dynamical systems. For example, they play an essential role in the study of interval maps. See [133] for example. Theorem 1.2.3 was essentially obtained in [76].

1.3   Partly motivated by Kameyama[80], the notion of the self-similar structure was introduced in [83].

1.4   Hutchinson has given the definition of self-similar measures in [76].

1.5 See Rogers[157] for details on the Hausdorff measures. See also [124] for results from the view point of geometric measure theory.
1.6 Theorem 1.6.2 was obtained in [64]. The results after Proposition 1.6.6 are new.

Exercises

Exercise 1.1. Let $f : \mathbb{R}^n \to \mathbb{R}^n$ be a similitude with Lipschitz constant r. Show that there exist $a \in \mathbb{R}^n$ and $U \in O(n)$ such that $f(x) = rUx + a$ for all $x \in \mathbb{R}^n$.

(Hint: if $g(x) = (f(x) - f(0))/r$, one can see that $|g(x) - g(y)| = |x - y|$. This implies that the natural inner product of $\mathbb{R}^n$ is invariant under g. Also one should show that g is a linear map.)

Exercise 1.2. Let (X, d) be a complete metric space and let $f_i : X \to X$ be a contraction for $i = 1, 2, \ldots, N$. For $A \subseteq X$, define $F(A) = \cup_{1 \le i \le N} f_i(A)$. Let K be the self-similar set with respect to $\{f_1, f_2, \ldots, f_N\}$. Then
(1) Suppose $A \ne \emptyset$. Show that $A \supseteq F(A)$ implies $\overline{A} \supseteq K$.
(2) Show that for any $x \in X$, $B_r(x) \supseteq F(B_r(x))$ for sufficiently large r.

Exercise 1.3. Define $F_i(z) = \frac{1}{2}(z - p_i) + p_i$ for $i \in \{1, 2, 3, 4\}$, where $p_1 = 0, p_2 = 1, p_3 = (1 + \sqrt{-1})$ and $p_4 = \sqrt{-1}$. Let K be the self-similar set with respect to $\{F_1, F_2, F_3, F_4\}$. Prove that V_0 coincides with the topological boundary of K.

Exercise 1.4. Let $K = [0, 1]$ and let $S = \{1, 2, \ldots, N\}$. Set $F_i(x) = a_i x + b_i$ for $i \in S$. Assume that $0 < a_i < 1$ for any $i \in S$ and that $K = \cup_{i \in S} F_i(K)$. Prove that $(K, S, \{F_i\}_{i \in S})$ is minimal if and only if $\sum_{i=1}^{N} a_i = 1$.

Exercise 1.5. Define $f_1(x) = x/3$ and $f_2(x) = x/3 + 2/3$. Let K be the self-similar set with respect to $\{f_1, f_2\}$. (K is the Cantor middle third set.) Set $g_i = f_i \circ f_i$ for $i = 1, 2$. Let K' be the self-similar set with respect to $\{g_1, g_2\}$. The natural map from $\Sigma(\{1, 2\}) \to K$ (resp. $\Sigma(S) \to K'$) is denoted by π (resp. π'). Note that both π and π' are homeomorphisms. Set $f_3 = f_1 \circ \pi' \circ \pi^{-1}$.
(1) Show that f_3 is a contraction on K.
(2) Let $\mathcal{L} = (K, \{1, 2, 3\}, \{f_1, f_2, f_3\})$. Show that $\text{int}(\mathcal{C}_{\mathcal{L}}) \ne \emptyset$ and $\text{int}(C_{\mathcal{L}, K}) = \emptyset$.

Exercise 1.6. Prove that $\mathcal{P}_{\mathcal{L}(\Lambda)} = \mathcal{P}_{\mathcal{L}}$ for any partition Λ for the self-similar structures corresponding to Example 1.2.7, 1.2.8, 1.2.9 in the last section.

Exercise 1.7. Let $S = \{1, 2, 3\}$. Set $\omega \sim \tau$ if and only if $\{\omega, \tau\} \subseteq \{w121\dot{2}, w3\dot{1}\}$ for some $w \in W_*(S)$ or $\omega = \tau$.
(1) Let $K = \Sigma(S)/\sim$ with the quotient topology. Also define $F_i : K \to K$ by $F_i(x) = \pi(\sigma_i(\pi^{-1}(x)))$ for $x \in K$. Then prove that $\mathcal{L} = (K, S, \{F_i\}_{i \in S})$ is a self-similar structure.
(2) Let $\Lambda = \{1, 21, 22, 23, 3\}$. Prove that $\mathcal{P}_{\mathcal{L}(\Lambda)}$ is a proper subset of $\mathcal{P}_{\mathcal{L}}$.

Exercise 1.8. Let $\mathcal{L} = (K, S, \{F_i\}_{i \in S})$ be a self-similar structure and let Λ be a partition of $\Sigma(S)$. Show that $\mathcal{L}$ is post critically finite if and only if $\mathcal{L}(\Lambda)$ is post critically finite.

Exercise 1.9. Evaluate the Hausdorff dimensions of the self-similar sets introduced in Examples 1.2.6–1.2.9 under the Euclidean metric.

2

Analysis on Limits of Networks

In this chapter, we will discuss limits of discrete Laplacians (or equivalently Dirichlet forms) on a increasing sequence of finite sets. The results in this chapter will play a fundamental role in constructing a Laplacian (or equivalently a Dirichlet form) on certain self-similar sets in the next chapter, where we will approximate a self-similar set by an increasing sequence of finite sets and then construct a Laplacian on the self-similar set by taking a limit of the Laplacians on the finite sets.

More precisely, we will define a Dirichlet form and a Laplacian on a finite set in 2.1. The key idea is that every Dirichlet form on a finite set can be associated with an electrical network consisting of resistors. From such a point of view, we will introduce the important notion of effective resistance. In 2.2, we will study a limit of a "compatible" sequence of Dirichlet forms on increasing finite sets. Roughly speaking, the word "compatible" means that the Dirichlet forms appearing in the sequence induce the same effective resistance on the union of the increasing finite sets. In 2.3 and 2.4, we will present further properties of limits of compatible sequences of Dirichlet forms.

2.1 Dirichlet forms and Laplacians on a finite set

In this section, we present some fundamental notions of analysis on a finite set, namely, Dirichlet forms, Laplacians and effective resistance.

Notation. For a set V, we define $\ell(V) = \{f : f \text{ maps } V \text{ into } \mathbb{R}\}$. If V is a finite set, $\ell(V)$ is considered to be equipped with the standard inner product $(\cdot, \cdot)$ defined by $(u, v) = \sum_{p \in V} u(p) v(p)$ for any $u, v \in \ell(V)$.

First we give a definition of Dirichlet forms on a finite set V. In B.3, one

41

can find a definition of Dirichlet forms for general locally compact metric spaces.

Definition 2.1.1 (Dirichlet forms). Let V be a finite set. A symmetric bilinear form on $\ell(V)$, $\mathcal{E}$ is called a Dirichlet form on V if it satisfies

(DF1) $\mathcal{E}(u,u) \geq 0$ for any $u \in \ell(V)$,

(DF2) $\mathcal{E}(u,u) = 0$ if and only if u is constant on V

and

(DF3) For any $u \in \ell(V)$, $\mathcal{E}(u,u) \geq \mathcal{E}(\bar{u},\bar{u})$, where $\bar{u}$ is defined by

$$\bar{u}(p) = \begin{cases} 1 & \text{if } u(p) \geq 1, \\ u(p) & \text{if } 0 < u(p) < 1, \\ 0 & \text{if } u(p) \leq 0. \end{cases}$$

We use $\mathcal{DF}(V)$ to denote the collection of Dirichlet forms on V. Also $\widetilde{\mathcal{DF}}(V)$ is the collection of all symmetric bilinear forms on $\ell(V)$ with (DF1) and (DF2).

Obviously $\mathcal{DF}(V) \subset \widetilde{\mathcal{DF}}(V)$. Condition (DF3) is called the Markov property.

This definition is a special case of Definition B.3.2 when X is a finite set V and the measure μ is the discrete measure on V.

Notation. Let V be a finite set. The characteristic function χ_U^V of a subset $U \subseteq V$ is defined by

$$\chi_U^V(q) = \begin{cases} 1 & \text{if } q \in U, \\ 0 & \text{otherwise.} \end{cases}$$

If no confusion can occur, we write χ_U instead of χ_U^V. If $U = \{p\}$ for a point $p \in V$, we write χ_p instead of $\chi_{\{p\}}$. If $H : \ell(V) \to \ell(V)$ is a linear map, then we set $H_{pq} = (H\chi_q)(p)$ for $p, q \in V$. For $f \in \ell(V)$, $(Hf)(p) = \sum_{q \in V} H_{pq} f(q)$.

Definition 2.1.2 (Laplacians). A symmetric linear operator $H : \ell(V) \to \ell(V)$ is called a Laplacian on V if it satisfies

(L1) H is non-positive definite,

(L2) $Hu = 0$ if and only if u is a constant on V,

and

(L3) $H_{pq} \geq 0$ for all $p \neq q \in V$.

We use $\mathcal{LA}(V)$ to denote the collection of Laplacians on V. Also $\widetilde{\mathcal{LA}}(V)$ is the collection of symmetric linear operators from $\ell(V)$ to itself with (L1) and (L2).

Obviously $\mathcal{LA}(V) \subset \widetilde{\mathcal{LA}}(V)$.

There is a natural correspondence between $\mathcal{DF}(V)$ and $\mathcal{LA}(V)$. For a symmetric linear operator $H : \ell(V) \to \ell(V)$, we can define a symmetric quadratic form $\mathcal{E}_H(\cdot, \cdot)$ on $\ell(V)$ by $\mathcal{E}_H(u, v) = -(u, Hv)$ for $u, v \in \ell(V)$. If we write $\pi(H) = \mathcal{E}_H$, it is easy to see that π is a bijective mapping between symmetric linear operators and symmetric quadratic forms.

This correspondence between Dirichlet forms and non-negative symmetric operators is a special case of the correspondence described in Theorem B.3.4.

Proposition 2.1.3. π *is a bijective mapping between* $\widetilde{\mathcal{LA}}(V)$ *and* $\widetilde{\mathcal{DF}}(V)$. *Moreover,* $\pi(\mathcal{LA}(V)) = \mathcal{DF}(V)$.

Proof. It is routine to show $\pi(\widetilde{\mathcal{LA}}(V)) = \widetilde{\mathcal{DF}}(V)$. To show $\pi(\mathcal{LA}(V)) = \mathcal{DF}(V)$, first note that $\mathcal{E}_H(u, u) = \frac{1}{2} \sum_{p,q \in V} H_{pq}(u(p) - u(q))^2$. By this expression, it is easy to see that $\pi(\mathcal{LA}(V)) \subseteq \mathcal{DF}(V)$. Now suppose $H \in \widetilde{\mathcal{LA}}(V) \backslash \mathcal{LA}(V)$. So there exist $p \neq q \in V$ with $H_{pq} < 0$. We can assume that $H_{pq} = -1$ without loss of generality. Set $u(p) = x, u(q) = y$ and $u(a) = z$ for all $a \in V \backslash \{p, q\}$. Then we have $\mathcal{E}_H(u, u) = \alpha(x - z)^2 + \beta(y - z)^2 - (x - y)^2$. As $\mathcal{E}_H$ is non-negative definite, α and β should be non-negative. If $x = 1$, $z = 0$ and $y < 0$, then $\mathcal{E}_H(u, u) = \alpha - 1 + 2y + (\beta - 1)y^2$ and $\mathcal{E}_H(\bar{u}, \bar{u}) = \alpha - 1$. If $|y|$ is small, we have $\mathcal{E}_H(u, u) < \mathcal{E}_H(\bar{u}, \bar{u})$. Hence $\mathcal{E}_H \notin \mathcal{DF}(V)$. This shows that $\pi(H) \in \mathcal{DF}(V)$ if and only if $H \in \mathcal{LA}(V)$. □

Example 2.1.4. Let V be a set with three elements, say, p_1, p_2, p_3. Set
$$H = \begin{pmatrix} -(1+\epsilon) & 1 & \epsilon \\ 1 & -2 & 1 \\ \epsilon & 1 & -(1+\epsilon) \end{pmatrix}. \text{ Then } \mathcal{E}_H(u, u) = (x - y)^2 + (y - z)^2 +$$
$\epsilon(x - z)^2$, where $x = u(p_1), y = u(p_2)$ and $z = u(p_3)$. Letting $X = x - y$ and $Y = y - z$, we have

$$\mathcal{E}_H(u, u) = X^2 + Y^2 + \epsilon(X + Y)^2$$
$$= (1 + 2\epsilon)(X^2 + Y^2) - \epsilon(X - Y)^2$$

So it is clear that if $\epsilon > -\frac{1}{2}$, then $H \in \widetilde{\mathcal{LA}}(V)$ and if $\epsilon \geq 0$, then $H \in \mathcal{LA}(V)$.

If V is a finite set and H is a Laplacian on V, the pair (V, H) is called a resistance network (an r-network, for short). In fact, we can relate an r-network to an actual electrical network as follows. For an r-network (V, H), we will attach a resistor of resistance $r_{pq} = H_{pq}^{-1}$ to the terminals p and q for $p, q \in V$. Also the plus-side of a battery is connected to every terminal p while its minus-side is grounded so that we can put any electrical potential on each terminal. For a given electric potential $v \in \ell(V)$, the current i_{pq} between p and q is given by $i_{pq} = H_{pq}(v(p) - v(q))$. So the total current $i(p)$ from a terminal p to the ground is obtained by $i(p) = (Hv)(p)$.

Let (V, H) be an r-network and let U be a proper subset of V. We next discuss the proper way of restricting H onto U from the analytical point of view.

Lemma 2.1.5. *Let V be a finite set and let U be a proper subset of V. For $H \in \widetilde{\mathcal{LA}}(V)$, we define $T_U : \ell(U) \to \ell(U), J_U : \ell(U) \to \ell(V \backslash U)$ and $X_U : \ell(V \backslash U) \to \ell(V \backslash U)$ by*

$$H = \begin{pmatrix} T_U & {}^t J_U \\ J_U & X_U \end{pmatrix},$$

where ${}^t J_U$ is the transpose matrix of J_U. (When no confusion can occur, we use T, J and X instead of T_U, J_U and X_U.) Then, $X = X_U$ is negative definite and, for any $u \in \ell(V)$,

$$\mathcal{E}_H(u, u) = \mathcal{E}_X(u_1 + X^{-1}Ju_0, u_1 + X^{-1}Ju_0) + \mathcal{E}_{T - {}^t JX^{-1}J}(u_0, u_0),$$

$$(2.1.1)$$

where $u_0 = u|_U$ and $u_1 = u|_{V \backslash U}$.

Proof. For $v \in \ell(V \backslash U)$, define $\tilde{v}$ by $\tilde{v}|_U = 0$ and $\tilde{v}|_{V \backslash U} = v$. Then $\mathcal{E}_X(v, v) = \mathcal{E}_H(\tilde{v}, \tilde{v}) \geq 0$. By (L2), we see that if $\mathcal{E}_X(v, v) = 0$ then $\tilde{v}$ should be a constant on V. This implies $v = 0$. Hence $\mathcal{E}_X$ is positive definite. By the definition of $\mathcal{E}_X$, X is negative definite. Now (2.1.1) can be obtained by an easy calculation. □

Theorem 2.1.6. *Assume the same situation as in Lemma 2.1.5. For $u \in \ell(U)$, define $h(u) \in \ell(V)$ by $h(u)|_U = u$ and $h(u)|_{V \backslash U} = -X^{-1}Ju$. Then $h(u)$ is the unique element that attains $\min_{v \in \ell(V), v|_U = u} \mathcal{E}_H(v, v)$. Also define $P_{V,U}(H) = T - {}^t JX^{-1}J$. Then, $P_{V,U} : \widetilde{\mathcal{LA}}(V) \to \widetilde{\mathcal{LA}}(U)$ and*

$$\mathcal{E}_{P_{V,U}(H)}(u, u) = \mathcal{E}_H(h(u), h(u)) = \min_{v \in \ell(V), v|_U = u} \mathcal{E}_H(v, v). \qquad (2.1.2)$$

Moreover, if $H \in \mathcal{LA}(V)$, then $P_{V,U}(H) \in \mathcal{LA}(U)$.

Proof. By (2.1.1), $\min_{v \in \ell(V), v|_U = u} \mathcal{E}_H(v, v)$ is attained if and only if $v|_{V \backslash U} + X^{-1}Ju = 0$. Hence we have the first part of the theorem.

Next we show that $P_{V,U}(H) = T - {}^t JX^{-1}J \in \widetilde{\mathcal{LA}}(V)$. By (2.1.1), we can verify (2.1.2). Hence, $\mathcal{E}_{P_{V,U}(H)}$ is non-negative definite. By (2.1.2), $\mathcal{E}_{P_{V,U}(H)}(u, u) = 0$ implies that $h(u)$ is a constant on V and therefore u is a constant on U. Thus we can show that $P_{V,U}(H) \in \widetilde{\mathcal{LA}}(U)$.

Finally, if $H \in \mathcal{LA}(V)$, then

$$\mathcal{E}_{P_{V,U}(H)}(u,u) = \mathcal{E}_H(h(u),h(u)) \geq \mathcal{E}_H(\overline{h(u)},\overline{h(u)}).$$

As $\overline{h(u)}|_U = \bar{u}$, we obtain $\mathcal{E}_H(\overline{h(u)},\overline{h(u)}) \geq \mathcal{E}_{P_{V,U}(H)}(\bar{u},\bar{u})$. Hence $\mathcal{E}_{P_{V,U}(H)}$ has the Markov property. By Proposition 2.1.3, $P_{V,U}(H) \in \mathcal{LA}(U)$. □

The linear operator $P_{V,U}(H)$ may be thought of as the proper restriction of H onto U from the viewpoint of electrical circuits. In fact, H and $P_{V,U}(H)$ give exactly the same effective resistance (which will be defined in Definition 2.1.9) on U. When no confusion can occur, we write $[H]_U$ in place of $P_{V,U}(H)$.

Remark. In general, $P_{V,U}$ is not injective. For example, set $V = \{p_1, p_2, p_3\}$ and $U = \{p_1, p_2\}$. If $H_\epsilon = \begin{pmatrix} -(1+\epsilon) & 1 & \epsilon \\ 1 & -1 & 0 \\ \epsilon & 0 & -\epsilon \end{pmatrix}$ for $\epsilon > 0$, then $H_\epsilon \in \mathcal{LA}(V)$ and $[H_\epsilon]_U = \begin{pmatrix} -1 & 1 \\ 1 & -1 \end{pmatrix}$.

Note that $h(u)$ is the unique solution of $(Hv)|_{V\setminus U} = 0$ and $v|_U = u$. Therefore, if we think of U as a boundary of V, $h(u)$ is called the harmonic function with boundary value $u \in \ell(U)$. For a Laplacian H on V, we obtain the following maximum principle for harmonic functions.

Proposition 2.1.7 (Maximum principle). *Let V be a finite set and let $H \in \mathcal{LA}(V)$. Also let U be a subset of V. For $p \in V\setminus U$, set*

$$U_p = \{q \in U : \text{There exist } p_1, p_2, \ldots, p_m \in V\setminus U \text{ with } p_1 = p$$
$$\text{such that } H_{p_i p_{i+1}} > 0 \text{ for } i = 1, 2, \ldots, m-1 \text{ and } H_{p_m q} > 0\}.$$

Then if $(Hu)|_{V\setminus U} = 0$, then, for any $p \in V\setminus U$,

$$\min_{q \in U_p} u(q) \leq u(p) \leq \max_{q \in U_p} u(q).$$

Moreover, $u(p) = \max_{q \in U_p} u(q)$ (or $u(p) = \min_{q \in U_p} u(q)$) if and only if u is constant on U_p.

Proof. For $p \in V$, set $N_p = \{q : H_{pq} > 0\}$. Also define

$$W_p = \{q \in V\setminus U : \text{there exist } p_1, p_2, \ldots, p_m \in V\setminus U \text{ with } p_1 = p$$
$$\text{and } p_m = q \text{ such that } H_{p_i p_{i+1}} > 0 \text{ for } i = 1, 2, \ldots, m-1\}$$

and $V_p = W_p \cup U_p$. First assume $u(p_*) = \max_{q \in V_p} u(q)$ for $p_* \in W_p$. Then $N_{p_*} \subseteq V_p$ and $(Hu)(p_*) = \sum_{q \in N_{p_*}} H_{p_* q}(u(q) - u(p_*)) = 0$. Since

$H_{p_*q} > 0$ and $u(q) - u(p_*) \leq 0$ for any $q \in N_{p_*}$, we have $u(q) = u(p_*)$ for all $q \in N_{p_*}$. Iterating this argument, we see that u is constant on V_p. Using the same argument, it follows that if there exists $p_* \in W_p$ such that $u(p_*) = \min_{q \in V_p} u(q)$, then u is constant on V_p. Hence,

$$\min_{q \in U_p} u(q) = \min_{q \in V_p} u(q) \leq u(p) \leq \max_{q \in V_p} u(q) = \max_{q \in U_p} u(q).$$

The rest of the statement is now obvious.   □

The following corollary of the maximum principle is called the Harnack inequality.

Corollary 2.1.8 (Harnack inequality). *Let V be a finite set and let $H \in \mathcal{LA}(V)$. Also let U be a subset of V. Assume that $A \subseteq V \backslash U$ and that $V_p = V_q$ for any $p, q \in A$. Then there exists a positive constant c such that*

$$\max_{p \in A} u(p) \leq c \min_{p \in A} u(p)$$

for any non-negative $u \in \ell(V)$ with $(Hu)|_{V \backslash U} = 0$.

Proof. Let $V' = V_p$ for some $p \in A$. (Note that V' is independent of the choice of $p \in A$.) Set $\mathcal{A} = \{u : (Hu)_{V \backslash U} = 0, \min_{p \in V} u(p) \geq 0, \max_{p \in V'} u(p) = 1\}$. By Proposition 2.1.7, $\min_{p \in A} u(p) > 0$ for $u \in \mathcal{A}$. If $\mathcal{A}_0 = \{u|_{V'} : u \in \mathcal{A}\}$, then $\mathcal{A}_0$ is a compact subset of $\ell(V')$. Therefore, $c = \inf\{\min_{p \in A} u(p) : u \in \mathcal{A}_0\} > 0$. By the definition of $\mathcal{A}_0$, it follows that $c = \inf\{\min_{p \in A} u(p) : u \in \mathcal{A}\}$. This immediately implies the Harnack inequality.   □

Next, we define effective resistances associated with a Laplacian or, equivalently, a Dirichlet form. From the viewpoint of electrical circuits, the effective resistance between two terminals is the actual resistance considering all the resistors in the circuit.

Definition 2.1.9 (Effective resistance). Let V be a finite set and let $H \in \widetilde{\mathcal{LA}}(V)$. For $p \neq q \in V$, we define

$$R_H(p,q) = \left(\min\{\mathcal{E}_H(u,u) : u \in \ell(V), u(p) = 1, u(q) = 0\}\right)^{-1}. \quad (2.1.3)$$

Also we define $R_H(p,p) = 0$ for all $p \in V$. $R_H(p,q)$ is called the effective resistance between p and q with respect to H.

By Theorem 2.1.6, if $U = \{p,q\}$, then it follows that

$$[H]_U = \frac{1}{R_H(p,q)} \begin{pmatrix} -1 & 1 \\ 1 & -1 \end{pmatrix}. \quad (2.1.4)$$

Definition 2.1.10. Let V_i be a finite set and let $H_i \in \widetilde{\mathcal{LA}}(V_i)$ for $i = 1, 2$. We write $(V_1, H_1) \leq (V_2, H_2)$ if and only if $V_1 \subseteq V_2$ and $P_{V_2,V_1}(H_2) = H_1$.

The next proposition is obvious by the above definitions.

Proposition 2.1.11. *Let V_i be a finite set and let $H_i \in \widetilde{\mathcal{LA}}(V_i)$ for $i = 1, 2$. If $(V_1, H_1) \leq (V_2, H_2)$, then $R_{H_1}(p, q) = R_{H_2}(p, q)$ for any $p, q \in V_1$.*

In fact, the converse of the above proposition is also true if both H_1 and H_2 satisfy (L3). This fact is a corollary of the following theorem, which says that a Laplacian is completely determined by its associated effective resistances.

Theorem 2.1.12. *Let V be a finite set. Suppose $H_1, H_2 \in \mathcal{LA}(V)$. Then $H_1 = H_2$ if and only if $R_{H_1}(p, q) = R_{H_2}(p, q)$ for any $p, q \in V$.*

Proof. We need to show the "if" part. We use an induction on $\#(V)$. When $\#(V) = 2$, the theorem follows immediately by (2.1.4). Now suppose that the statement holds if $\#(V) < n$. Let $V = \{p_1, p_2, \ldots, p_n\}$. We write $h_{ij} = (H_1)_{p_i p_j}$ and $H_{ij} = (H_2)_{p_i p_j}$. Also let $V_i = V \backslash \{p_i\}$ and let

$$D_k^i = T_k^i - {}^t J_k^i (X_k^i)^{-1} J_k^i$$

for $k = 1, 2$, where $T_k^i : \ell(V_i) \to \ell(V_i), J_k^i : \ell(V_i) \to \ell(\{p_i\})$ and $X_k^i : \ell(\{p_i\}) \to \ell(\{p_i\})$ are defined by

$$H_k = \begin{pmatrix} T_k^i & {}^t J_k^i \\ J_k^i & X_k^i \end{pmatrix}.$$

Since $(V_i, D_k^i) \leq (V, H_k)$, $R_{D_1^i}(p, q) = R_{D_2^i}(p, q)$ for all $p, q \in V_i$. By the induction hypothesis, $D_1^i = D_2^i$. Now define $D^i = D_1^i = D_2^i$ and $d_{kl}^i = (D^i)_{p_k p_l}$. Calculating directly and then using the fact that $h_{ik} = h_{ki}$ and $H_{ik} = H_{ki}$, we obtain

$$d_{kl}^i = h_{kl} - h_{ik}h_{il}/h_{ii} = H_{kl} - H_{ik}H_{il}/H_{ii}.$$

In particular,

$$d_{kk}^i = h_{kk} - h_{ik}^2/h_{ii} = H_{kk} - H_{ik}^2/H_{ii}. \tag{2.1.5}$$

Exchanging k and i, we see that $d_{ii}^k/d_{kk}^i = h_{ii}/h_{kk} = H_{ii}/H_{kk}$. Therefore, there exists $t > 0$ such that $H_{ii} = t\,h_{ii}$ for $i = 1, 2, \ldots, N$. Again, by (2.1.5), we have

$$(h_{ik})^2 = h_{kk}h_{ii} - d_{kk}^i h_{ii}$$

and

$$(H_{ik})^2 = H_{kk}H_{ii} - d^i_{kk}H_{ii} = t^2 h_{kk}h_{ii} - t d^i_{kk}h_{ii}.$$

As $-h_{kk} = \sum_{i:i\neq k} h_{ik}$ and $-th_{kk} = -H_{kk} = \sum_{i:i\neq k} H_{ik}$, we obtain

$$-h_{kk} = \sum_{i:i\neq k} \sqrt{h_{kk}h_{ii} - d^i_{kk}h_{ii}} = \sum_{i:i\neq k} \sqrt{h_{kk}h_{ii} - d^i_{kk}h_{ii}/t}.$$

As a function of t, the right-hand side of the above equation is monotonically increasing. Hence the above equality holds only for $t = 1$. Therefore we obtain $H_1 = H_2$. $\qquad\square$

Corollary 2.1.13. *Let V_i be a finite set and let $H_i \in \mathcal{LA}(V_i)$ for $i = 1, 2$. Then $(V_1, H_1) \leq (V_2, H_2)$ if and only if $R_{H_1}(p,q) = R_{H_2}(p,q)$ for any $p, q \in V_1$.*

Remark. It is reasonable to expect that Theorem 2.1.12 remains true even if we only assume $H_1, H_2 \in \widetilde{\mathcal{LA}}(V)$. However, the above proof cannot be extended to such a case, because it uses the fact that $H_{pq} \geq 0$. Unfortunately, we don't know whether such an extension is true or not.

One reason why effective resistance is important is that it is a metric on V if $H \in \mathcal{LA}(V)$. This metric, called the effective resistance metric, will play a crucial role in the theory of Laplacians and Dirichlet forms on (post critically finite) self-similar sets.

Theorem 2.1.14. *Let V be a finite set and let $H \in \mathcal{LA}(V)$. Then $R_H(\cdot,\cdot)$ is a metric on V. This metric R_H is called the effective resistance metric on V associated with H.*

Remark. Not every metric on a finite set V corresponds to an effective resistance metric with respect to a Laplacian $H \in \mathcal{LA}(V)$. See Exercise 2.1 and Exercise 2.2.

We need the following well-known formula about electrical networks to show Theorem 2.1.14.

Lemma 2.1.15 (Δ–Y transform). *Let $U = \{p_1, p_2, p_3\}$ and let $V = \{p_0\} \cup U$. Set $R_{ij} = H^{-1}_{p_ip_j}$ for $H \in \mathcal{LA}(U)$, where we assume that $H_{p_ip_j} > 0$. Define*

$$R_1 = \frac{R_{12}R_{31}}{R_{12} + R_{23} + R_{31}}, R_2 = \frac{R_{23}R_{12}}{R_{12} + R_{23} + R_{31}}, R_3 = \frac{R_{31}R_{23}}{R_{12} + R_{23} + R_{31}}.$$

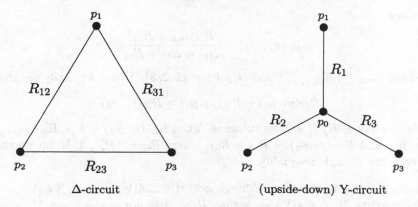

Fig. 2.1. Δ–Y transform

If $H' \in \mathcal{LA}(V)$ is defined by

$$H'_{p_i p_j} = \begin{cases} R_j^{-1} & \text{if } i = 0, \\ 0 & \text{othewise,} \end{cases}$$

for $i < j$, then $[H']_U = H$.

A direct calculation shows this formula.

As we mentioned before, we can associate an actual electrical circuit with a Laplacian. In the above lemma, the circuit associated with $H \in \mathcal{LA}(U)$ has three terminals $\{p_1, p_2, p_3\}$ and the terminals p_i and p_j are connected by a resistor of resistance R_{ij}. Let us call this circuit a Δ-circuit, which reflects the triangular shape of the circuit. At the same time, the circuit associated with $H' \in \mathcal{LA}(V)$ consists of four terminals $\{p_0, p_1, p_2, p_3\}$ and each terminal p_i is connected only to p_0 by a resistor of resistance R_i for $i = 1, 2, 3$. p_0 is a focal point of the circuit. Let us call this circuit a Y-circuit because of its "upside-down Y" shape. The Δ–Y transform says that the Δ-circuit and the Y-circuit are equivalent to each other as electrical networks. See Figure 2.1.

Proof of Theorem 2.1.14. Definition 2.1.9 and (2.1.4) imply that $R_H(p, q) \geq 0$ and that $R_H(p, q) = 0$ if and only if $p = q$. Next we must show the triangle inequality. We may assume that $\#(V) \geq 3$. For $U = \{p_1, p_2, p_3\} \subset V$, let $H' = [H]_U$. By Proposition 2.1.11, we have $R_{H'}(p_i, p_j) = R_H(p_i, p_j)$.

First assume that $H'_{p_m p_n} > 0$ for any $m \neq n$. Then the Δ–Y transform shows

$$R_{H'}(p_i, p_j) = \frac{R_{ij}(R_{ik} + R_{kj})}{R_{12} + R_{23} + R_{31}}, \qquad (2.1.6)$$

where $R_{mn} = (H'_{p_m p_n})^{-1}$ and $\{i, j, k\} = \{1, 2, 3\}$. Hence we easily see that

$$R_H(p_1, p_2) + R_H(p_2, p_3) \geq R_H(p_1, p_3).$$

Next, suppose $H'_{p_1 p_3} = 0$ for instance. Then $R_{H'}(p_1, p_2) = R_{12}, R_{H'}(p_2, p_3) = R_{23}$ and $R_{H'}(p_1, p_3) = R_{12} + R_{23}$, where $R_{ij} = (H'_{p_i p_j})^{-1}$. So we can verify the triangle inequality. □

$R_H(\cdot, \cdot)$ is not a metric on V for general $H \in \widetilde{\mathcal{LA}}(V)$. In fact, if $\#(V) > 3$, there exists $H \notin \mathcal{LA}(V)$ such that $R_H(\cdot, \cdot)$ is not a metric on V. (See Exercise 2.4 and Exercise 2.5.) As we will see, however, $\sqrt{R_H(\cdot, \cdot)}$ is always a metric on V for all $H \in \widetilde{\mathcal{LA}}(V)$.

The following is an alternative expression for the effective resistance.

Proposition 2.1.16. *Let V be a finite set and let $H \in \widetilde{\mathcal{LA}}(V)$. Then, for any $p, q \in V$,*

$$R_H(p, q) = \max\{\frac{|u(p) - u(q)|^2}{\mathcal{E}_H(u, u)} : u \in \ell(V), \mathcal{E}_H(u, u) \neq 0)\}. \qquad (2.1.7)$$

Proof. Note that $|u(p) - u(q)|^2 / \mathcal{E}_H(u, u) = |v(p) - v(q)|^2 / \mathcal{E}_H(v, v)$ if $v = \alpha u + \beta$ for any $\alpha, \beta \in \mathbb{R}$ with $\alpha \neq 0$. For given $u \in \ell(V)$ with $u(p) \neq u(q)$, there exist α and β such that $v(p) = 1$ and $v(q) = 0$, where $v = \alpha u + \beta$. Hence the right-hand side of (2.1.7) equals

$$\max\{\frac{1}{\mathcal{E}_H(v, v)} : v \in \ell(V), v(p) = 1, v(q) = 0\}.$$

Now (2.1.3) immediately implies (2.1.7). □

Applying (2.1.7), we can obtain an inequality between $|u(p) - u(q)|$, $R_H(p, q)$ and $\mathcal{E}_H(u, u)$.

Corollary 2.1.17. *Let V be a finite set and let $H \in \widetilde{\mathcal{LA}}(V)$. For any $p, q \in V$ and any $u \in \ell(V)$,*

$$|u(p) - u(q)|^2 \leq R_H(p, q)\mathcal{E}_H(u, u). \qquad (2.1.8)$$

This estimate will play an important role when we discuss the limit of a sequence of r-networks in the following sections.

As another application of Proposition 2.1.16, we can show that $\sqrt{R_H(\cdot, \cdot)}$ is a metric on V.

Theorem 2.1.18. *Let V be a finite set and let $H \in \widetilde{\mathcal{LA}}(V)$. Set $R_H^{1/2}(p,q)$ $= \sqrt{R_H(p,q)}$. Then $R_H^{1/2}(\cdot,\cdot)$ is a metric on V.*

Proof. We only need to show the triangle inequality. By (2.1.7),

$$R_H^{1/2}(p,q) = \max\{\frac{|u(p) - u(q)|}{\sqrt{\mathcal{E}_H(u,u)}} : u \in \ell(V), \mathcal{E}_H(u,u) \neq 0\}.$$

This immediately implies the triangle inequality for $R_H^{1/2}(\cdot,\cdot)$. $\qquad\square$

2.2 Sequence of discrete Laplacians

In this section, we will discuss the limit of r-networks on an increasing sequence of finite sets that satisfy a certain compatible condition, namely:

Definition 2.2.1. Let V_m be a finite set and let $H_m \in \widetilde{\mathcal{LA}}(V)$ for each $m \geq 0$. $\{(V_m, H_m)\}_{m \geq 0}$ is called a *compatible sequence* if $(V_m, H_m) \leq (V_{m+1}, H_{m+1})$ for all $m \geq 0$. Let $\mathcal{S} = \{(V_m, H_m)\}_{m \geq 0}$ be a compatible sequence. Set $V_* = \cup_{m \geq 0} V_m$ and define

$$\mathcal{F}(\mathcal{S}) = \{u : u \in \ell(V_*), \lim_{m \to \infty} \mathcal{E}_{H_m}(u|_{V_m}, u|_{V_m}) < +\infty\}, \qquad (2.2.1)$$

$$\mathcal{E}_{\mathcal{S}}(u,v) = \lim_{m \to \infty} \mathcal{E}_{H_m}(u|_{V_m}, v|_{V_m}), \qquad (2.2.2)$$

for $u, v \in \mathcal{F}(\mathcal{S})$. Also, for $p, q \in V_*$, define the effective resistance associated with $\mathcal{S}$ by

$$R_{\mathcal{S}}(p,q) = R_{H_m}(p,q), \qquad (2.2.3)$$

where m is chosen so that $p, q \in V_m$.

In the next chapter, we will approximate a self-similar set by a sequence of increasing finite sets. Then we will construct Dirichlet forms and Laplacians on the self-similar set by taking a limit of a compatible sequence of r-networks.

Throughout this section, $\mathcal{S} = \{(V_m, H_m)\}_{m \geq 0}$ is assumed to be a compatible sequence.

Let us regard V_m as a boundary of V_*. Then for any $u \in \ell(V_m)$, we consider a minimizing problem of $\mathcal{E}_{\mathcal{S}}(\cdot, \cdot)$ under the fixed boundary value u as follows.

Lemma 2.2.2. *There exists a linear map $h_m : \ell(V_m) \to \mathcal{F}(\mathcal{S})$ such that, for any $u \in \ell(V_m)$, $h_m(u)|_{V_m} = u$ and*

$$\mathcal{E}_{H_m}(u,u) = \mathcal{E}_{\mathcal{S}}(h_m(u), h_m(u)) = \min_{v \in \mathcal{F}(\mathcal{S}), v|_{V_m} = u} \mathcal{E}_{\mathcal{S}}(v,v). \qquad (2.2.4)$$

Moreover, if $v \in \mathcal{F}(\mathcal{S})$ with $v|_{V_m} = u$ attains the above minimum then $v = h_m(u)$.

Proof. As $[H_n]_{V_m} = H_m$ for $n > m$, we can apply Theorem 2.1.6 with $V = V_n, U = V_m$ and $H = H_n$. Set $h_{n,m} = h$ where h is the linear map $\ell(U) \to \ell(V)$ defined in Theorem 2.1.6. Then define $h_m(u)|_{V_n} = h_{n,m}(u)$. For any $n > m$, this definition is compatible and $h_m(u) \in \ell(V_*)$ is well-defined. By (2.1.2), we have

$$\mathcal{E}_{H_m}(u, u) = \mathcal{E}_{H_n}(h_m(u)|_{V_n}, h_m(u)|_{V_n})$$

for all $n > m$. Therefore $h_m(u) \in \mathcal{F}(\mathcal{S})$. Also (2.1.2) implies (2.2.4) immediately. $\qquad\square$

Let us fix m. Then $h_m(u)$ is also characterized as the unique solution of

$$\begin{cases} (H_n v_n)|_{V_n \setminus V_m} = 0 & \text{for all } n > m, \\ v|_{V_m} = u, \end{cases}$$

where $v \in \ell(V_*)$ and $v_n = v|_{V_n}$. So $h_m(u)$ may be thought of as the harmonic function with boundary values $u \in V_m$. If $H_m \in \mathcal{LA}(V_m)$ for all $m \geq 0$, we can show the following maximum principle for harmonic functions.

We will sometimes regard $\ell(V_m)$ as a subset of $\mathcal{F}(\mathcal{S})$ by identifying $\ell(V_m)$ with $h_m(\ell(V_m))$ through the injective map h_m. With this identification, one can write $\mathcal{E}_{H_m}(u, u) = \mathcal{E}_{\mathcal{S}}(u, u)$ for any $u \in \ell(V_m) \subset \mathcal{F}(\mathcal{S})$.

Lemma 2.2.3 (Maximum principle). *Assume $H_m \in \mathcal{LA}(V_m)$ for all $m \geq 0$. If $v \in \ell(V_*)$ satisfies $(H_n v_n)|_{V_n \setminus V_m} = 0$ for all $n > m$, where $v_n = v|_{V_n}$, then*

$$\min_{q \in V_m} v(q) \leq v(p) \leq \max_{q \in V_m} v(q)$$

for any $p \in V_*$.

Proof. This follows immediately by the maximum principle for harmonic functions on a finite set, Proposition 2.1.7. $\qquad\square$

Next we discuss the effective resistance $\mathcal{R}_{\mathcal{S}}(\cdot, \cdot)$. As in the case for finite sets, $\sqrt{R_{\mathcal{S}}(\cdot, \cdot)}$ is a metric on V_*.

Proposition 2.2.4. *If $R_{\mathcal{S}}^{1/2}(\cdot, \cdot) = \sqrt{R_{\mathcal{S}}(\cdot, \cdot)}$, then $R_{\mathcal{S}}^{1/2}$ is a metric on V_*. Moreover if $H_m \in \mathcal{LA}(V_m)$ for all $m \geq 0$, then $R_{\mathcal{S}}$ is a metric on V_*.*

Proof. This is an easy corollary of Theorem 2.1.14 and Theorem 2.1.18 along with Proposition 2.1.11. $\qquad\square$

The following lemma follows immediately from its counterpart, Proposition 2.1.16.

Lemma 2.2.5. *For any $p, q \in V_*$,*

$$R_S(p,q) = \left(\min\{\mathcal{E}_S(u,u) : u \in \mathcal{F}(S), u(p) = 1, u(q) = 0\}\right)^{-1}$$

$$= \max\left\{\frac{|u(p) - u(q)|^2}{\mathcal{E}_S(u,u)} : u \in \mathcal{F}(S), \mathcal{E}_S(u,u) > 0\right\}. \quad (2.2.5)$$

This lemma implies that

$$|u(p) - u(q)|^2 \leq R_S(p,q)\mathcal{E}_S(u,u) \quad (2.2.6)$$

for any $u \in \mathcal{F}(S)$ and $p, q \in V_*$. By (2.2.6), $\mathcal{F}(S) \subset C(V_*, R_S^{1/2})$. For a metric space (X, d), $C(X, d)$ is the collection of real-valued functions on X that are uniformly continuous on (X, d) and bounded on every bounded subset of (X, d).

Next we present important results on the limit of compatible sequences. In the following chapter, the results will be applied in constructing Dirichlet forms and Laplacians on a self-similar set.

Theorem 2.2.6.
(1) $\mathcal{F}(S) \subset C(V_*, R_S^{1/2})$.
(2) $\mathcal{E}_S$ is a non-negative symmetric form on $\mathcal{F}(S)$. Moreover, $\mathcal{E}_S(u,u) = 0$ if and only if u is a constant on V_*.
(3) Define an equivalence relation $\sim$ on $\mathcal{F}(S)$ by letting $u \sim v$ if and only if $u - v$ is constant on V_*. Then $\mathcal{E}_S$ is a naturally defined positive definite symmetric form on $\mathcal{F}(S)/\sim$ and $(\mathcal{F}(S)/\sim, \mathcal{E}_S)$ is a Hilbert space.
(4) Assume that $H_m \in \mathcal{LA}(V)$ for all $m \geq 0$. If $\bar{u}$ is defined as in (DF3) for any $u \in \mathcal{F}(S)$, then $\bar{u} \in \mathcal{F}(S)$ and $\mathcal{E}_S(\bar{u}, \bar{u}) \leq \mathcal{E}_S(u, u)$.

Proof. Every statement but (3) follows easily from the results and discussions in this and the previous section. We will use $\mathcal{E}$ and $\mathcal{F}$ in place of $\mathcal{E}_S$ and $\mathcal{F}(S)$ respectively. To show (3), first note that $\mathcal{E}(u, u) = \mathcal{E}(v, v)$ if $u \sim v$. Hence $\mathcal{E}$ is a well-defined positive definite symmetric form on $\mathcal{F}/\sim$. Choose any $p \in V_*$ and set $\mathcal{F}_p = \{u : u \in \mathcal{F}, u(p) = 0\}$. Then $(\mathcal{F}/\sim, \mathcal{E})$ is naturally isomorphic to $(\mathcal{F}_p, \mathcal{E})$. Hence it suffices to show that $(\mathcal{F}_p, \mathcal{E})$ is a Hilbert space. Now let $\{v_n\}_{n \geq 0}$ be a Cauchy sequence in $(\mathcal{F}_p, \mathcal{E})$ and let $v_n^m = h_m(v_n|_{V_m})$. Then by Lemma 2.2.2

$$\mathcal{E}(v_k^m - v_l^m, v_k^m - v_l^m) \leq \mathcal{E}(v_k - v_l, v_k - v_l).$$

Note that $p \in V_m$ for sufficiently large m. Hence $\mathcal{E}$ is an inner product on $\mathcal{F}_p \cap \ell(V_m)$, where $\ell(V_m)$ is identified with $h_m(\ell(V_m))$. So there exists

$v^m \in \mathcal{F}_p \cap \ell(V_m)$ such that $v_n^m \to v^m$ as $n \to \infty$. As $v^{m+1}|_{V_m} = v^m$, there exists $v \in \ell(V_*)$ such that $v|_{V_m} = v^m$.

On the other hand, let $C = \sup_{n \geq 0} \mathcal{E}(v_n, v_n)$. Then we have $\mathcal{E}(v^m, v^m) \leq \sup_{n,m} \mathcal{E}(v_n^m, v_n^m) = C$. Hence $v \in \mathcal{F}$.

Now, we fix $\epsilon > 0$. Then, we can choose n so that $\mathcal{E}(v_n - v_k, v_n - v_k) < \epsilon$ for all $k > n$. Also, we can choose m so that

$$|\mathcal{E}(v_n - v, v_n - v) - \mathcal{E}(v_n^m - v^m, v_n^m - v^m)| < \epsilon.$$

Furthermore, we can choose k so that $k > n$ and

$$|\mathcal{E}(v_n^m - v_k^m, v_n^m - v_k^m) - \mathcal{E}(v_n^m - v^m, v_n^m - v^m)| < \epsilon.$$

As $\mathcal{E}(v_n^m - v_k^m, v_n^m - v_k^m) \leq \mathcal{E}(v_n - v_k, v_n - v_k) < \epsilon$, we have $\mathcal{E}(v_n - v, v_n - v) < 3\epsilon$. Thus we have completed the proof of (3). $\qquad\square$

Finally we show two examples. The first one is related to one of the most basic examples in probability theory.

Example 2.2.7 (Simple random walk on $\mathbb{Z}$). Let $V_m = \{i \in \mathbb{Z} : |i| \leq m\}$ and let $H_m \in \mathcal{LA}(V_m)$ be defined by $(H_m)_{ij} = 1$ if $|i - j| = 1$, $(H_m)_{ij} = 0$ if $|i - j| > 1$. Then $\mathcal{S} = \{(V_m, H_m)\}_{m \geq 1}$ is a compatible sequence. We can easily see that $V_* = \mathbb{Z}$, $R_\mathcal{S}(i, j) = |i - j|$ and

$$\mathcal{E}_\mathcal{S}(u, v) = \sum_{i \in \mathbb{Z}} (u(i + 1) - u(i))(v(i + 1) - v(i)).$$

Also we can see that $(\mathcal{E}_\mathcal{S}, \mathcal{F}(\mathcal{S}) \cap L^2(\mathbb{Z}, \mu))$ is a regular Dirichlet form on $L^2(\mathbb{Z}, \mu)$ for every Borel measure μ on $\mathbb{Z}$ that satisfies $0 < \mu(\{i\}) < \infty$ for all $i \in \mathbb{Z}$. (See Definition B.3.2 for the definition of a regular Dirichlet form.)

Define a linear operator Δ_μ on $L^2(\mathbb{Z}, \mu)$ by

$$(\Delta_\mu u)(i) = \mu(i)^{-1}(u(i + 1) + u(i - 1) - 2u(i)).$$

Then Δ_μ is a non-positive self-adjoint operator on $L^2(\mathbb{Z}, \mu)$. Also $\mathrm{Dom}(\Delta_\mu) \subset \mathcal{F}(\mathcal{S}) \cap L^2(\mathbb{Z}, \mu)$ and

$$\mathcal{E}_\mathcal{S}(u, v) = -\int_\mathbb{Z} u \Delta_\mu v \, d\mu$$

for any $u, v \in \mathrm{Dom}(\Delta_\mu)$. From this fact, Δ_μ is identified as the self-adjoint operator associated with the closed form $(\mathcal{E}_\mathcal{S}, \mathcal{F}(\mathcal{S}) \cap L^2(\mathbb{Z}, \mu))$ on $L^2(\mathbb{Z}, \mu)$. (See B.1 for more information about closed forms and their associated self-adjoint operators.)

Now if $\nu(i) = 1$ for all $i \in \mathbb{Z}$, then Δ_ν is the self-adjoint operator associated with a simple random walk on $\mathbb{Z}$ in the following sense. Let $u_0 \in \text{Dom}(\Delta_\nu)$ and think about the following evolution equation with discrete time $n = 0, 1, 2, \dots$:

$$u_{n+1} - u_n = \Delta_\nu u/2.$$

One can easily see that $u_n = (I + \Delta_\nu/2)^n u_0$ for any n. For $i \in \mathbb{Z}$, if $u_0(i) = 1$ and $u_0(k) = 0$ for any $k \neq i$, then $u_n(j)$ is the transition probability from i at time 0 to j at time n for the simple random walk on $\mathbb{Z}$.

The next example is an extreme case where $R_{\mathcal{S}}$ is the trivial metric on V_*.

Example 2.2.8 (Discrete topology). Let $V_m = \{1, 2, \dots, m\}$ and let $H_m \in \mathcal{LA}(V_m)$ be defined by $(H_m)_{ij} = 2/m$ for $i \neq j$. Then $\mathcal{S} = \{(V_m, H_m)\}_{m \geq 2}$ is a compatible sequence. We can easily verify that $V_* = \mathbb{N}$ and $R_{\mathcal{S}}(i, j) = 1$ for $i \neq j$. This metric $R_{\mathcal{S}}$ induces the discrete topology on $\mathbb{N}$. As $\chi_i \in \mathcal{F}(\mathcal{S})$ for all $i \in \mathbb{N}$, $(\mathcal{E}_{\mathcal{S}}, \mathcal{F}(\mathcal{S}))$ is a regular Dirichlet form on $L^2(\mathbb{N}, \mu)$ for every Borel measure μ on $\mathbb{N}$ that satisfies $0 < \mu(\{i\}) < \infty$ for all $i \in \mathbb{N}$. In particular, if $\mu(\{i\}) = 1$ for all $i \in \mathbb{N}$, then $L^2(\mathbb{N}, \mu) = \ell^2(\mathbb{N})$. We see that $\ell^2(\mathbb{N}) \cap \mathcal{F}(\mathcal{S}) = \ell^2(\mathbb{N})$ and, for all $u, v \in \ell^2(\mathbb{N})$,

$$\mathcal{E}_{\mathcal{S}}(u, v) = 2 \int_{\mathbb{N}} uv \, d\mu.$$

2.3 Resistance form and resistance metric

In the previous section, we constructed a quadratic form $(\mathcal{E}_{\mathcal{S}}, \mathcal{F}(\mathcal{S}))$ and a metric $R_{\mathcal{S}}$ from a compatible sequence of r-networks $\mathcal{S} = \{(V_m, H_m)\}_{m \geq 0}$. In this section, we will give characterizations of the form $(\mathcal{E}_{\mathcal{S}}, \mathcal{F}(\mathcal{S}))$ and the metric $R_{\mathcal{S}}$ and show that there is an one-to-one correspondence between such forms and metrics.

First we give a characterization of quadratic forms.

Definition 2.3.1 (Resistance form). Let X be a set. A pair $(\mathcal{E}, \mathcal{F})$ is called a resistance form on X if it satisfies the following conditions (RF1) through (RF5).

(RF1) $\mathcal{F}$ is a linear subspace of $\ell(X)$ containing constants and $\mathcal{E}$ is a nonnegative symmetric quadratic form on $\mathcal{F}$. $\mathcal{E}(u, u) = 0$ if and only if u is constant on X.

(RF2) Let $\sim$ be an equivalent relation on $\mathcal{F}$ defined by $u \sim v$ if and only if $u - v$ is constant on X. Then $(\mathcal{F}/\sim, \mathcal{E})$ is a Hilbert space.

(RF3) For any finite subset $V \subset X$ and for any $v \in \ell(V)$, there exists $u \in \mathcal{F}$ such that $u|_V = v$.

(RF4) For any $p, q \in X$,

$$\sup\{\frac{|u(p) - u(q)|^2}{\mathcal{E}(u, u)} : u \in \mathcal{F}, \mathcal{E}(u, u) > 0\}$$

is finite. The above supremum is denoted by $M(p, q)$.

(RF5) If $u \in \mathcal{F}$, then $\bar{u} \in \mathcal{F}$ and $\mathcal{E}(\bar{u}, \bar{u}) \leq \mathcal{E}(u, u)$, where $\bar{u}$ is defined in the same way as (DF3) in Definition 2.1.1.

We use $\mathcal{RF}(X)$ to denote the collection of resistance forms on X. Also we define

$$\widetilde{\mathcal{RF}}(X) = \{(\mathcal{E}, \mathcal{F}) : (\mathcal{E}, \mathcal{F}) \text{ satisfies the condtions (RF1) through (RF4)}\}.$$

The condition (RF5) is called the Markov property.

Let V be a finite set. Then $(\mathcal{E}, \ell(V)) \in \widetilde{\mathcal{RF}}(V)$ (or $(\mathcal{E}, \ell(V)) \in \mathcal{RF}(V)$) if and only if $\mathcal{E} \in \widetilde{\mathcal{DF}}(V)$ (or $\mathcal{E} \in \mathcal{DF}(V)$ respectively.) Also, immediately from Theorem 2.2.6, $(\mathcal{E}_S, \mathcal{F}(S))$ belongs to $\widetilde{\mathcal{RF}}(V_*)$ for any compatible sequence $S = \{(V_m, H_m)\}_{m \geq 0}$. Moreover, if $H_m \in \mathcal{LA}(V_m)$ for all m, then $(\mathcal{E}_S, \mathcal{F}(S))$ is a resistance form on V_*.

Next we consider a characterization of metrics.

Definition 2.3.2 (Resistance metric). Let X be a set. A function $R : X \times X \to [0, +\infty)$ is called a resistance metric on X if and only if, for any finite subset $V \subset X$, there exists $H_V \in \mathcal{LA}(V)$ such that $R|_{V \times V} = R_{H_V}$, where R_{H_V} is the effective resistance with respect to H_V. The collection of resistance metrics on X is denoted by $\mathcal{RM}(X)$. Also we define

$$\widetilde{\mathcal{RM}}(X) = \{R : X \times X \to \mathbb{R}_+ : \text{for any finite subset } V \subset X, \text{ there exists}$$
$$H_V \in \widetilde{\mathcal{LA}}(V) \text{ with } R|_{V \times V} = R_{H_V} \text{ and } H_{V_1} = [H_{V_2}]_{V_1} \text{ if } V_1 \subseteq V_2\}.$$

Remark. Recall that $[H_{V_2}]_{V_1} = P_{V_2, V_1}(H_{V_2})$ by definition. Notice that by Corollary 2.1.13, the condition $H_{V_1} = [H_{V_2}]_{V_1}$ is satisfied for a resistance metric R. If we could extend Theorem 2.1.12 to $\widetilde{\mathcal{LA}}(V)$, which is quite likely, then we could remove the assumption $H_{V_1} = [H_{V_2}]_{V_1}$ from the definition of $\widetilde{\mathcal{RM}}(X)$.

Since R_{H_V} is a metric on V, a resistance metric R is a distance on X. Also, for $R \in \widetilde{\mathcal{RM}}(X)$, $\sqrt{R(\cdot, \cdot)}$ is a distance on X.

Let V be a finite set. Then $R \in \widetilde{\mathcal{RM}}(V)$ (or $R \in \mathcal{RM}(V)$) if and only if $R = R_H$ for some $H \in \widetilde{\mathcal{LA}}(V)$ (or $H \in \mathcal{LA}(V)$ respectively). Furthermore, it is natural to expect that R_S is a resistance metric for a compatible sequence S. More precisely, we have the following proposition.

Proposition 2.3.3. *If* $\mathcal{S} = \{(V_m, H_m)\}_{m \geq 0}$ *is a compatible sequence, then* $R_{\mathcal{S}} \in \widetilde{\mathcal{RM}}(V_*)$. *In particular, if* $H_m \in \mathcal{LA}(V)$ *for all* m, *then* $R_{\mathcal{S}}$ *is a resistance metric.*

Proof. Let V be a finite subset of V_*. Then $V \subseteq V_m$ for sufficiently large m. If $H_V = [H_m]_V$, then $R_{H_V} = R_{\mathcal{S}}|_{V \times V}$. The rest of the conditions are obvious. □

There is a natural one-to-one correspondence between resistance forms and resistance metrics. First we will construct a resistance metric from a resistance form.

Theorem 2.3.4. *If* $(\mathcal{E}, \mathcal{F}) \in \widetilde{\mathcal{RF}}(X)$, *then*

$$\min\{\mathcal{E}(u,u) : u \in \mathcal{F}, u(p) = 1, u(q) = 0\}$$

exists for any $p, q \in X$ *with* $p \neq q$. *If we define* $R(p,q)^{-1}$ *to be equal to the minimum value, then* $R \in \widetilde{\mathcal{RM}}(X)$ *and*

$$R(p,q) = \max\{\frac{|u(p) - u(q)|^2}{\mathcal{E}(u,u)} : u \in \mathcal{F}, \mathcal{E}(u,u) > 0\}. \tag{2.3.1}$$

Moreover, if $(\mathcal{E}, \mathcal{F}) \in \mathcal{RF}(X)$, *then* $R \in \mathcal{RM}(X)$.

To prove the above theorem, we need the following lemma.

Lemma 2.3.5. *If* $(\mathcal{E}, \mathcal{F}) \in \widetilde{\mathcal{RF}}(X)$ *and* V *is a finite subset of* X, *then there exists a linear map* $h_V : \ell(V) \to \mathcal{F}$ *such that* $h_V(u)|_V = u$ *and*

$$\mathcal{E}(h_V(u), h_V(u)) = \min_{v \in \mathcal{F}, v|_V = u} \mathcal{E}(v,v). \tag{2.3.2}$$

Furthermore, $h_V(u)$ *is the unique element that attains the above minimum. Also set* $\mathcal{E}^V(u_1, u_2) = \mathcal{E}(h_V(u_1), h_V(u_2))$. *Then* $\mathcal{E}^V \in \widetilde{\mathcal{DF}}(V)$. *Moreover, if* $(\mathcal{E}, \mathcal{F})$ *is a resistance form, then* $\mathcal{E}^V \in \mathcal{DF}(V)$.

Proof. For $p \in V$, let $\mathcal{F}^p = \{u : u \in \mathcal{F}, u|_{V \setminus \{p\}} = 0\}$. Then by (RF3), $\mathcal{F}^p$ is not trivial. By (RF2), $(\mathcal{F}^p, \mathcal{E})$ is a Hilbert space. Define $\Phi_p : \mathcal{F}^p \to \mathbb{R}$ by $\Phi_p(u) = u(p)$. Then by (RF4) we have, for $q \in V \setminus \{p\}$, $|\Phi_p(u)|^2 \leq M(p,q)\mathcal{E}(u,u)$ for all $u \in \mathcal{F}^p$. Hence Φ_p is a continuous linear functional on $(\mathcal{F}^p, \mathcal{E})$. Therefore there exists $g_p \in \mathcal{F}^p$ such that for all $u \in \mathcal{F}^p$, $\mathcal{E}(g_p, u) = \Phi_p(u) = u(p)$. Noting that $\mathcal{E}(g_p, g_p) = g_p(p) > 0$, we define $\psi_p^V = g_p/g_p(p)$.

Now for any $u \in \ell(V)$, define $h_V(u) = \sum_{p \in V} u(p)\psi_p^V$. If $v \in \mathcal{F}$ with

$v|_V = u$, set $\tilde{v} = v - h_V(u)$. Then

$$\mathcal{E}(v,v) = \mathcal{E}(\tilde{v} + h_V(u), \tilde{v} + h_V(u))$$
$$= \mathcal{E}(\tilde{v}, \tilde{v}) + 2\mathcal{E}(\tilde{v}, h_V(u)) + \mathcal{E}(h_V(u), h_V(u)).$$

As $\mathcal{E}(\tilde{v}, h_V(u)) = \sum_{p \in V} u(p)\tilde{v}(p)/g_p(p) = 0$, we have

$$\mathcal{E}(v,v) = \mathcal{E}(\tilde{v}, \tilde{v}) + \mathcal{E}(h_V(u), h_V(u)) \geq \mathcal{E}(h_V(u), h_V(u)).$$

Equality holds only when $\tilde{v}$ is constant on X and so $\tilde{v} \equiv 0$ on X. It is easy to see that $\mathcal{E}^V \in \widetilde{\mathcal{DF}}(V)$. Also the Markov property (RF5) of $(\mathcal{E}, \mathcal{F})$ implies the Markov property (DF3) of $\mathcal{E}^V$. Thus we have completed the proof. $\qquad\square$

Proof of Theorem 2.3.4. By Lemma 2.3.5,

$$\min\{\mathcal{E}(u,u) : u \in \mathcal{F}, u(p) = 1, u(q) = 0\}$$

exists for any $p, q \in X$ with $p \neq q$. Now define $H_V \in \widetilde{\mathcal{LA}}(V)$ by $\mathcal{E}^V = \mathcal{E}_{H_V}$. If $V_1 \subset V_2$, then

$$\mathcal{E}_{[H_{V_2}]_{V_1}}(u,u) = \min_{v \in \ell(V_2), v|_{V_1} = u} \mathcal{E}(h_{V_2}(v), h_{V_2}(v))$$
$$= \min_{v' \in \mathcal{F}, v'|_{V_1} = u} \mathcal{E}(v', v') = \mathcal{E}_{H_{V_1}}(u,u).$$

Hence $[H_{V_2}]_{V_1} = H_{V_1}$. This fact also implies that $R_{H_V} = R|_{V \times V}$. Therefore we see that $R \in \widetilde{\mathcal{RM}}(X)$. If $(\mathcal{E}, \mathcal{F}) \in \mathcal{RF}(X)$, then $H_V \in \mathcal{DF}(V)$ and hence $R \in \mathcal{RM}(X)$. The same argument as in the proof of Proposition 2.1.16 implies (2.3.1). $\qquad\square$

Theorem 2.3.4 says that each $(\mathcal{E}, \mathcal{F}) \in \widetilde{\mathcal{RF}}(X)$ is associated with $R \in \widetilde{\mathcal{RM}}(X)$. So we can define a map $FM_X : \widetilde{\mathcal{RF}}(X) \to \widetilde{\mathcal{RM}}(X)$, which is called the "form to metric" map, by $R = FM_X((\mathcal{E}, \mathcal{F}))$. This form to metric map is, in fact, bijective: we can construct the inverse of FM_X.

Theorem 2.3.6. *For $R \in \widetilde{\mathcal{RM}}(X)$, there exists a unique $(\mathcal{E}, \mathcal{F}) \in \widetilde{\mathcal{RF}}(X)$ that satisfies (2.3.1). Moreover, if $R \in \mathcal{RM}(X)$, then $(\mathcal{E}, \mathcal{F}) \in \mathcal{RF}(X)$.*

Assuming the above theorem, we can define the "metric to form" map $MF_X : \widetilde{\mathcal{RM}}(X) \to \widetilde{\mathcal{RF}}(X)$. It is easy to see that MF_X is the inverse of FM_X.

We will only present the proof of a special case of Theorem 2.3.6, namely Theorem 2.3.7. Theorem 2.3.6 can be proven by using routine and tedious arguments about limiting procedures similar to the special case.

If $R \in \widetilde{\mathcal{RM}}(X)$, $R^{1/2}(\cdot, \cdot) = \sqrt{R(\cdot, \cdot)}$ is a metric on X. Assume that the metric space $(X, R^{1/2})$ is separable. Equivalently, there exists a family of finite subsets $\{V_m\}_{m \geq 0}$ of X that satisfy $V_m \subset V_{m+1}$ for $m \geq 0$ and $V_* = \cup_{m \geq 0} V_m$ is dense in X. Set $H_m = H_{V_m}$. Then $H_m \in \widetilde{\mathcal{LA}}(V_m)$ and $[H_{m+1}]_{V_m} = H_m$ by definition. Hence $(V_m, H_m) \leq (V_{m+1}, H_{m+1})$ and so $\mathcal{S} = \{(V_m, H_m)\}_{m \geq 0}$ is a compatible sequence. We know that $(\mathcal{E}_{\mathcal{S}}, \mathcal{F}(\mathcal{S})) \in \widetilde{\mathcal{RF}}(V_*)$. Also it is obvious that $R = R_{\mathcal{S}}$ on V_*. Now as $\mathcal{F}(\mathcal{S}) \in C(V_*, R_{\mathcal{S}}^{1/2})$, $u \in \mathcal{F}(\mathcal{S})$ has a natural extension to a function in $C(X, R^{1/2})$. We will think of $\mathcal{F}(\mathcal{S})$ as a subset of $C(X, R^{1/2})$ in this way. Then it is easy to see that $(\mathcal{E}_{\mathcal{S}}, \mathcal{F}(\mathcal{S}))$ satisfies (RF1) and (RF2). This $(\mathcal{E}_{\mathcal{S}}, \mathcal{F}(\mathcal{S}))$ is the candidate for $(\mathcal{E}, \mathcal{F})$ in Theorem 2.3.6. The problem is to show (RF3) and (2.3.1) for any $p, q \in X$. (We already know that (2.3.1) holds for $p, q \in V_*$ by Lemma 2.2.5.) We do this in the next theorem.

Theorem 2.3.7. *For $R \in \widetilde{\mathcal{RM}}(X)$, assume that $(X, R^{1/2})$ is separable. Let $\{V_m\}_{m \geq 0}$ be a family of finite subsets of X such that $V_m \subset V_{m+1}$ for any $m \geq 0$ and that $V_* = \cup_{m \geq 0} V_m$ is dense in X. Set $\mathcal{S} = \{(V_m, H_m)\}_{m \geq 0}$ where $H_m = H_{V_m}$. Then $(\mathcal{E}_{\mathcal{S}}, \mathcal{F}(\mathcal{S})) \in \widetilde{\mathcal{RF}}(X)$ and*

$$R(p, q) = \max\{\frac{|u(p) - u(q)|^2}{\mathcal{E}_{\mathcal{S}}(u, u)} : u \in \mathcal{F}(\mathcal{S}), \mathcal{E}_{\mathcal{S}}(u, u) > 0\} \qquad (2.3.3)$$

for all $p, q \in X$. Moreover, $(\mathcal{E}_{\mathcal{S}}, \mathcal{F}(\mathcal{S}))$ is independent of the choice of $\{V_m\}_{m \geq 0}$. Also if $R \in \mathcal{RM}(X)$, then $(\mathcal{E}_{\mathcal{S}}, \mathcal{F}(\mathcal{S})) \in \mathcal{RF}(X)$.

Before proving the theorem, we need two lemmas.

Lemma 2.3.8. *Let $(\mathcal{E}, \mathcal{F}) \in \widetilde{\mathcal{RF}}(X)$ and let $\{V_m\}$ be a sequence of finite subsets of X such that $V_m \subset V_{m+1}$ for $m \geq 0$ and that $V_* = \cup_{m \geq 0} V_m$ is dense in $(X, R^{1/2})$, where $R = FM_X((\mathcal{E}, \mathcal{F}))$. If $\mathcal{S} = \{(V_m, H_m)\}_{m \geq 0}$ where $H_m = H_{V_m}$, then $(\mathcal{E}_{\mathcal{S}}, \mathcal{F}(\mathcal{S})) = (\mathcal{E}, \mathcal{F})$.*

Remark. In this lemma, again we regard $\mathcal{F}(\mathcal{S})$ as a subset of $C(X, R^{1/2})$ because $R = R_{\mathcal{S}}$ on V_* and $\mathcal{F}(\mathcal{S}) \subset C(V_*, R_{\mathcal{S}}^{1/2})$.

Proof. First we show that $\mathcal{F}(\mathcal{S}) \subseteq \mathcal{F}$ and $\mathcal{E}_{\mathcal{S}}(u, u) = \mathcal{E}(u, u)$ for $u \in \mathcal{F}(\mathcal{S})$. Let $u \in \mathcal{F}(\mathcal{S})$. Set $u_m = h_{V_m}(u|_{V_m})$, where h_{V_m} is defined in Lemma 2.3.5. As $\mathcal{E}_{H_m}(u|_{V_m}, u|_{V_m}) = \mathcal{E}(u_m, u_m)$, we obtain $\mathcal{E}(u_m, u_m) \leq \mathcal{E}(u_{m+1}, u_{m+1}) \leq \mathcal{E}_{\mathcal{S}}(u, u)$. Now without loss of generality we may assume that $u(p) = 0$ for some $p \in V_0$. (We can just replace u by $u - u(p)$.) Note that $(u_m - u_n)|_{V_n} = 0$ for $m \geq n$. Then, recalling the definition of h_V in the proof of Lemma 2.3.5, it follows that

$$\mathcal{E}(u_m - u_n, u_n) = \sum_{p \in V_n} (u_m(p) - u_n(p))u(p)/g_p(p) = 0$$

for $m \geq n$. Hence $\mathcal{E}(u_m - u_n, u_m - u_n) = \mathcal{E}(u_m, u_m) - \mathcal{E}(u_n, u_n) \to 0$ as $m, n \to \infty$. Therefore $\{u_m\}_{m \geq 0}$ is a Cauchy sequence in $(\mathcal{F}_p, \mathcal{E})$. As $(\mathcal{F}_p, \mathcal{E})$ is complete by (RF2), there exists $u_* \in \mathcal{F}_p$ such that $\mathcal{E}(u_* - u_m, u_* - u_m) \to 0$ as $m \to \infty$. So $\mathcal{E}(u_*, u_*) = \lim_{m \to \infty} \mathcal{E}(u_m, u_m) = \mathcal{E}_S(u, u)$. For $q \in V_*$, we have

$$|u_*(q) - u_m(q)|^2 \leq R(p, q)\mathcal{E}(u_* - u_m, u_* - u_m).$$

Letting $m \to \infty$, we obtain $u|_{V_*} = u_*|_{V_*}$. As u and u_* are continuous on X with respect to $R^{1/2}$, it follows that $u = u_*$. Thus we have shown that $u \in \mathcal{F}$ and $\mathcal{E}(u, u) = \mathcal{E}_S(u, u)$.

Secondly, for $u \in \mathcal{F}$, define u_m exactly the same as before. Then

$$\mathcal{E}(u_m, u_m) = \min_{v \in \mathcal{F}, v|_{V_m} = u|_{V_m}} \mathcal{E}(v, v) \leq \mathcal{E}(u, u).$$

Hence $u \in \mathcal{F}(S)$. Now using the argument of the latter half of this proof, we see that $\mathcal{E}_S(u, u) = \mathcal{E}(u, u)$. $\square$

Lemma 2.3.9. *Let $(\mathcal{E}, \mathcal{F}) \in \widetilde{RF}(Y)$ and let $(\overline{Y}, \overline{R}^{1/2})$ be the completion of $(Y, R^{1/2})$, where $R = FM_Y((\mathcal{E}, \mathcal{F}))$. Then, for any $p, q \in \overline{Y}$,*

$$\overline{R}(p, q) = \max\{\frac{|u(p) - u(q)|^2}{\mathcal{E}(u, u)} : u \in \mathcal{F}, \mathcal{E}(u, u) > 0\}.$$

Proof. First we will show that

$$\overline{R}(p, q) = \sup\{\frac{|u(p) - u(q)|^2}{\mathcal{E}(u, u)} : u \in \mathcal{F}, \mathcal{E}(u, u) > 0\}. \qquad (2.3.4)$$

We will denote the right-hand side of (2.3.4) by $M(p, q)$. Choose $\{p_n\}, \{q_n\} \subset Y$ so that $p_n \to p$ and $q_n \to q$ as $n \to \infty$. Note that,

$$R(p_n, q_n) = \max\{\frac{|u(p_n) - u(q_n)|^2}{\mathcal{E}(u, u)} : u \in \mathcal{F}, \mathcal{E}(u, u) > 0\}. \qquad (2.3.5)$$

Hence, we have $|u(p_n) - u(q_n)|^2 \leq R(p_n, q_n)\mathcal{E}(u, u)$ for any $u \in \mathcal{F}$. Letting $n \to \infty$, we obtain $|u(p) - u(q)|^2 \leq \overline{R}(p, q)\mathcal{E}(u, u)$. Hence $M(p, q) \leq \overline{R}(p, q)$.

Suppose $M(p, q) < \overline{R}(p, q)$. Then we can choose ϵ so that for all $u \in \mathcal{F}$,

$$|u(p) - u(q)| < (\sqrt{\overline{R}(p, q)} - 5\epsilon)\sqrt{\mathcal{E}(u, u)}.$$

On the other hand, since $R(p_n, q_n) \to \overline{R}(p, q)$ as $n \to \infty$, using (2.3.5), there exists $\{u_n\}$ such that $\mathcal{E}(u_n, u_n) = 1$ and $|u_n(p_n) - u_n(q_n)|^2 \to \overline{R}(p, q)$ as $n \to \infty$. For sufficiently large n, we have

$$|u_n(p_n) - u_n(q_n)| > (\sqrt{\overline{R}(p, q)} - \epsilon)$$

and $R(p_n, p_m), R(q_n, q_m) < \epsilon^2$ for all $m > n$. Furthermore, we can choose m so that $m > n$ and

$$|u_n(p_m) - u_n(p)| < \epsilon \quad \text{and} \quad |u_n(q_m) - u_n(q)| < \epsilon.$$

Now we have

$$|u_n(p_n) - u_n(q_n)| \leq |u_n(p_n) - u_n(p_m)| + |u_n(p_m) - u_n(p)|$$
$$+ |u_n(p) - u_n(q)| + |u_n(q) - u_n(q_m)| + |u_n(q_m) - u_n(q_n)|$$
$$\leq |u_n(p_n) - u_n(p_m)| + |u_n(q_n) - u_n(q_m)| + (\sqrt{R(p,q)} - 3\epsilon).$$

Hence $|u_n(p_n) - u_n(p_m)| \geq \epsilon$ or $|u_n(q_n) - u_n(q_m)| \geq \epsilon$. This contradicts the fact that $R(p_n, p_m), R(q_n, q_m) < \epsilon^2$. Therefore we have shown (2.3.4).

Now using the same argument as in the proof of Lemma 2.3.5, it follows that there exists $\psi \in \mathcal{F}$ such that $\psi(p) = 1$, $\psi(q) = 0$ and ψ attains the supremum in (2.3.4). $\qquad\square$

Proof of Theorem 2.3.7. As we mentioned before, $(\mathcal{E}_\mathcal{S}, \mathcal{F}(\mathcal{S}))$ (recall that we think of $\mathcal{F}(\mathcal{S})$ as a subset of $C(X, R^{1/2})$) satisfies (RF1) and (RF2). To show (RF3), set $V'_m = V_m \cup V$ for a finite set $V \subset X$. Let $H'_m = H_{V'_m}$ and let $\mathcal{S}' = \{(V'_m, H'_m)\}$. Then for any $u \in \ell(V)$, there exists $v \in \mathcal{F}(\mathcal{S}')$ such that $v|_V = u$. As $(V_m, H_m) \leq (V'_m, H'_m)$, $\mathcal{E}_{H_m}(v|_{V_m}, v|_{V_m}) \leq \mathcal{E}_{H'_m}(v|_{V'_m}, v|_{V'_m}) \leq \mathcal{E}_{\mathcal{S}'}(v, v)$. Hence $\lim_{m \to \infty} \mathcal{E}_{H_m}(v|_{V_m}, v|_{V_m}) \leq \mathcal{E}_{\mathcal{S}'}(v, v)$. Therefore $v \in \mathcal{F}(\mathcal{S})$. This shows (RF3).

Next, applying Lemma 2.3.9 to the case that $Y = V_*$, we obtain (2.3.3) because $X \subset \overline{Y}$. This implies (RF4). Thus we have shown $(\mathcal{E}_\mathcal{S}, \mathcal{F}(\mathcal{S})) \in \widetilde{\mathcal{RF}}(X)$. Furthermore, (2.3.3) also implies $R = FM_X((\mathcal{E}_\mathcal{S}, \mathcal{F}(\mathcal{S})))$.

Let $\{U_m\}$ be a sequence of finite subsets of X that satisfies the same condition as $\{V_m\}$ and let $\mathcal{S}_1$ be the compatible sequence associated with U_m. Then, applying Lemma 2.3.8, we can see that $(\mathcal{E}_\mathcal{S}, \mathcal{F}(\mathcal{S})) = (\mathcal{E}_{\mathcal{S}_1}, \mathcal{F}(\mathcal{S}_1))$. Hence $(\mathcal{E}_\mathcal{S}, \mathcal{F}(\mathcal{S}))$ is independent of the choice of $\{V_m\}$.

Finally, if $R \in \mathcal{RM}(X)$, then $H_{V_m} \in \mathcal{LA}(V_m)$. Hence $(\mathcal{E}_\mathcal{S}, \mathcal{F}(\mathcal{S}))$ has the Markov property. $\qquad\square$

Using the discussions in this section, we can show another important fact about resistance forms and resistance metrics. If $\mathcal{S} = \{(V_m, H_m)\}_{m \geq 0}$ is a compatible sequence, then $(\mathcal{E}_\mathcal{S}, \mathcal{F}(\mathcal{S})) \in \widetilde{\mathcal{RF}}(V_*)$ and $R_\mathcal{S} \in \widetilde{\mathcal{RM}}(V_*)$. The space V_* is merely a countable set. So if we constructed analytical objects like Laplacians or Dirichlet forms from $(\mathcal{E}_\mathcal{S}, \mathcal{F}(\mathcal{S}))$, we would end up with analysis on a countable set. That is hardly what we want! One way of overcoming this difficulty is to consider the completion of

V_* with respect to the metric $R_S^{1/2}$. Let $(\Omega_S, R_S^{1/2})$ be the completion of $(V_*, R_S^{1/2})$. Then $(\Omega_S, R_S^{1/2})$ might be an interesting uncountably infinite set. As we mentioned before, $\mathcal{F}(S)$ can be naturally regarded as a subset of $C(\Omega_S, R_S^{1/2})$. Hence $(\mathcal{E}_S, \mathcal{F}(S))$ can be considered as a quadratic form on $(\Omega_S, R_S^{1/2})$. There is, however, a delicate question about this completion procedure. Is the extended R_S in $\widetilde{\mathcal{RM}}(\Omega_S)$? Equivalently, do we have $(\mathcal{E}_S, \mathcal{F}(S)) \in \widetilde{\mathcal{RF}}(X)$? This is not a trivial problem. In fact, this is not true in general. See Exercise 2.7 for a counterexample. Fortunately, if we assume the Markov property, i.e., $H_m \in \mathcal{LA}(V_m)$ for all $m \geq 0$, then it follows that $R_S^{1/2} \in \mathcal{RM}(\Omega_S)$ and $(\mathcal{E}_S, \mathcal{F}(S)) \in \mathcal{RF}(\Omega_S)$ by virtue of the next theorem.

Theorem 2.3.10. *Let $(\mathcal{E}, \mathcal{F}) \in \mathcal{RF}(X)$. If $(\overline{X}, R)$ is the completion of (X, R), where $R = FM_X((\mathcal{E}, \mathcal{F}))$, then $(\mathcal{E}, \mathcal{F}) \in \mathcal{RF}(\overline{X})$ and $R \in \mathcal{RM}(\overline{X})$.*

Proof. (RF1), (RF2) and (RF5) follow immediately. Also (RF4) is an obvious consequence of Lemma 2.3.9. Instead of (RF3), we will show the following (RF3*).

(RF3*) For each finite subset $V \subset \overline{X}$ and for each $p \in V$, there exists $u \in \mathcal{F}$ such that $u|_V = \chi_p$, where χ_p is the characteristic function of the one point set $\{p\}$.

We use an induction on $\#(V)$ to prove (RF3*). If $\#(V) = 2$, say $V = \{p, q\}$, then by Lemma 2.3.9, there exists $u \in \mathcal{F}$ such that $u(p) \neq u(q)$. If we set $f = (u - u(q))/(u(p) - u(q))$, we have $f|_V = \chi_p$.

Next suppose (RF3*) holds for $\#(V) < n$. Let $V = \{p_1, p_2, \ldots, p_n\}$. Then by the induction hypothesis, there exists $u \in \mathcal{F}$ such that $u(p_1) = 1$ and $u(p_i) = 0$ for $i \geq 3$.

Case 1. If $u(p_2) < u(p_1)$, then for some $\alpha, \beta \in \mathbb{R}$, $v = \alpha u + \beta$ satisfies $v(p_1) = 1$ and $v(p_j) \leq 0$ for $j \geq 2$. Define $\bar{v}$ as in (DF3) of Definition 2.1.1. Then by the Markov property (RF5), we have $\bar{v} \in \mathcal{F}$. Obviously $\bar{v}|_V = \chi_{p_1}$.

Case 2. If $u(p_2) = u(p_1)$, then choose $f \in \mathcal{F}$ that satisfies $f(p_1) > f(p_2)$ and $|f(p_i)| < 1/2$ for all $i = 1, 2, \ldots, n$. For some $\alpha, \beta \in \mathbb{R}$, $v = \alpha(u+f)+\beta$ has the same properties as v in Case 1.

Case 3. If $u(p_1) < u(p_2)$, then using the same argument as in Case 1, we can find $v \in \mathcal{F}$ that satisfies $v|_V = \chi_{p_2}$. Thus if $f = u - u(p_2)v$, then $f|_V = \chi_{p_1}$.

Thus we have shown that (RF3*) holds for $\#(V) = n$. $\square$

2.4 Dirichlet forms and Laplacians on limits of networks

In the last section, we studied relations among a compatible sequence of r-networks, a resistance form and a resistance metric. In this section, we will take a first step to establish an "analysis" on limits of networks. In particular, we are interested in constructing a counterpart of the Laplacian defined as a differential operator in classical calculus. By the results in the last section, it is reasonable to start from a compatible sequence of r-networks $\mathcal{S} = \{(V_m, H_m)\}$. (We are not concerned with how to obtain a compatible sequence of r-networks in this section.) Then we obtain a resistance form $(\mathcal{E}, \mathcal{F})$ and a resistance metric R by taking a limit of $\mathcal{S}$. Naturally, the resistance form $(\mathcal{E}, \mathcal{F})$ and the resistance metric R are important elements in our "analysis". However, those are not enough. We need to introduce integration, that is, to introduce a measure on the space. The following is a general result concerning a resistance form, a resistance metric and a measure.

Theorem 2.4.1. *Let $R \in \widetilde{\mathcal{RM}}(X)$ and suppose that $(X, R^{1/2})$ is separable. Set $(\mathcal{E}, \mathcal{F}) = MF_X(R)$. Also let μ be a σ-finite Borel measure on $(X, R^{1/2})$. Define*

$$\mathcal{E}_1(u, v) = \mathcal{E}(\dot{u}, v) + \int_X u(x)v(x)\mu(dx)$$

for $u, v \in L^2(X, \mu) \cap \mathcal{F}$. Then $(L^2(X, \mu) \cap \mathcal{F}, \mathcal{E}_1)$ is a Hilbert space. Moreover, if $\mu(X) < \infty$ and $\int_X R(p, p_*)\mu(dp) < \infty$ for some $p_* \in X$, then the identity map from $L^2(X, \mu) \cap \mathcal{F}$ with $\mathcal{E}_1$-norm to $L^2(X, \mu)$ with L^2-norm is a compact operator.

Proof. Let $\{u_n\}_{n \geq 0}$ be a Cauchy sequence in $(L^2(X, \mu) \cap \mathcal{F}, \mathcal{E}_1)$ and let $v_n = u_n - u_n(p)$ for $p \in X$. Then by (RF2), there exists $v \in \mathcal{F}_p$ such that $\mathcal{E}(v_n - v, v_n - v) \to 0$ as $n \to \infty$. By (RF4), we have

$$|v_n(q) - v(q)|^2 \leq R(p, q)\mathcal{E}(v_n - v, v_n - v). \tag{2.4.1}$$

Since μ is σ-finite, there exists $\{K_m\}_{m \geq 0}$ such that $K_m \subset X$ is bounded, $0 < \mu(K_m) < \infty$ and $\cup_{m \geq 0} K_m = X$. By (2.4.1), we see that $v_n \to v$ as $n \to \infty$ in $L^2(K_m, \mu|_{K_m})$. Also $\{u_n|_{K_m}\}_{n \geq 0}$ is a Cauchy sequence in $L^2(K_m, \mu|_{K_m})$. As $u_n(p) = (u_n - v_n)|_{K_m}$, there exists $c \in \mathbb{R}$ such that $u_n(p) \to c$ as $n \to \infty$. If we let $u = v + c$, then $\mathcal{E}(u - u_n, u - u_n) \to 0$ as $n \to \infty$. Also, $u_n|_{K_m} \to u|_{K_m}$ as $n \to \infty$ in $L^2(K_m, \mu|_{K_m})$.

On the other hand, $\{u_n\}_{n \geq 0}$ is a Cauchy sequence in $L^2(X, \mu)$ and so there exists $u^* \in L^2(X, \mu)$ such that $u_n \to u^*$ as $n \to \infty$ in $L^2(X, \mu)$. As

$u^*|_{K_m} = u|_{K_m}$ in $L^2(K_m, \mu|_{K_m})$, $u = u^*$ in $L^2(X, \mu)$. Hence $u_n \to u$ as $n \to \infty$ in $(L^2(X, \mu) \cap \mathcal{F}, \mathcal{E}_1)$.

Now suppose $\mu(X) < \infty$. Let $\mathcal{U}$ be a bounded subset of $(L^2(X, \mu) \cap \mathcal{F}, \mathcal{E}_1)$. If $C = \sup_{u \in \mathcal{U}} \mathcal{E}_1(u, u)$, then

$$|u(p) - u(q)|^2 \le C R(p, q) \qquad (2.4.2)$$

for all $u \in \mathcal{U}$ and all $p, q \in X$. Let V be a countable dense subset of X. Note that $\{u(p)\}_{u \in \mathcal{U}}$ is bounded for any $p \in V$ by (2.4.2). Hence, by standard diagonal construction, we can find $v \in \ell(V)$ and $\{v_n\} \subset \mathcal{U}$ satisfying $v_n(p) \to v(p)$ as $n \to \infty$ for all $p \in V$. Using (2.4.2), we see that v satisfies (2.4.2) for $p, q \in V$. Therefore v extends naturally to a function $v \in C(X, R^{1/2})$ and it satisfies (2.4.2) for all $p, q \in X$ as well. For any $p \in X$, choose $\{p_n\} \subset V$ so that $p_n \to p$ as $n \to \infty$. Then

$$|v_k(p) - v(p)| \le |v_k(p) - v_k(p_n)| + |v_k(p_n) - v(p_n)| + |v(p_n) - v(p)|$$
$$\le 2\sqrt{C R(p, p_n)} + |v_k(p_n) - v(p_n)|.$$

Hence $v_n(p) \to v(p)$ as $n \to \infty$ for all $p \in X$. By (RF4),

$$|(v_k(p) - v_l(p)) - (v_k(p_*) - v_l(p_*))|^2 \le \mathcal{E}(v_k - v_l, v_k - v_l) R(p, p_*).$$

As $\mathcal{E}(v_n, v_n) \le C$, the above inequality implies

$$|v_k(p) - v_l(p)| \le \sqrt{4C R(p, p_*)} + |v_k(p_*) - v_l(p_*)|.$$

Letting $l \to \infty$, we have $|v_k(p) - v(p)|^2 \le 4C R(p, p_*) + 1$ for large k. Since $\int_X R(p, p_*)\mu(dp) < \infty$, it follows that $v_k \to v$ in $L^2(X, \mu)$ as $k \to \infty$ by Lebesgue's dominated convergence theorem. $\qquad \square$

Now we have collected enough facts to use some abstract theory from functional analysis. In fact, by Theorem 2.4.1, we can apply the well-developed theory of closed forms and self-adjoint operators, which is introduced in Appendix B.1.

Theorem 2.4.2. *Let $R \in \widetilde{\mathcal{RM}}(X)$ and suppose that $(X, R^{1/2})$ is separable. Set $(\mathcal{E}, \mathcal{F}) = MF_X(R)$. Also let μ be a σ-finite Borel measure on $(X, R^{1/2})$. Also assume that $L^2(X, \mu) \cap \mathcal{F}$ is dense in $L^2(X, \mu)$ with respect to the L^2-norm. Then there exists a non-negative self-adjoint operator H on $L^2(K, \mu)$ such that $\mathrm{Dom}(H^{1/2}) = \mathcal{F}$ and $\mathcal{E}(u, v) = (H^{1/2}u, H^{1/2}v)$ for all $u, v \in \mathcal{F}$. Moreover, if $\mu(X) < \infty$ and $\int_X R(p, p_*)\mu(dp) < \infty$ for some $p_* \in X$, then H has compact resolvent.*

Proof. Set $\mathcal{H} = L^2(X, \mu)$, $Q(\cdot, \cdot) = \mathcal{E}$ and $\mathrm{Dom}(Q) = \mathcal{F}$. Then Theorem 2.4.1 along with Theorem B.1.6 immediately implies the required results. $\qquad \square$

Assume that $R \in \mathcal{RM}(X)$. Also, in addition to the assumptions of Theorem 2.4.2, let us assume that X is a locally compact metric space. Then $(\mathcal{E}, \mathcal{F} \cap L^2(X, \mu))$ is a Dirichlet form on $L^2(X, \mu)$. Moreover, if $\mathcal{F} \cap L^2(X, \mu) \cap C_0(X)$ is dense in $C_0(X)$, then $(\mathcal{E}, \mathcal{F} \cap L^2(X, \mu))$ is a regular Dirichlet form. (See B.2 for the definition of (regular) Dirichlet forms and $C_0(X)$.) In fact, if $(\mathcal{E}, \mathcal{F})$ comes from a regular harmonic structure, which is defined in 3.1, we can verify all the conditions above and get a Dirichlet form and a Laplacian immediately from the theorems in this section. See the next chapter for details.

From an abstract point of view, the self-adjoint operator $-H$ should be our Laplacian. However, this abstract construction is too general to study detailed information about our Laplacian. For example, it is quite difficult to get concrete expressions for harmonic functions and Green's function from this abstract definition. So, we also need to construct a Laplacian on a self-similar set in a classical way, specifically as a direct limit of discrete Laplacians. See Chapter 3, in particular 3.7.

Example 2.4.3. Let K be any closed subset of $\mathbb{R}$. We can always find an increasing sequence of finite sets $\{V_m\}_{m \geq 0}$ that satisfy $V_m \subseteq V_{m+1}$ and $\overline{\cup_{m \geq 0} V_m} = K$. If $V_m = \{p_{m,i}\}_{i=1}^{n_m}$ and $p_{m,i} < p_{m,i+1}$ for all i, then we define $H_m \in \mathcal{LA}(V_m)$ by

$$(H_m)_{p_{m,i} p_{m,j}} = \begin{cases} |p_{m,i} - p_{m,j}|^{-1} & \text{if } |i - j| = 1, \\ 0 & \text{otherwise,} \end{cases}$$

for $i \neq j$. Then $\{(V_m, H_m)\}_{m \geq 0}$ is a compatible sequence. Also if R is the effective resistance defined on $\cup_{m \geq 0} V_m$, then R coincides with the restriction of the Euclidean metric. Let $(\mathcal{E}, \mathcal{F})$ be the corresponding resistance form and let μ be a σ-finite Borel regular measure on K. First note that $f|_K$ belongs to $\mathcal{F}$ for any piecewise linear function f on $\mathbb{R}$ with $\text{supp}(f)$ compact. ($\text{supp}(f)$ denotes the support of f.) By this fact, it follows that $\mathcal{F} \cap L^2(K, \mu)$ is dense in $L^2(K, \mu)$ with respect to the L^2-norm. Set $\mathcal{F}_1 = \mathcal{F} \cap L^2(K, \mu)$. Then $(\mathcal{E}, \mathcal{F}_1)$ is a local regular Dirichlet form on $L^2(K, \mu)$.

This example contains many interesting cases. The most obvious one is the case where $K = \mathbb{R}$. In this case, $\mathcal{F}_1$ coincides with $H^1(\mathbb{R})$, which is the completion of

$$\{u \in C^1(\mathbb{R}) : \int_{\mathbb{R}} u'(x)^2 dx < \infty, \text{supp}(u) \text{ is compact.}\}$$

with respect to the H^1-norm $|| \cdot ||_1$ defined by

$$||u||_1 = \sqrt{\int_{\mathbb{R}} (u(x)^2 + u'(x)^2)dx}.$$

Also $\mathcal{E}(u,v) = \int_{\mathbb{R}} u'(x)v'(x)dx$. If μ is the Lebesgue measure on $\mathbb{R}$, then the non-negative self-adjoint operator H coincides with the standard $-\Delta = -d^2/dx^2$.

One of the other interesting cases is the Cantor set. Let K be the Cantor set defined in Example 1.2.6. Let μ be a self-similar measure on K. (See 1.4 for the definition of self-similar measures.) Then $(\mathcal{E}, \mathcal{F})$ is a local regular Dirichlet form on $L^2(K, \mu)$. By Theorem 2.4.2, we obtain a non-negative self-adjoint operator H. Set $\Delta_\mu = -H$. Then Δ_μ may be thought of as a Laplacian on the Cantor set K. In [39, 40], Fujita studied the spectrum of Δ_μ and the asymptotic behavior of the associated (generalized) diffusion process on the Cantor set.

Notes and references

Most of the contents in this chapter are taken from [87]. New aspects are the introduction of $\widetilde{\mathcal{DF}}(\cdot)$, $\widetilde{\mathcal{LA}}(\cdot)$, $\widetilde{\mathcal{RM}}(\cdot)$ and $\widetilde{\mathcal{RF}}(\cdot)$. Indeed, in [87], condition (DF3) was forgotten in the definition of the Dirichlet forms. (The author thanks V. Metz for having pointed this out.) This motivated the author to study what would happen without (DF3).

The relation between electrical networks and Dirichlet forms (or Laplacians) on graphs is a classical subject. See, for example, Doyle & Snell [31].
2.1   The operator $P_{U,V}$ appearing in Theorem 2.1.6 is known as the trace of Dirichlet forms. In 2.1, we only treat the special case of Dirichlet forms on finite sets. See [43, §6.2] for the definition of the trace of Dirichlet forms for general cases. This operation has been known in various areas. For example, it is called the shorted operator in electrical network theory and is an example of a Shur complement. See Metz [127] and Barlow [6, Remark 4.26] for details.

Exercises

Exercise 2.1. Show that every metric on V coincides with an effective resistance metric associated with a Laplacian $H \in \mathcal{LA}(V)$ if $\#(V) = 3$.

Exercise 2.2. Let $V = \{p_1, p_2, p_3, p_4\}$ and let d be a metric on V defined by

$$d(p_i, p_j) = \begin{cases} 1 & \text{if } (i,j) \neq (1,2) \text{ and } i \neq j, \\ x & \text{if } (i,j) = (1,2), \\ 0 & \text{if } i = j, \end{cases}$$

for some x with $0 < x \leq 2$. Show that there exists a Laplacian $H \in \mathcal{LA}(V)$ such that $R_H = d$ if and only if $x \leq 3/2$.

Exercise 2.3. Verify that the Δ–Y transform remains true even if $H \in \widetilde{\mathcal{LA}}(V)$.

Exercise 2.4. Show that if $\#(V) = 3$, $H \in \mathcal{LA}(V)$ if and only if $R_H(\cdot, \cdot)$ is a metric on V.

Exercise 2.5. Let $V = \{p_1, p_2, p_3, p_4\}$. For $i \neq j$, set

$$H_{p_i p_j} = \begin{cases} 1 & \text{if } (i,j) \neq (1,4), \\ -\epsilon & \text{if } (i,j) = (1,4), \end{cases}$$

where $\epsilon > 0$. Show that if ϵ is sufficiently small, $H \in \widetilde{\mathcal{LA}}(V)$ and $R_H(\cdot, \cdot)$ becomes a metric on V.

Exercise 2.6. Let $V = \{p_1, p_2, p_3\}$. Define

$$H_m = \begin{pmatrix} -(1+m) & 1+2m & -m \\ 1+2m & -2(1+2m) & 1+2m \\ -m & 1+2m & -(1+m) \end{pmatrix}.$$

Show that $R_{H_m}(p_i, p_j)$ converges as $m \to \infty$. Also show that there exists no $H \in \widetilde{\mathcal{LA}}(V)$ such that $R = R_H$, where $R(p_i, p_j) = \lim_{m \to \infty} R_{H_m}(p_i, p_j)$.

Exercise 2.7. Let $X = \{a, b\} \cup \{p_m\}_{m \geq 1}$. Define $R(a, b) = 2$, $R(a, p_m) = R(b, p_m) = (1+m)/(1+2m)$ and $R(p_j, p_k) = |k - j|/(1+2k)(1+2j)$.
(1) Show that $R \in \widetilde{\mathcal{RM}}(X)$.
(2) Let $(\overline{X}, R^{1/2})$ be the completion of $(X, R^{1/2})$. Show that $R \notin \widetilde{\mathcal{RM}}(\overline{X})$.
   (Hint: see Exercise 2.6.)

3

Construction of Laplacians on P. C. F. Self-Similar Structures

In this chapter, we will construct the analysis associated with Laplacians on connected post critically finite self-similar structures. In this chapter, $\mathcal{L} = (K, S, \{F_i\}_{i \in S})$ is a post critically finite (p. c. f. for short) self-similar structure and K is assumed to be connected. (Also in this chapter, we always set $S = \{1, 2, \ldots, N\}$.) Recall that a condition for K being connected was given in 1.6.

The key idea of constructing a Laplacian (or a Dirichlet form) on K is finding a "self-similar" compatible sequence of r-networks on $\{V_m\}_{m \geq 0}$, where $V_m = V_m(\mathcal{L})$ was defined in Lemma 1.3.11. Note that $\{V_m\}_{m \geq 0}$ is a monotone increasing sequence of finite sets. We will formulate such a self-similar compatible sequence in 3.1. Once we get such a sequence, we can use the general theory in the last chapter and construct a resistance form $(\mathcal{E}, \mathcal{F})$ and a resistance metric R on V_*, where $V_* = \cup_{m \geq 0} V_m$.

If the closure of V_* with respect to the metric R were always identified with K, then we could apply Theorem 2.4.2 and see that $(\mathcal{E}, \mathcal{F})$ is a regular local Dirichlet form on $L^2(K, \mu)$ for any self-similar measure μ on K. Consequently, we could immediately obtain a Laplacian associated with the Dirichlet form $(\mathcal{E}, \mathcal{F})$ on $L^2(K, \mu)$. Unfortunately, as we will see in 3.3, in general, the closure of V_* with respect to R is merely a proper subset of K in certain cases. In spite of this difficulty, we will give a condition on probability measure μ which is sufficient for $(\mathcal{E}, \mathcal{F})$ to be a regular local Dirichlet form on $L^2(K, \mu)$ in 3.4.

As was mentioned in 2.4, there is an abstract way of constructing the Laplacian from a Dirichlet form $(\mathcal{E}, \mathcal{F})$ on $L^2(K, \mu)$. (See B.1 for details.) However, we will develop our analysis on a p. c. f. self-similar set K in a classical and explicit way similar to that of the ordinary Laplacian d^2/dx^2 on the unit interval. In 3.5, the Green's function will be given in a constructive manner. In the following sections, we will study some counterparts of classical analysis on Euclidean spaces, for example, Green's operator in 3.6 and Gauss–Green's formula in Theorem 3.7.8. Finally, we will define a Laplacian as a scaling limit of discrete Laplacians on V_m in 3.7.

Throughout this chapter, d is a metric on K which gives the original topology of K as a compact metric space. Also we write $C(K) = C(K, d)$. Since (K, d) is compact, $C(K)$ is the collection of all real-valued continuous functions on K.

3.1 Harmonic structures

In this section, we start constructing Dirichlet forms and Laplacians on K. As was mentioned above, the basic idea is to find a "self-similar" compatible sequence of r-networks on $\{V_m\}_{m \geq 0}$ and to take a limit. (Recall that $V_m \subseteq V_{m+1}$ by Lemma 1.3.11.)

For any initial $D \in \mathcal{LA}(V_0)$, we can construct a sequence of self-similar Laplacians $H_m \in \mathcal{LA}(V_m)$ as follows.

Definition 3.1.1. If $D \in \mathcal{LA}(V_0)$ and $\mathbf{r} = (r_1, r_2, \ldots, r_N)$, where $r_i > 0$ for $i \in S$, we define $\mathcal{E}^{(m)} \in \mathcal{DF}(V_m)$ by

$$\mathcal{E}^{(m)}(u, v) = \sum_{w \in W_m} \frac{1}{r_w} \mathcal{E}_D(u \circ F_w, v \circ F_w)$$

for $u, v \in \ell(V_m)$, where $r_w = r_{w_1} \ldots r_{w_m}$ for $w = w_1 w_2 \ldots w_m \in W_m$. Also $H_m \in \mathcal{LA}(V_m)$ is characterized by $\mathcal{E}^{(m)} = \mathcal{E}_{H_m}$.

It is easy to see that

$$\mathcal{E}^{(m+1)}(u, v) = \sum_{i=1}^{N} \frac{1}{r_i} \mathcal{E}^{(m)}(u \circ F_i, v \circ F_i) \tag{3.1.1}$$

for $u, v \in \ell(V_m)$. Also $H_m = \sum_{w \in W_m} \frac{1}{r_w}{}^t R_w D R_w$, where $R_w : \ell(V_m) \to \ell(V_0)$ is defined by $R_w f = f \circ F_w$ for $w \in W_m$. We write $\mathcal{E}_m = \mathcal{E}^{(m)}$ hereafter.

Considering (3.1.1), we may regard (V_m, H_m) as a self-similar sequence of r-networks. If it is also a compatible sequence, then it is possible to construct a Laplacian on K using the theory in the previous chapter.

Definition 3.1.2 (Harmonic structures). $(D, \mathbf{r})$ is called a harmonic structure if and only if $\{(V_m, H_m)\}_{m \geq 0}$ is a compatible sequence of r-networks. Also, a harmonic structure $(D, \mathbf{r})$ is said to be regular if $0 < r_i < 1$ for all $i \in S$.

Once we get a harmonic structure, we can use the general framework in Chapter 2 (in particular, Theorem 2.2.6, Theorem 2.4.1 and Theorem 2.4.2) to construct a quadratic form $(\mathcal{E}, \mathcal{F})$ on $V_* = \cup_{m \geq 0} V_m$ and an associated non-negative self-adjoint operator H on $L^2(\Omega, \mu)$, where Ω is the completion of V_* under the resistance metric associated with $(\mathcal{E}, \mathcal{F})$ and μ is a given σ-finite Borel regular measure. This H should be our Laplacian. It seems easy but there remains a "slight" problem: the topology on V_* given by the resistance metric may be different from the original topology of K. In such a case, Ω does not coincide with K. In fact, we will see in the next section that $\Omega = K$ if and only if the harmonic structure is regular.

Another important problem is whether there exists any harmonic structure on a p. c. f. self-similar structure. By virtue of the self-similar construction of H_m, we can simplify the condition for harmonic structures as follows.

Proposition 3.1.3. $(D, \mathbf{r})$ *is a harmonic structure if and only if* $(V_0, D) \leq (V_1, H_1)$.

Proof. Assume that $(V_{m-1}, H_{m-1}) \leq (V_m, H_m)$. Then, for any $u \in \ell(V_m)$, we have $\mathcal{E}_{m-1}(u \circ F_i, u \circ F_i) = \min\{\mathcal{E}_m(v \circ F_i, v \circ F_i) : v \in \ell(V_{m+1}), v|_{V_m} = u\}$. Hence, by (3.1.1), $\mathcal{E}_m(u, u) = \min\{\mathcal{E}_{m+1}(v, v) : v \in \ell(V_{m+1}), v|_{V_m} = u\}$. Therefore $(V_m, H_m) \leq (V_{m+1}, H_{m+1})$. So, by induction, if $(V_0, D) \leq (V_1, H_1)$, then $(V_m, H_m) \leq (V_{m+1}, H_{m+1})$ for any $m \geq 0$. The converse is obvious. $\qquad \square$

For given $\mathbf{r} = (r_1, r_2, \ldots, r_N)$, define $\mathcal{R}_{\mathbf{r}} : \mathcal{LA}(V_0) \to \mathcal{LA}(V_0)$ by

$$\mathcal{R}_{\mathbf{r}}(D) = [H_1]_{V_0},$$

where $H_1 \in \mathcal{LA}(V_1)$ is given by Definition 3.1.1. $\mathcal{R}_{\mathbf{r}}$ is called a renormalization operator on $\mathcal{LA}(V_0)$. By the above proposition, D is a harmonic structure if and only if D is a fixed point of $\mathcal{R}_{\mathbf{r}}$. Also, it is easy to see that $\mathcal{R}_{\lambda\mathbf{r}}(\alpha D) = (\lambda)^{-1}\alpha\mathcal{R}_{\mathbf{r}}(D)$ for any $\alpha, \lambda > 0$. Hence, if D is an eigenvector of $\mathcal{R}_{\mathbf{r}}$, i.e., $\mathcal{R}_{\mathbf{r}}(D) = \lambda D$, then D is a fixed point of $\mathcal{R}_{\lambda\mathbf{r}}$. So, the existence problem of harmonic structures is reduced to a fixed point problem (or eigenvalue problem) for the non-linear homogeneous map $\mathcal{R}_r$. In general, this problem is not easy and we do not fully understand the situation yet. For example, it is not known whether there exists at least one harmonic structure on a p. c. f. self-similar set. The only general result on existence of a harmonic structure is the theory of nested fractals by Lindstrøm [116]. Nested fractals are highly symmetric self-similar structures. (See 3.8 for the definition.) We will present a slightly extended version of Lindstrøm's result on the existence of a harmonic structure on nested fractals in 3.8.

Example 3.1.4 (Interval). Set $F_1(x) = x/2$ and $F_2(x) = x/2 + 1/2$. Then $\mathcal{L} = ([0,1], \{1,2\}, \{F_1, F_2\})$ is a p.c.f. self-similar structure. We see that $V_m = \{i/2^m\}_{i=0,1,\dots,2^m}$. Let us define $D \in \mathcal{L}\mathcal{A}(V_0)$ by

$$D = \begin{pmatrix} -1 & 1 \\ 1 & -1 \end{pmatrix}.$$

Then $(D, \mathbf{r})$ is a harmonic structure on $\mathcal{L}$ if $\mathbf{r} = (r_1, r_2)$ satisfies $r_1 + r_2 = 1$ and $0 < r_i < 1$ for $i = 1, 2$. Also, it is easy to see that those are all the harmonic structures on $\mathcal{L}$.

Example 3.1.5 (Sierpinski gasket). Recall Examples 1.2.8 and 1.3.15. The Sierpinski gasket is a p.c.f. self-similar set with $V_0 = \{p_1, p_2, p_3\}$. Define $D \in \mathcal{L}\mathcal{A}(V_0)$ by

$$D = \begin{pmatrix} -2 & 1 & 1 \\ 1 & -2 & 1 \\ 1 & 1 & -2 \end{pmatrix}.$$

Also set $\mathbf{r} = (3/5, 3/5, 3/5)$. Then we see that $(D, \mathbf{r})$ is a harmonic structure on the Sierpinski gasket. $(D, \mathbf{r})$ is called the standard harmonic structure on the Sierpinski gasket. There are other harmonic structures on the Sierpinski gasket if we weaken the symmetry. See Exercise 3.1.

Example 3.1.6 (Hata's tree-like set). Let $\mathcal{L}$ be the self-similar structure associated with Hata's tree-like set appearing in Examples 1.2.9 and 1.3.16. Then $V_0 = \{c, 0, 1\}$ as in the previous example. Define $D \in \mathcal{L}\mathcal{A}(V_0)$ by

$$D = \begin{pmatrix} -h & h & 0 \\ h & -(h+1) & 1 \\ 0 & 1 & -1 \end{pmatrix}.$$

and define $\mathbf{r} = (r, 1 - r^2)$ for $r \in (0, 1)$. If $rh = 1$, then $(D, \mathbf{r})$ is a regular harmonic structure on $\mathcal{L}$.

So far we have presented examples of regular harmonic structure. Of course, there are many examples of non-regular harmonic structures.

Example 3.1.7. Set $F_1(z) = z/2$, $F_2(z) = z/2 + 1/2$ and $F_3(z) = \sqrt{-1}z/3 + 1/2$. Let K be the self-similar set with respect to $\{F_1, F_2, F_3\}$ and let $\mathcal{L} = (K, \{1, 2, 3\}, \{F_1, F_2, F_3\})$. Then $\mathcal{L}$ is a p.c.f. self-similar structure. In fact, $\mathcal{C_L} = \{1\dot{2}, 2\dot{1}, 3\dot{1}\}, \mathcal{P_L} = \{\dot{1}, \dot{2}\}$ and $V_0 = \{0, 1\}$. If $D = \begin{pmatrix} -1 & 1 \\ 1 & -1 \end{pmatrix}$ and $\mathbf{r} = (r, 1 - r, s)$ for $r \in (0, 1)$ and $s > 0$, then $(D, \mathbf{r})$ is a harmonic structure on $\mathcal{L}$. Obviously, $(D, \mathbf{r})$ is not regular for $s \geq 1$.

See Example 3.2.3 and Exercise 3.2 for a more natural example of non-regular harmonic structures.

Proposition 3.1.8. *For $w \in W_*$, let $\dot{w}$ denote the periodic sequence in Σ defined by $\dot{w} = www\ldots$ Let $(D, \mathbf{r})$ be a harmonic structure and let $\dot{w} \in \mathcal{P}$ for $w \in W_*$. Then $r_w < 1$. In particular, there exists $i \in S$ such that $r_i < 1$.*

Remark. If $(K, S, \{F_i\}_{i \in S})$ is a p. c. f. self-similar structure, then the post critical set $\mathcal{P}$ consists of eventually periodic points: for any $\omega \in \mathcal{P}$, there exist $w \in W_*$ and $m \geq 0$ such that $\sigma^m \omega = \dot{w}$.

Corollary 3.1.9. *Let $(D, \mathbf{r})$ be a harmonic structure. If $r_1 = \cdots = r_N$, then $r_i < 1$ for any $i \in S$. In particular, $(D, \mathbf{r})$ is a regular harmonic structure.*

To prove Proposition 3.1.8, we need the following lemma.

Lemma 3.1.10. *Let V be a finite set and let $H \in \mathcal{LA}(V)$. Suppose $U \subsetneq V$ and $p \in U$. If there exists $q_* \in V \backslash U$ such that $H_{pq_*} \neq 0$, then $-h_{pp} < -H_{pp}$, where $(h_{kl})_{k,l \in U} = [H]_U$.*

Proof. H can be expressed as $\begin{pmatrix} T & {}^t J \\ J & X \end{pmatrix}$, where $T : \ell(U) \to \ell(U), J : \ell(U) \to \ell(V \backslash U)$ and $X : \ell(V \backslash U) \to \ell(V \backslash U)$. Then $[H]_U = T - {}^t J X^{-1} J$. Now let $\psi_p = -X^{-1} J \chi_p^U$, where $\chi_p^U(x) = 1$ if $x = p$ and $\chi_p^U(x) = 0$ if $x \neq p$ on U. It follows that

$$h_{pp} = H_{pp} + \sum_{q \in V \backslash U} H_{pq} \psi_p(q).$$

As $H_{pq_*} \neq 0$, the maximum principle (Proposition 2.1.7) implies that $\psi_p(q_*) > 0$. Therefore $\sum_{q \in V \backslash U} H_{pq} \psi_p(q) > 0$. $\square$

Proof of Proposition 3.1.8. First we assume that

$$\#(F_i(V_0) \cap V_0) \leq 1 \text{ for all } i \in S. \tag{3.1.2}$$

As $H_1 = \sum_{i=1}^N \frac{1}{r_i} {}^t R_i D R_i$, we have $(H_1)_{pp} = \sum_{(q,i):q \in V_0, F_i(q)=p} \frac{1}{r_i} D_{qq}$. Set $p = \pi(\dot{w})$, where $w = w_1 w_2 \ldots w_m \in W_*$ and $\dot{w} \in \mathcal{P}$. By Lemma 1.3.14, $\pi^{-1}(\pi(\dot{w})) = \{\dot{w}\}$. Hence, $\{(q, i) : q \in V_0, F_i(q) = p\} = \{(\pi(\sigma\dot{w}), w_1)\}$.

Therefore, $(H_1)_{pp} = \frac{1}{r_{w_1}} D_{qq}$, where $q = \pi(\sigma \dot{w})$. So, letting $p_i = \pi(\sigma^{i-1}\dot{w})$ for $i = 1, 2, \ldots, m+1$, then $(H_1)_{p_i p_i} = \frac{1}{r_{w_i}} D_{p_{i+1}p_{i+1}}$, where we set $w_{m+1} = w_1$. Now by (3.1.2), we can apply Lemma 3.1.10 and obtain $-D_{p_{i+1}p_{i+1}} < -(H_1)_{p_{i+1}p_{i+1}}$. So we have $\prod_{i=1}^{m} -(H_1)_{p_i p_i} < r_w^{-1} \prod_{i=1}^{m} -(H_1)_{p_{i+1}p_{i+1}}$. Hence $r_w < 1$.

If (3.1.2) is not satisfied, we replace the original self-similar structure $\mathcal{L} = (K, S, \{F_i\}_{i \in S})$ by $\mathcal{L}_m = (K, W_m, \{F_w\}_{w \in W_m})$. Then, by Proposition 1.3.12, $\mathcal{P}_{\mathcal{L}} = \mathcal{P}_{\mathcal{L}_m}$. Also, it is easy to see that $(D, \mathbf{r}_m)$, where $\mathbf{r}_m = (r_w)_{w \in W_m}$, is a harmonic structure on $\mathcal{L}_m$. For sufficiently large m, $\mathcal{L}_m$ satisfies (3.1.2) and hence we can apply the above argument to the harmonic structure $(D, \mathbf{r}_m)$. Therefore $(r_w)^m < 1$. Thus we obtain that $r_w < 1$. $\qquad \square$

3.2 Harmonic functions

Let $(D, \mathbf{r})$ be a harmonic structure on a connected p. c. f. self-similar structure $\mathcal{L} = (K, S, \{F_i\}_{i \in S})$, where $S = \{1, 2, \ldots, N\}$. Then $\{(V_m, H_m)\}_{m \geq 0}$ is a compatible sequence of r-networks. So we can construct $(\mathcal{E}, \mathcal{F})$ as in (2.2.1) and (2.2.2). By Theorem 2.2.6, $(\mathcal{E}, \mathcal{F}) \in \mathcal{RF}(V_*)$, where $V_* = \cup_{m \geq 0} V_m$. In this section, we consider harmonic functions associated with $(\mathcal{E}, \mathcal{F})$. The arguments in the last chapter, in particular Lemma 2.2.2, imply the following result.

Proposition 3.2.1. *For any $\rho \in \ell(V_0)$, there exists a unique $u \in \mathcal{F}$ such that $u|_{V_0} = \rho$ and $\mathcal{E}(u, u) = \min\{\mathcal{E}(v, v) : v \in \mathcal{F}, v|_{V_0} = \rho\}$. Furthermore, u is the unique solution of*

$$\begin{cases} (H_m v)|_{V_m \backslash V_0} = 0 & \text{for all } m \geq 1, \\ v|_{V_0} = \rho. \end{cases} \tag{3.2.1}$$

The function u obtained in the above theorem is called a harmonic function with boundary value ρ.

Corollary 3.2.2. *Let u be a harmonic function. Then, for any $p \in V_0$ and any $m \geq 0$, $(H_m u)(p) = (Du)(p)$.*

Proof. By Lemma 2.2.2, $\mathcal{E}(u, u) = \mathcal{E}_{H_m}(u, u)$ for any $m \geq 0$. Hence, if v is a harmonic function, then $\mathcal{E}(u, v) = \mathcal{E}_{H_m}(u, v)$ for any $m \geq 0$. Now let $v|_{V_0} = \chi_p$. Then $\mathcal{E}(u, v) = \mathcal{E}_{H_m}(u, v) = -\sum_{q \in V_m} v(q)(H_m u)(q) = (H_m u)(p)$. $\qquad \square$

Example 3.2.3. Let $\{p_1, p_2, p_3\}$ be the vertices of an equilateral triangle in $\mathbb{C}$ and let

$$p_4 = \frac{1}{2}(p_2 + p_3), p_5 = \frac{1}{2}(p_1 + p_3), p_6 = \frac{1}{2}(p_1 + p_2), p_7 = \frac{1}{3}(p_1 + p_2 + p_3).$$

Let $S = \{1, 2, \dots, 7\}$. For $i \in S$, let $F_i(z) = \beta_i(z - p_i) + p_i$, where $\beta_1 = \beta_2 = \beta_3 = \beta$, $\beta_4 = \beta_5 = \beta_6 = 1 - 2\beta$, $\beta_7 = 1 - 3\beta$ for $1/3 < \beta < 1/2$.

If K is the self-similar set with respect to $\{F_i\}_{i \in S}$, then the self-similar structure $(K, S, \{F_i\}_{i \in S})$ is independent of the value of β and is post critically finite. In fact

$$\mathcal{C} = \bigcup_{k=1}^{3} \{7\dot{k}, (k+3)\dot{k}\} \ \cup \bigcup_{\substack{k<l<m \\ k+l+m=9}} \{k\dot{l}, m\dot{k}, m\dot{l}, l\dot{k}\},$$

$$\mathcal{P} = \{\dot{1}, \dot{2}, \dot{3}\} \quad \text{and} \quad p_k = \pi(\dot{k})$$

for $k = 1, 2, 3$. Define $q_k = \pi(7\dot{k}) = \pi((k+3)\dot{k})$ for $k = 1, 2, 3$. Also let $q_{lm} = \pi(m\dot{l}) = \pi(l\dot{k}), q_{km} = \pi(k\dot{l}) = \pi(m\dot{k})$ for (k, l, m) such that $k < l < m$ and $k + l + m = 9$. See Figure 3.1.

Define

$$D = \begin{pmatrix} -2 & 1 & 1 \\ 1 & -2 & 1 \\ 1 & 1 & -2 \end{pmatrix}.$$

and $\mathbf{r} = (1, 1, 1, 1, 1, 1, t)$ for $t > 0$. Considering the symmetry of $(D, \mathbf{r})$ and K, we see that $\mathcal{R}_{\mathbf{r}}(D) = \lambda D$ for some λ. (Note that $\mathcal{R}_{\mathbf{r}}$ preserves the symmetry of D.) So $(D, \lambda \mathbf{r})$ is a harmonic structure. Now let us calculate the value of λ by using Corollary 3.2.2. Let u be the harmonic function that satisfies $u(p_2) = u(p_3) = 0$ and $u(p_1) = 1$. (Note that the harmonic function u is independent of the value of λ because it is the solution of (3.2.1).) Taking the symmetry into account, we deduce that $u(q_{16}) = u(q_{15}), u(q_{26}) = u(q_{35}), u(q_2) = u(q_3), u(q_{24}) = u(q_{34})$. By (3.2.1),

$$u(q_{16}) = \frac{4t + 8}{7t + 15}, u(q_{26}) = \frac{2t + 4}{7t + 15}, u(q_2) = \frac{3t + 5}{7t + 15}$$

$$u(q_1) = \frac{t + 5}{7t + 15}, u(q_{24}) = \frac{t + 3}{7t + 15}.$$

By Corollary 3.2.2, $(Du)(p_1) = -2 = \lambda^{-1}2(u(q_{16}) - 1) = (H_1 u)(p_1)$. Therefore, we obtain $\lambda = (3t + 7)/(7t + 15)$. Note that $0 < \lambda < 1$ for any $t > 0$. However, $\lambda t \geq 1$ if $t \geq \sqrt{5}$. So the harmonic structure $(D, \lambda \mathbf{r})$ is regular if and only if $0 < t < \sqrt{5}$. Note that $(K, S, \{F_i\}_{i \in S})$ is an affine nested fractal. (See 3.8 for the definition of an affine nested fractal.)

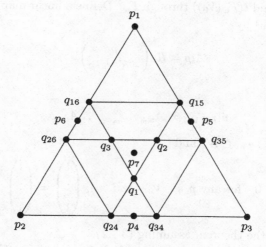

Fig. 3.1. p_i, q_i and q_{ij}

Let R be the resistance metric on V_* associated with $(\mathcal{E}, \mathcal{F})$. Then, by (2.2.6), $u \in C(V_*, R)$. Note that V_* is a countable subset of K and the topology of (V_*, R) may be different from that of V_* with the relative topology from the original metric on K. We will see, however, that a harmonic function has a unique extension to a continuous function on K.

Now recall that d is a metric on K which is compatible with the original topology of K.

Theorem 3.2.4. *Let u be a harmonic function. Then there exists a unique $\tilde{u} \in C(K)$ such that $u|_{V_*} = \tilde{u}|_{V_*}$.*

Remark. As is shown in 3.3, the closure of (V_*, R) equals K with the original topology if and only if $(D, \mathbf{r})$ is regular harmonic structure. In such a case, the above theorem is obvious.

Proof. Let u be a harmonic function with boundary value ρ. Set $H_1 = \begin{pmatrix} T & {}^t J \\ J & X \end{pmatrix}$, where $T : \ell(V_0) \to \ell(V_0), J : \ell(V_0) \to \ell(V_1 \backslash V_0)$ and $X : \ell(V_1 \backslash V_0) \to \ell(V_1 \backslash V_0)$. Then it follows that

$$(u \circ F_i)|_{V_0} = R_i(u|_{V_1}) = R_i \begin{pmatrix} \rho \\ -X^{-1} J \rho \end{pmatrix}.$$

As F_w is a bijective mapping between V_0 and $F_w(V_0)$ for $w \in W_*$, we will identify $\ell(V_0)$ and $\ell(F_w(V_0))$ through F_w. Define a linear map $A_i : \ell(V_0) \rightarrow \ell(F_i(V_0)) \cong \ell(V_0)$ by

$$A_i \rho = R_i \begin{pmatrix} \rho \\ -X^{-1} J \rho \end{pmatrix}. \tag{3.2.2}$$

Then

$$u|_{F_w(V_0)} = A_{w_m} A_{w_{m-1}} \ldots A_{w_1} \rho \tag{3.2.3}$$

for $w = w_1 w_2 \ldots w_m \in W_*$ and

$$(A_i)_{pq} \geq 0 \quad \text{for any } p, q \in V_0 \quad \text{and} \quad A_i \begin{pmatrix} 1 \\ \vdots \\ 1 \end{pmatrix} = \begin{pmatrix} 1 \\ \vdots \\ 1 \end{pmatrix}. \tag{3.2.4}$$

First we prove the theorem assuming (3.1.2).

Claim 1. Set $v(f) = \max_{p,q \in V_0} |f(p) - f(q)|$ for $f \in \ell(V_0)$. Then $v(A_i f) < v(f)$ if $v(f) \neq 0$.

Proof of Claim 1: If $H_1 g|_{V_1 \setminus V_0} = 0$ and $g|_{V_0} = f$, then $A_i f = g|_{F_i(V_0)}$. Applying the maximum principle (Proposition 2.1.7) and taking (3.1.2) into account, we can see that $\max_{q \in F_i(V_0)} g(q) - \min_{q \in F_i(V_0)} g(q) < v(f)$. Hence $v(A_i f) < v(f)$.

Claim 2. There exists c_i such that $0 < c_i < 1$ and $v(A_i f) \leq c_i v(f)$ for any $f \in \ell(V_0)$.

Proof of Claim 2: Define $Q : \ell(V_0) \rightarrow \ell(V_0)$ by

$$(Qf)(p) = f(p) - \#(V_0)^{-1} \sum_{q \in V_0} f(q),$$

then $v(f) = v(Qf)$ and $v(A_i f) = v(A_i Q f)$ for any $f \in \ell(V_0)$. Hence

$$\sup\{ \frac{v(A_i f)}{v(f)} : f \in \ell(V_0), v(f) \neq 0 \} = \sup\{ \frac{v(A_i Q f)}{v(Qf)} : f \in \ell(V_0), v(f) \neq 0 \}$$

$$= \sup\{ \frac{v(A_i f)}{v(f)} : f \in \ell(V_0), \sum_{q \in V_0} f(q) = 0, v(f) = 1 \}.$$

As $\{ f \in \ell(V_0) : \sum_{q \in V_0} f(q) = 0, v(f) = 1 \}$ is compact, the above supremum is less than 1.

Now by Claim 2 and the maximum principle,

$$v_w(u) = \sup\{ |f(p) - f(q)| : p, q \in K_w \cap V_* \} \leq c^m v(\rho)$$

for any $w \in W_*$, where $c = \max_{i \in S} c_i$. Hence, if $\{p_i\}_{i \geq 1}$ is a Cauchy sequence with respect to a metric on K which is compatible with the original topology of K, then $\{u(p_i)\}_{i \geq 1}$ is convergent as $i \to \infty$. Using this limit, we can extend u to a continuous function $\tilde{u}$ on K.

Next if (3.1.2) is not satisfied, we can exchange the harmonic structure as in the proof of Proposition 3.1.8. Then again we can use the result under a new self-similar structure $\mathcal{L}_m$ and harmonic structure $(D, \mathbf{r}_m)$. Note that harmonic functions remain the same after we replace the harmonic structure. $\qquad\qquad\square$

Hereafter, we identify u with its extension $\tilde{u}$ and think of a harmonic function as a continuous function on K. Immediately, by Lemma 2.2.3, we have the maximum principle for harmonic functions.

Theorem 3.2.5 (Maximum principle: weak version). *Let u be harmonic. Then, for any $w \in W_*$,*

$$\min_{p \in F_w(V_0)} u(p) \leq u(x) \leq \max_{p \in F_w(V_0)} u(p)$$

for any $x \in K_w$.

This maximum principle is a little weak because we say nothing about the case when $u(x)$ attains the maximum. We will give a stronger version of the maximum principle in Theorem 3.2.14.

Example 3.2.6 (Sierpinski gasket). Let us calculate the stochastic matrices $\{A_i\}_{i \in S}$ for the standard harmonic structure on the Sierpinski gasket given in Example 3.1.5. Recall that $V_0 = \{p_1, p_2, p_3\}$. See Figure 1.2. It follows that $V_1 = \{p_i, q_i\}_{i \in S}$, where $S = \{1, 2, 3\}$. Now let $f(p_1) = a, f(p_2) = b$ and $f(p_3) = c$ and solve the linear equation $(H_1 f)(q_i) = 0$ for $i \in S$. Then we get $f(q_1) = (2b + 2c + a)/5, f(q_2) = (2c + 2a + b)/5$ and $f(q_3) = (2a + 2b + c)/5$. By this result,

$$A_1 = \frac{1}{5} \begin{pmatrix} 5 & 0 & 0 \\ 2 & 2 & 1 \\ 2 & 1 & 2 \end{pmatrix} \quad A_2 = \frac{1}{5} \begin{pmatrix} 2 & 2 & 1 \\ 0 & 5 & 0 \\ 1 & 2 & 2 \end{pmatrix} \quad A_3 = \frac{1}{5} \begin{pmatrix} 2 & 1 & 2 \\ 1 & 2 & 2 \\ 0 & 0 & 5 \end{pmatrix}.$$

It is easy to see that the eigenvalues of A_i are $1, 3/5, 1/5$. Note that the second eigenvalue $3/5$ is equal to r_i. In fact, this is not a coincidence. (Recall that $\mathbf{r} = (3/5, 3/5, 3/5)$.) In A.1, we will give a general result on the second eigenvalue of A_i.

See Exercise 3.3 for more examples.

The stochastic matrices $\{A_i\}_{i \in S}$ determine the harmonic functions through (3.2.3). The behavior of a harmonic function around a point $\pi(\omega)$ for $\omega \in \Sigma(S)$ is given by the asymptotic behavior of (3.2.3) for $m \to \infty$.

This is the problem of random iterations of matrices and, in general, it is very difficult. Even in the above example, we do not know how to calculate the behavior of $A_{\omega_m} A_{\omega_{m-1}} \cdots A_{\omega_1}$ as $m \to \infty$ unless the sequence ω is (eventually) periodic. Kusuoka used $\{A_i\}_{i \in S}$ to construct Dirichlet forms on finitely ramified self-similar sets in Kusuoka [107] and obtain some result about almost sure behavior of the random iteration of $\{A_i\}_{i \in S}$.

An important property of harmonic functions is the Harnack inequality, which follows from the discrete version, Corollary 2.1.8.

Proposition 3.2.7 (Harnack inequality). *If X is a compact subset of K that is contained in a connected component of $K \backslash V_0$, then there exists a constant $c > 0$ such that $\max_{x \in X} u(x) \leq c \min_{x \in X} u(x)$ for any non-negative harmonic function u on K.*

Proof. Set $X_m = \cup_{w \in W_m : K_w \cap X \neq \emptyset} K_w$. Then we can choose m so that $X_m \cap V_0 = \emptyset$. Now set $V = V_m$, $U = V_0$, $H = H_m$ and $A = X_m \cap V_m$. The Harnack inequality (Corollary 2.1.8) implies that there exists $c > 0$ such that $\max_{p \in A} u(p) \leq c \min_{p \in A} u(p)$ for any non-negative harmonic function u on K. By using the maximum principle, it follows that $\max_{x \in X} u(x) \leq \max_{p \in A} u(p)$ and $\min_{p \in A} u(p) \leq \min_{x \in X} u(x)$. Hence we have shown the required inequality.   □

For $p \in V_0$, let ψ_p be the harmonic function that satisfies $\psi_p|_{V_0} = \chi_p^{V_0}$. It is easy to see that $\{\psi_p\}_{p \in V_0}$ is a partition of unity on K: $\sum_{p \in V_0} \psi_p(x) = 1$ for any $x \in K$.

Theorem 3.2.8. *Let $p \in V_0$ and let $x \in K \backslash V_0$. Then $\psi_p(x) > 0$ if and only if there exists $C \in J(p, V_0)$ which contains x.*

Recall that $J(p, V_0)$ is the collection of all connected components of $K \backslash V_0$ whose closure contains p.

Definition 3.2.9. (1) For $p, q \in V_m$, $\{p_i\}_{i=1}^n$ is called an H_m-path between p and q if $p_i \in V_m \backslash V_0$ for any $i = 1, 2, \ldots, n$, $(H_m)_{pp_1} > 0$, $(H_m)_{p_n q} > 0$ and $(H_m)_{p_i p_{i+1}} > 0$ for any $i = 1, 2, \ldots, n-1$.
(2) For $p, q \in V_m \backslash V_0$, we write $p \underset{m}{\sim} q$ if and only if there exists an H_m-path between p and q.

The notion $\underset{m}{\sim}$ will not appear later in this section. It plays, however, an important role in 3.5, where we will show that $p \underset{m}{\sim} q$ if and only if p and q belong to the same connected component of $K \backslash V_0$ by using the results in this section.

Note that $(H_m)_{pq} = \sum_{w:p,q\in F_w(V_0)}(r_w)^{-1}D_{(F_w)^{-1}(p)(F_w)^{-1}(q)}$. Since D is a Laplacian on V_0, we may immediately verify the following lemma.

Lemma 3.2.10. *Let* $w \in W_m$ *and let* $p, q \in F_w(V_0)$.
(1) *If* $F_w(V_0) \cap V_0 = \emptyset$, *then there exists an* H_m-*path between* p *and* q *contained in* $F_w(V_0)$.
(2) *If* $F_w(V_0) \cap V_0 = \{q\}$ *and* $p \neq q$, *then there exists an* H_m-*path between* p *and* q *contained in* $F_w(V_0)$.

Proof of Theorem 3.2.8. First assume that $x \in C \in J(p, V_0)$. Then, by Proposition 1.6.6, there exists a path $\gamma : [0,1] \to C \cup \{p\}$ between x and p. Choose m so that

$$\sup_{t\in[0,1]} \operatorname{diam}(K_{m,\gamma(t)}) < \operatorname{dist}(\gamma([0,1]), V_0\backslash\{p\}),$$

where $\operatorname{dist}(A, B) = \inf_{x\in A, y\in B} d(x,y)$. Let $\{p_1,\dots,p_n\} = V_m \cap \gamma([0,1])$, where $p_i = p(t_i)$ and $t_1 < t_2 < \cdots < t_n = 1$. Note that $p_n = p$. Then there exist $w(0),\dots,w(n-1) \in W_m$ such that $x, p_1 \in K_{w(0)}$ and $p_i, p_{i+1} \in K_{w(i)}$ for all $i = 1,2,\dots,n-1$. Note that $K_{w(i)} \cap V_0 \subseteq \{p\}$ for any i. Let $j = \min\{i : K_{w(i)} \cap V_0 = \{p\}\}$.

Suppose $\psi_p(x) = 0$. Then, by Theorem 3.2.5, there exists $q \in F_{w(0)}(V_0)$ such that $\psi_p(q) = 0$. Let $q = p_0$. Then, by Lemma 3.2.10, there exists an H_m-path between p_i and p_{i+1} for any $i = 0,1,\dots,j-1$. Also, by Lemma 3.2.10, there exists an H_m-path between p_j and p. Joining those H_m-paths, we may construct an H_m-path between q and p. Let $V = V_m$ and let $U = V_0$. Then the maximum principle, Proposition 2.1.7, implies that $\psi_p(q) > 0$. Therefore, $\psi_p(x) > 0$.

Next let C be the connected component of $K\backslash V_0$. Assume that $p \notin \overline{C}$. We show that $\psi_p(q) = 0$ for any $q \in C \cap V_*$. Let $q \in C \cap V_m$. Define $U_m = \cup_{y\in V_0\cap\overline{C}}(K_{m,y} \cap V_m)$. Since ψ_p is continuous and $\psi_p(y) = 0$ for any $p \in V_0 \cap \overline{C}$, it follows that $\max_{z\in U_m} \psi_p(z) \to 0$ as $m \to \infty$. Also $q \notin U_m$ and $p \notin U_m$ for sufficiently large m. Now apply the maximum principle, Proposition 2.1.7 with $V = V_m$, $U = U_m \cup V_0$ and $H = H_m$. Since $p \notin \overline{C}$, any H_m-path between p and q must intersect U_m. Hence $U_q \subseteq U_m$, where U_q is defined in Proposition 2.1.7. So $0 \leq \psi_p(q) \leq \max_{z\in U_m} \psi_p(z)$. Letting $m \to \infty$, we see that $\psi_p(q) = 0$. Since $C \cap V_*$ is dense in C, it follows that $\psi_p|_C \equiv 0$. $\square$

By the above theorem, we get a relation between the topological properties of K and the positivity of the elements of D, where $(D, \mathbf{r})$ is the harmonic structure.

Theorem 3.2.11. *Let $(D, \mathbf{r})$ be a harmonic structure. Then $D_{pq} > 0$ if and only if $J(p, V_0) \cap J(q, V_0) \neq \emptyset$.*

The condition that $J(p, V_0) \cap J(q, V_0) \neq \emptyset$ is equivalent to the one that there exists a connected component of $K \backslash V_0$ whose closure contains both p and q.

Proof: Part I. Let $(\mathcal{E}, \mathcal{F})$ be the resistance form associated with $(D, \mathbf{r})$. Then it follows that $D_{pq} = -\mathcal{E}_0(\psi_p|_{V_0}, \psi_q|_{V_0}) = -\mathcal{E}_m(\psi_p|_{V_m}, \psi_q|_{V_m})$. Hence

$$D_{pq} = (H_m \psi_p)(q) = (H_m \psi_q)(p). \tag{3.2.5}$$

Assume that $J(p, V_0) \cap J(q, V_0) = \emptyset$. Since $K_{m,q} \subseteq \cup_{C \in J(q, V_0)} C$ for sufficiently large m, Theorem 3.2.8 implies that $\psi_p|_{K_{m,q}} \equiv 0$. Hence, by (3.2.5), $D_{pq} = 0$. $\qquad\square$

To prove the rest of this theorem, we need the following lemmas.

Lemma 3.2.12. *Let p and q belong to V_0. If $D_{pq} > 0$, then there exists a path $\gamma : [0, 1] \to K$ between p and q that satisfies $\gamma((0, 1)) \subseteq K \backslash V_0$.*

Proof. By the above proof of Theorem 3.2.11, if $D_{pq} > 0$, then $J(p, V_0) \cap J(q, V_0) \neq \emptyset$. Let $C \in J(p, V_0) \cap J(q, V_0)$. Then, by Proposition 1.6.6, there exists a path $\gamma : [0, 1] \to C \cup \{p, q\}$ with $\gamma(0) = p$ and $\gamma(1) = q$. $\qquad\square$

Lemma 3.2.13. *Let p and q belong to V_m. Let $\{p_i\}_{i=1,2,\ldots,n}$ be an H_m-path between p and q. Then there exists a connected component of $K \backslash V_0$, C, such that $p_i \in C$ for all $i = 1, 2, \ldots, n$, $p \in \overline{C}$ and $q \in \overline{C}$.*

Proof. For each $i \in \{1, 2, \ldots, n - 1\}$, there exists $w \in W_m$ such that $p_i, p_{i+1} \in K_w$ and $D_{(F_w)^{-1}(p_i), (F_w)^{-1}(p_{i+1})} > 0$. By Lemma 3.2.12, it follows that there exists a path γ_i between p_i and p_{i+1} that satisfies $\gamma_i((0, 1)) \subseteq K \backslash V_m$. Hence there exists a connected component of $K \backslash V_0$ which contains both p_i and p_{i+1}. Inductively, we see that the connected component, say C, contains all p_i. By the same argument as p_i and p_{i+1}, we may find a path γ between p and p_1 that satisfies $\gamma((0, 1)) \subseteq K \backslash V_m$. So it follows that $\gamma((0, 1)) \subseteq C$. Hence $p \in \overline{C}$. Exactly the same argument implies that $q \in \overline{C}$ as well. $\qquad\square$

Part II of the proof of Theorem 3.2.11. Assume that $J(p, V_0) \cap J(q, V_0) \neq \emptyset$. Let $C \in J(p, V_0) \cap J(q, V_0)$. Then, by Proposition 1.6.6, there exists a path $\gamma : [0, 1] \to C \cup \{p, q\}$ with $\gamma(0) = p$ and $\gamma(1) = q$. Now by the same argument as in the proof of Theorem 3.2.8, we may construct an H_m-path between p and q from the continuous path γ. Let $\{p_i\}_{i=1}^n$ be the H_m-path between p and q. Then, by Lemma 3.2.13, p_i belongs to C for all $i = 1, 2, \dots, n$. Since $(H_m)_{p_nq} > 0$ and $\psi_p(p_n) > 0$, (3.2.5) implies that $D_{pq} = (H_m\psi_p)(q) \geq (H_m)_{p_nq}(\psi_p(p_n) - \psi_p(q)) > 0$. $\qquad\square$

By the above discussion, we also obtain a stronger version of the maximum principle.

Theorem 3.2.14 (Maximum principle: strong version).
Let $x \in K\backslash V_0$ and let C be the connected component of $K\backslash V_0$ with $x \in C$. Then, for any harmonic function u,

$$\min_{p \in V_0 \cap \overline{C}} u(p) \leq u(x) \leq \max_{p \in V_0 \cap \overline{C}} u(p).$$

Moreover, if any of the equalities holds, then u is constant on $\overline{C}$.

Proof. First assume that $x \in V_m$ for some m. Then the same arguments as in the proof of Theorem 3.2.8 imply that, for any $p \in \overline{C} \cap V_0$, there exists an H_m-path between x and p. Conversely, by Lemma 3.2.13, if there exists an H_m-path between x and $p \in V_0$, then $p \in \overline{C} \cap V_0$. Hence, let $V = V_m$, $U = V_0$ and $H = H_m$ and apply Proposition 2.1.7. Then $U_x = \overline{C} \cap V_0$. Therefore, the desired results are immediate from the conclusion of Proposition 2.1.7.

Next if $x \notin V_*$, then choose $w \in W_*$ so that $x \in K_w \subseteq C$. Then, by the weak version of the maximum principle (Theorem 3.2.5),

$$\min_{q \in F_w(V_0)} u(q) \leq u(x) \leq \max_{q \in F_w(V_0)} u(q).$$

Also, if $u(x) = \max_{q \in F_w(V_0)} u(q)$, then $u(x) = u(q)$ for some $q \in F_w(V_0)$. Since $F_w(V_0) \subseteq C \cap V_m$, where $m = |w|$, the first part of this proof implies the desired results for this case as well. $\qquad\square$

Next we define the notion of piecewise harmonic functions. The following is an immediate corollary of Theorem 3.2.4.

Corollary 3.2.15. *For $\rho \in \ell(V_m)$, there exists a unique continuous function u on K such that $u|_{V_m} = \rho$ and $\mathcal{E}(u|_{V_*}, u|_{V_*}) = \min\{\mathcal{E}(v, v) : v \in \mathcal{F}, v|_{V_m} = \rho\}$.*

u in the above corollary is called an m-harmonic function with boundary value ρ. Another characterization of m-harmonic functions is that u is an m-harmonic function if and only if $u \circ F_w$ is a harmonic function for any $w \in W_m$. For $p \in V_m$, define ψ_p^m to be the m-harmonic function with boundary value $\chi_p^{V_m}$. Then any m-harmonic function u is a linear combination of $\{\psi_p^m\}$. In fact, $u = \sum_{p \in V_m} u(p)\psi_p^m$. For $u \in \ell(V_*)$, we define $P_m u$ by

$P_m u = \sum_{p \in V_m} u(p) \psi_p^m$. If we write $u_m = P_m u$, then $\mathcal{E}_m(u|_{V_m}, u|_{V_m}) = \mathcal{E}(u_m, u_m)$.

In the rest of this section, we will give an expansion of $u \in \ell(V_*)$ in a piecewise harmonic basis $\{\psi_p\}_{p \in V_*}$, where $\psi_p = \psi_p^m$ if $p \in V_m \backslash V_{m-1}$.

Lemma 3.2.16. *Let u be an m-harmonic function. Then $\mathcal{E}(u, f) = 0$ for $f \in \mathcal{F}$ if $f|_{V_m} = 0$.*

Proof. For $n > m$, we have $(H_n u)(p) = 0$ if $p \in V_n \backslash V_m$ and $f(p) = 0$ if $p \in V_m$. Hence $\mathcal{E}_n(u, f) = - \sum_{p \in V_n} f(p)(H_n u)(p) = 0$. □

Lemma 3.2.17. *For $u \in \mathcal{F}$, $\mathcal{E}(u - u_m, u - u_m) \to 0$ as $m \to \infty$, where $u_m = P_m u$.*

Proof. By Lemma 3.2.16, $\mathcal{E}(u - u_m, u_m) = 0$. Hence, $\mathcal{E}(u - u_m, u - u_m) = \mathcal{E}(u, u) - \mathcal{E}(u_m, u_m) = \mathcal{E}(u, u) - \mathcal{E}_m(u, u)$. □

Definition 3.2.18. *Let $u \in \ell(V_*)$. For $p \in V_m \backslash V_{m-1}$, define $\alpha_p(u) = u_m(p) - u_{m-1}(p)$. ($\alpha_p(u) = u(p)$ if $p \in V_0$.) Also define $a_w(u) = u_{m+1} \circ F_w - u_m \circ F_w$ for $w \in W_m$.*

Note that $\alpha_p(u) = u(p) - u_{m-1}(p)$ for $p \in V_m \backslash V_{m-1}$. By the above definition,

$$u_m = \sum_{p \in V_m} \alpha_p(u) \psi_p \quad \text{and} \quad u = \sum_{p \in V_*} \alpha_p(u) \psi_p,$$

where the infinite sum in the latter equality is pointwise convergent. (In fact, it is a finite sum for each $p \in V_*$.) This is the expansion of a function on V_* in the piecewise basis $\{\psi_p\}_{p \in V_*}$.

Also, by Definition 3.2.18, $a_w(u)$ is a 1-harmonic function and $a_w(u)|_{V_0} = 0$. Using $\alpha_p(u)$, $a_w(u)$ is given by

$$a_w(u) = \sum_{p \in V_1 \backslash V_0} \alpha_{F_w(p)}(u) \psi_p. \tag{3.2.6}$$

Proposition 3.2.19. *For $u \in \ell(V_*)$, $u \in \mathcal{F}$ if and only if*

$$\mathcal{E}_0(u, u) + \sum_{m \geq 0} \sum_{w \in W_m} \frac{1}{r_w} \mathcal{E}_1(a_w(u), a_w(u)) < \infty. \tag{3.2.7}$$

Moreover, if $u \in \mathcal{F}$, then the sum in (3.2.7) is equal to $\mathcal{E}(u, u)$.

Proof. By Lemma 3.2.16,

$$\mathcal{E}_m(u, u) = \sum_{k=0}^{m-1} \mathcal{E}_{k+1}(u_{k+1} - u_k, u_{k+1} - u_k) + \mathcal{E}_0(u, u).$$

By the self-similarity of $\mathcal{E}_{k+1}$, we obtain

$$\mathcal{E}_{k+1}(u_{k+1} - u_k, u_{k+1} - u_k) = \sum_{w \in W_k} \frac{1}{r_w} \mathcal{E}_1(a_w(u), a_w(u)).$$

Combining the above two equalities, we easily verify the statement of the proposition. $\qquad\square$

3.3 Topology given by effective resistance

As in the last section, we assume that $(D, \mathbf{r})$ is a harmonic structure on $\mathcal{L} = (K, S, \{F_i\}_{i \in S})$, where $S = \{1, 2, \ldots, N\}$. Let $\{(V_m, H_m)\}_{m \geq 0}$ be the compatible sequence induced by $(D, \mathbf{r})$. Then, there exist $(\mathcal{E}, \mathcal{F}) \in \mathcal{RF}(V_*)$ and $R \in \mathcal{RM}(V_*)$ associated with $\{(V_m, H_m)\}_{m \geq 0}$. Moreover, if (Ω, R) is the completion of (V_*, R), the results in 2.3 and 2.4 imply that $(\mathcal{E}, \mathcal{F}) \in \mathcal{RF}(\Omega)$.

In this section, we study the difference between Ω and K, in other words, the difference between two topologies on V_*. One is given by the resistance metric R and the other is given by the relative topology from the original metric space (K, d). Roughly speaking, it turns out that Ω can be always identified with a subset of K and that $\Omega = K$ if and only if $(D, \mathbf{r})$ is regular.

First, we observe an important property of $(\mathcal{E}, \mathcal{F})$, that is, the self-similarity of $(\mathcal{E}, \mathcal{F})$.

Proposition 3.3.1. *For any $u \in \mathcal{F}$ and $i \in S$, $u \circ F_i \in \mathcal{F}$. Moreover,*

$$\mathcal{E}(u, v) = \sum_{i=1}^{N} \frac{1}{r_i} \mathcal{E}(u \circ F_i, v \circ F_i)$$

for any $u, v \in \mathcal{F}$.

The above proposition is easily verified by (3.1.1). Moreover, if Λ is a partition of $\Sigma = S^{\mathbb{N}}$, then an inductive argument on $\#(\Lambda)$ shows that

$$\mathcal{E}(u, v) = \sum_{w \in \Lambda} \frac{1}{r_w} \mathcal{E}(u \circ F_w, v \circ F_w). \tag{3.3.1}$$

Now we start to study the difference between Ω and K.

Proposition 3.3.2. *There exists a continuous injective map $\theta : \Omega \to K$ which is the identity on V_*.*

Lemma 3.3.3. *If $F_w(V_*) \cap F_v(V_*) = \emptyset$ for $w, v \in W_*$, then $\inf\{R(p, q) : p \in F_w(V_*), q \in F_v(V_*)\} > 0$.*

Proof. Choose $m \geq 0$ so that $w, v \in W_m$. Define $\rho \in \ell(V_m)$ by

$$\rho(q) = \begin{cases} 1 & \text{if } q \in F_w(V_*) \cap V_m, \\ 0 & \text{otherwise.} \end{cases}$$

Then $\rho|_{F_v(V_*) \cap V_m} = 0$. Hence, if u is the m-harmonic function with boundary value ρ, then $u|_{F_w(V_*)} \equiv 1$ and $u|_{F_v(V_*)} \equiv 0$. By (2.2.6), if $p \in F_w(V_*)$ and $q \in F_v(V_*)$, then

$$R(p, q) \geq \frac{|u(p) - u(q)|^2}{\mathcal{E}(u, u)} \geq \frac{1}{\mathcal{E}(u, u)}.$$

$\square$

Recall that d is a metric on K which is compatible with the original topology of K. Note that (K, d) is a compact metric space.

Proof of Proposition 3.3.2. Step 1: Construction of θ

Let $\{p_i\}$ be a Cauchy sequence in (V_*, R). By Lemma 3.3.3, defining δ_m by

$$\min_{w, v \in W_m, F_w(V_*) \cap F_v(V_*) = \emptyset} \left(\inf_{p \in F_w(V_*), q \in F_v(V_*)} R(p, q) \right)$$

then we have $\delta_m > 0$. Choose $n(m)$ so that $R(p_k, p_l) \leq \delta_m$ for $k, l \geq n(m)$, then there exists $w \in W_m$ such that $p_k \in \cup_{v \in W_m : F_w(V_*) \cap F_v(V_*) \neq \emptyset} F_v(V_*)$ for $k \geq n(m)$. As $\max_{w \in W_m} \operatorname{diam}(K_w, d) \to 0$ as $m \to \infty$, $\{p_i\}$ becomes a Cauchy sequence in (K, d). ($\operatorname{diam}(\cdot, d)$ means the diameter with respect to a metric d.) So let p be the limit point of $\{p_i\}$ in (V_*, R). Then we define $\theta(p)$ by the limit point of $\{p_i\}$ in (K, d). It is routine to show that θ is a well-defined continuous map from Ω to K and $\theta|_{V_*}$ is the identity.

Step 2: θ is injective.

Assume that $\theta(p) = \theta(q)$ for $p, q \in \Omega$. If $R(p_i, p) \to 0$ and $R(q_i, q) \to 0$ as $i \to \infty$, then the discussion in Step 1 shows that $d(p_i, \theta(p)) \to 0$ and $d(q_i, \theta(q)) \to 0$ as $i \to \infty$. Let v be an m-harmonic function on V_*. Since v can be extended as a continuous function on both (K, d) and (Ω, R), we see that $v(p) = \lim_{i \to \infty} v(p_i) = v(\theta(p)) = v(\theta(q)) = \lim_{i \to \infty} v(q_i) = v(q)$.

Now, for $u \in \mathcal{F}$, by (2.2.6)

$$|u_m(p) - u(p)|^2 \leq \mathcal{E}(u_m - u, u_m - u) R(p, p_0),$$

where $p_0 \in V_0$. (Recall that u_m is an m-harmonic function defined by $u_m = \sum_{x \in V_m} u(x) \psi_m^x$.) Hence, using Lemma 3.2.17, we have $\lim_{m \to \infty} u_m(p) = u(p)$. In the same manner, we can also show that $\lim_{m \to \infty} u_m(q) = u(q)$. Since u_m is a m-harmonic function, it follows that $u_m(p) = u_m(q)$. Hence $u(p) = u(q)$. Therefore, using (2.3.1), we can conclude that $p = q$. $\square$

By virtue of Proposition 3.3.2, we will identify Ω with $\theta(\Omega)$ and think of Ω as a subset of K.

Now we ask when Ω is equal to K. The answer is given by the following theorem.

Theorem 3.3.4. *The following are equivalent.*
(1) $\Omega = K$.
(2) (Ω, R) *is compact.*
(3) (Ω, R) *is bounded.*
(4) *For any $u \in \mathcal{F}$, $\sup_{p \in \Omega} |u(p)| < \infty$.*
(5) $(D, \mathbf{r})$ *is regular.*

Moreover, if (D, r) is regular, then R is a metric on K which is compatible with the original topology.

To prove the above theorem, we need the following lemmas.

Lemma 3.3.5. *For any $w \in W_*$ and for any $p, q \in \Omega$,*

$$r_w R(p, q) \geq R(F_w(p), F_w(q)). \tag{3.3.2}$$

Proof. By Proposition 3.3.1,

$$\mathcal{E}(u, u) = \sum_{w' \in W_m} r_{w'}^{-1} \mathcal{E}(u \circ F_{w'}, u \circ F_{w'}) \geq r_w^{-1} \mathcal{E}(u \circ F_w, u \circ F_w)$$

for $w \in W_m$. Therefore, we have

$$r_w \frac{|u \circ F_w(p) - u \circ F_w(q)|^2}{\mathcal{E}(u \circ F_w, u \circ F_w)} \geq \frac{|u(F_w(p)) - u(F_w(q))|^2}{\mathcal{E}(u, u)}.$$

Hence, by (2.3.1), we obtain (3.3.2). $\quad\square$

Lemma 3.3.6. *Let $\omega = \omega_1 \omega_2 \ldots \in \Sigma$. Assume that*

$$\limsup_{m \to \infty} \min_{\omega' \in \mathcal{P}} \delta(\sigma^m \omega, \omega') > 0, \tag{3.3.3}$$

where δ is any of the metrics δ_r on Σ defined in Theorem 1.2.2. If

$$\liminf_{m \to \infty} r_{\omega_1 \omega_2 \ldots \omega_m} > 0,$$

then $\pi(\omega) \notin \Omega$.

Proof. Set $\varphi = \sum_{p \in V_1 \setminus V_0} \psi_p$ and

$$\varphi_w(x) = \begin{cases} \varphi \circ F_w^{-1}(x) & \text{if } x \in K_w, \\ 0 & \text{otherwise,} \end{cases}$$

for $w \in W_*$. By (3.3.3), there exist $\tau \in \Sigma \setminus \mathcal{P}$ and $\{m_n\}_{n \geq 1} \subset \mathbb{N}$ such that $m_n < m_{n+1}$ for any $n \geq 1$ and $\sigma^{m_n}\omega \to \tau$ as $n \to \infty$. Set $w^n = \omega_1\omega_2\ldots\omega_{m_n}$ and define $u = \sum_{n\geq 1} n^{-1}\varphi_{w^n}$. Then

$$a_w(u) = \begin{cases} n^{-1}\varphi & \text{if } w = w^n, \\ 0 & \text{otherwise.} \end{cases}$$

Note that $\liminf_{n\to\infty} r_{w^n} > 0$. Hence

$$\mathcal{E}_0(u,u) + \sum_{m\geq 0} \sum_{w\in W_m} \frac{1}{r_w}\mathcal{E}_1(a_w(u), a_w(u))$$
$$= \mathcal{E}_0(u,u) + \sum_{n\geq 1} \frac{1}{n^2 r_{w^n}}\mathcal{E}_1(\varphi,\varphi) < \infty.$$

By Proposition 3.2.19, we see that $\mathcal{E}(u,u) < \infty$ and $u \in \mathcal{F}$.

Now we assume (3.1.2). Then, by Lemma 1.6.10, $\varphi(x) > 0$ for any $x \in K\backslash V_0$. Since $\tau \notin \mathcal{P}$, we can choose $m \geq 0$ so that $C = \min_{x\in K_{\tau_1\tau_2\ldots\tau_m}} \varphi(x) > 0$. Without loss of generality, we may assume that $\sigma^{m_n}\omega \in \Sigma_{\tau_1\tau_2\ldots\tau_m}$. So, if $w(n) = w^n\tau_1\tau_2\ldots\tau_m$, then $w(n) = \omega_1\omega_2\ldots\omega_{m_n+m}$ and

$$\min_{x\in K_{w(n)}} \varphi_{w^n}(x) = \min_{x\in K_{\tau_1\tau_2\ldots\tau_m}} \varphi(x) = C.$$

Hence $\inf\{u(x) : x \in K_{w(n)} \cap V_*\} \geq \sum_{k=1}^{n-1} C/k \to \infty$ as $n \to \infty$. Suppose that $\pi(\omega) \in \Omega$. Then there exists $\{p_i\}_{i\geq 1} \subset V_*$ such that $p_i \to \pi(\omega)$ as $i \to \infty$ in (Ω, R). Since θ is continuous, it follows that $p_i \to \pi(\omega)$ as $i \to \infty$ in (K,d) as well. Therefore, $p_i \in K_{w(n)}$ for sufficiently large i. Hence $u(p_i) \to \infty$ as $i \to \infty$. On the other hand, u is a continuous function on (Ω, R). Hence $\{u(p_i)\}_{i\geq 1}$ converges to a finite limit as $i \to \infty$. Thus we conclude that $\pi(\omega) \notin \Omega$.

If (3.1.2) is not satisfied, again we may repeat the same arguments as in the proofs of Proposition 3.1.8 and Theorem 3.2.4. $\square$

Lemma 3.3.7. *If $(D, \mathbf{r})$ is not regular, then $\Omega \neq K$. Also there exists $u \in \mathcal{F}$ such that $\sup_{p\in V_*} |u(p)| = \infty$.*

Proof. If $(D, \mathbf{r})$ is not regular, there exists $r_k \geq 1$. Let $\omega = \dot{k}$. Then Proposition 3.1.8 implies that $\omega \notin \mathcal{P}$. Hence ω satisfies the conditions in Lemma 3.3.6. So $\pi(\omega) \notin \Omega$ and $\Omega \neq K$. Also, by the proof of Lemma 3.3.6, we can find $u \in \mathcal{F}$ that is unbounded on V_*. $\square$

Proof of Theorem 3.3.4. Assume that $(D, \mathbf{r})$ is regular. Then Lemma 3.3.5 implies that $\{F_i\}_{i\in S}$ is a system of contractions on a complete metric space (Ω, R). Hence, by Theorem 1.1.4, there exists a non-empty compact subset $\tilde{K}$ of (Ω, R) such that $\tilde{K} = \cup_{i\in S} F_i(\tilde{K})$. As θ is continuous, $\tilde{K}$ is also a compact subset of (K, d). For $x \in \tilde{K}$, $\cup_{w \in W_*} F_w(x)$ is a subset of $\tilde{K}$. It is easy to see that $\cup_{w \in W_*} F_w(x)$ is dense in (K, d). Hence $K = \tilde{K}$. Therefore $\Omega = K$, (Ω, R) is compact, and θ is a homeomorphism between (Ω, R) and (K, d).

Obviously (2) implies (3). Also as $|u(p) - u(q)|^2 \leq \mathcal{E}(u, u)R(p, q)$ for $u \in F$, we can see that (3) implies (4).

The above discussion with Lemma 3.3.7 completes the proof of Theorem 3.3.4. $\qquad\square$

Next, we consider how large Ω is in K if $(D, \mathbf{r})$ is not regular. For instance, we will calculate $\mu(\Omega)$ for a self-similar measure μ on K.

Theorem 3.3.8. *Let μ be a self-similar measure on K with weight $(\mu_i)_{i \in S}$.*
(1) *If $\prod_{i \in S}(r_i)^{\mu_i} < 1$, then $\mu(\Omega) = 1$.*
(2) *If $\prod_{i \in S}(r_i)^{\mu_i} > 1$, then $\mu(\Omega) = 0$.*

Remark. A. Teplyaev showed that if $\prod_{i \in S}(r_i)^{\mu_i} = 1$, then $\mu(\Omega) = 0$.

This theorem is somewhat intriguing, in particular, in the second case where $\mu(\Omega) = 0$. As is shown in the next section, even in such a case, if $r_i \mu_i < 1$ for all $i \in S$, then $(\mathcal{E}, \mathcal{F})$ is a local regular Dirichlet form on $L^2(K, \mu)$. So $u \in \mathcal{F}$ is originally a continuous function on (V_*, R) and it can be extended to a continuous function on (Ω, R). This is not surprising. But, despite the fact that Ω is a null set in K, u can be extended to an L^2-class function on a far larger space K. This fact might suggest that some of the analysis on a large (higher dimensional) space could be determined by information on a small subset of the space.

Lemma 3.3.9. *Let $\omega = \omega_1\omega_2 \ldots \in \Sigma$. If $\sum_{m \geq 0} r_{\omega_1\omega_2\ldots\omega_m} < \infty$, then $\pi(\omega) \in \Omega$.*

Proof. Let $\tau \in \mathcal{P}$ and set $p_m = \pi(\omega_1\omega_2\ldots\omega_m\tau)$. Then $p_m \in V_m$, and, using Lemma 3.3.5, we see that

$$R(p_m, p_{m+1}) = R(F_{\omega_1\omega_2\ldots\omega_m}(p_0), F_{\omega_1\omega_2\ldots\omega_m}(F_{\omega_{m+1}}(p_0)))$$
$$\leq r_{\omega_1\omega_2\ldots\omega_m} \max_{x,y \in V_1} R(x, y).$$

Hence $\{p_m\}$ is a Cauchy sequence in (Ω, R). Let p be its limit. On the other hand, $d(p_m, \pi(\omega)) \to 0$ as $m \to \infty$. Hence $\theta(p) = \pi(\omega) \in \Omega$. $\qquad\square$

Proof of Theorem 3.3.8. Set $n_m(i, \omega) = \#\{j : \omega_j = i, 1 \leq j \leq m\}$ for $\omega \in \Sigma$, $i \in S$ and $m \geq 1$. Define

$$\mathcal{N} = \{\omega \in \Sigma : \lim_{m \to \infty} \frac{n_m(i, \omega)}{m} = \mu_i \quad \text{for all } i \in S\}.$$

Then the strong law of large number implies that $\mu(\mathcal{N}) = 1$. Note that $\frac{1}{m} \log r_{\omega_1\omega_2...\omega_m} = \sum_{i\in S} \frac{n_m(i,\omega)}{m} \log r_i$. Hence, for $\omega \in \mathcal{N}$,

$$\lim_{m\to\infty} |r_{\omega_1\omega_2...\omega_m}|^{1/m} = \prod_{i\in S} (r_i)^{\mu_i}.$$

Now, if $\prod_{i\in S}(r_i)^{\mu_i} < 1$, then $\sum_{m\geq 0} r_{\omega_1\omega_2...\omega_m} < \infty$ for $\omega \in \mathcal{N}$. Therefore, Lemma 3.3.9 implies $\pi(\omega) \in \Omega$ for any $\omega \in \mathcal{N}$. Hence $\mu(\Omega) = 1$.

Next, suppose $\prod_{i\in S}(r_i)^{\mu_i} > 1$. Let $\mathcal{P}_m = \{\omega_1\omega_2...\omega_m : \omega \in \mathcal{P}\}$. Then we can choose m so that $\#(\mathcal{P}_m) < \#(W_m)$. Set

$$B_k = \{\omega \in \Sigma : \omega_k\omega_{k+1}...\omega_{k+m-1} \in \mathcal{P}_m\}$$

and $A = \liminf_{k\to\infty} B_k$. If $\min_{\tau\in\mathcal{P}} \delta(\sigma^n\omega, \tau) \to 0$ as $n \to \infty$, then, for sufficiently large k, $\omega_k\omega_{k+1}...\omega_{k+m-1} \in \mathcal{P}_m$. Hence $\omega \in A$. Now, as B_{k+ml} for $l = 1, 2, \ldots$ are independent and $\mu(B_k) = \mu(B_{k+ml}) < 1$, we see that $\mu(\cap_{j\geq k}B_j) = 0$. Hence $\mu(A) = 0$. So $\mu(\mathcal{N} \cap A^c) = 1$. Assume that $\omega \in \mathcal{N} \cap A^c$. Then $\limsup_{m\to\infty} \min_{\tau\in\mathcal{P}} \delta(\sigma^m\omega, \tau) > 0$. Also, $\liminf_{m\to\infty} r_{\omega_1\omega_2...\omega_m} > 0$ because $\lim_{m\to\infty} |r_{\omega_1\omega_2...\omega_m}|^{1/m} = \prod_{i\in S}(r_i)^{\mu_i} > 1$. Hence, by Lemma 3.3.6, $\pi(\omega) \notin \Omega$. As $\mu(\mathcal{N} \cap A^c) = 1$, it follows that $\mu(\Omega) = 0$. $\qquad\square$

3.4 Dirichlet forms on p. c. f. self-similar sets

In this section, we continue with the assumptions of the previous section: $(D, \mathbf{r})$ is a harmonic structure on $(K, S, \{F_i\}_{i\in S})$. Then we have $(\mathcal{E}, \mathcal{F}) \in \mathcal{RF}(\Omega)$, where $(\mathcal{E}, \mathcal{F})$ is the natural limit of the compatible sequence $\{(V_m, H_m)\}_{m\geq 0}$. If the harmonic structure $(D, \mathbf{r})$ is regular, it was shown in Theorem 3.3.4 that $K = \Omega$. Then, results from Chapter 2, in particular, Theorems 2.2.6, 2.3.10 and 2.4.1, immediately imply that $(\mathcal{E}, \mathcal{F})$ is a local regular Dirichlet form on $L^2(K, \mu)$ for any Borel regular probability measure μ on K. See B.3 for the definition of Dirichlet forms.

On the other hand, by Theorem 3.3.4, Ω is a proper subset of K if $(D, \mathbf{r})$ is not regular. In this case, functions in $\mathcal{F}$ may not be extendable to continuous functions on K. (In fact, $\mathcal{F}$ turns out to contain an unbounded function by Theorem 3.3.4.) In this section, however, we will show that $\mathcal{F}$ can be embedded in $L^2(K, \mu)$ for a certain measure μ and that $(\mathcal{E}, \mathcal{F})$ is a local regular Dirichlet form on $L^2(K, \mu)$.

Let $M(K)$ be the collection of all Borel regular probability measures on

K. Also we define

$$\widetilde{M}(K) = \{\mu \in M(K) : \mu(V_*) = 0 \text{ and } \mu(O) > 0$$
$$\text{for any non-empty open set } O \subset K.\}$$

Of course, all self-similar measures belong to $\widetilde{M}(K)$.

The following sufficient condition for μ to be in $\widetilde{M}(K)$ is sometimes useful.

Lemma 3.4.1. *Let $\mu \in M(K)$. If there exists $\gamma > 0$ such that $\gamma \mu(K_w) \leq \mu(K_{wi})$ for any $(w, i) \in W_* \times S$, then $\mu \in \widetilde{M}(K)$.*

A measure satisfying the condition in the above theorem is called a γ-elliptic measure.

Proof. If we choose m large enough, then for any $w \in W_m$, there exists $w' \in W_m$ such that $K_w \cap K_{w'} = \emptyset$. So, changing the self-similar structure $\mathcal{L} = (K, S, \{F_i\}_{i \in S})$ to $\mathcal{L}_m = (K, W_m, \{F_w\}_{w \in W_m})$ if necessary, without loss of generality we may assume that for any $i \in S$, there exists $j \in S$ such that $K_i \cap K_j = \emptyset$. Then $1 = \mu(K) \geq \mu(K_i) + \mu(K_j) \geq 2\gamma$. Therefore $\gamma \leq 1/2$. Let $\delta = 1 - \gamma$. Then $0 < \delta < 1$. Now, for any $w \in W_*$ and any $i \in S$,

$$\mu(K_w) \geq \mu(K_{wi}) + \mu(K_{wj}) \geq \mu(K_{wi}) + \gamma \mu(K_w)$$

where we choose $j \in S$ so that $K_i \cap K_j = \emptyset$. This implies that $\mu(K_{wi}) \leq \delta \mu(K_w)$. Hence $\gamma^m \leq \mu(K_w) \leq \delta^m$ for any $w \in W_m$. This inequality implies that $\mu(p) = 0$ for any $p \in K$ and that $\mu(O) > 0$ for any non-empty open subset $O \subset K$. $\square$

As in 3.2, set $P_m : \mathcal{F} \to C(K)$ by $P_m u = \sum_{p \in V_m} u(p)\psi_p^m$. (In 3.2, $P_m u$ is denoted by u_m. We also use this notation in this section.) Recall that u_m is an m-harmonic function and it can be regarded as a continuous function on K. Also, define $\mathcal{H}_{1,0} = \{u : u \text{ is a 1-harmonic function and } u|_{V_0} \equiv 0\}$.

Lemma 3.4.2. *Then there exists $c > 0$ such that $(u, u)_\mu \leq c \mathcal{E}(u, u)$ for any $u \in \mathcal{H}_{1,0}$ and for any $\mu \in M(K)$, where $(u, v)_\mu = \int_K uv d\mu$.*

Proof. Note that both $(\cdot, \cdot)_\mu$ and $\mathcal{E}$ are inner-products on a finite dimensional vector space $\mathcal{H}_{1,0}$. As $u = \sum_{p \in V_1 \setminus V_0} u(p)\psi_p$ for any $u \in \mathcal{H}_{1,0}$, we have

$$(u, u)_\mu = \sum_{p,q \in V_1 \setminus V_0} a_{pq} u(p) u(q) \leq \frac{1}{2} \sum_{p,q \in V_1 \setminus V_0} (u(p)^2 + u(q)^2)$$

$$= \#(V_1 \setminus V_0) \sum_{p \in V_1 \setminus V_0} u(p)^2,$$

where $a_{pq} = \int_K \psi_p \psi_q d\mu$. (Since $0 \leq \psi_p(x) \leq 1$ for any $x \in K$, we have $0 \leq a_{pq} \leq 1$.) Also there exists $c > 0$ such that $\#(V_1 \setminus V_0) \sum_{p \in V_1 \setminus V_0} u(p)^2 \leq c\mathcal{E}(u, u)$ for any $u \in \mathcal{H}_{1,0}$. This proves the lemma. $\qquad\square$

Lemma 3.4.3. *Let $\mu \in \widetilde{M}(K)$. Define $R_m(\mu) = \max_{w \in W_m} r_w \mu(K_w)$. If $\sum_{m \geq 0} R_m(\mu) < \infty$, then $\{P_m u\}$ converges in $L^2(K, \mu)$ as $m \to \infty$ for any $u \in \mathcal{F}$.*

For ease of notation, we write R_m instead of $R_m(\mu)$ if no confusion can occur.

Proof. For $w \in W_*$, define μ^w by $\mu^w(A) = \mu(F_w(A))/\mu(K_w)$. Then $\mu^w \in \widetilde{M}(K)$. Note that $\int_K f \circ F_w d\mu^w = \frac{1}{\mu(K_w)} \int_{K_w} f d\mu$. Now, by Proposition 3.3.1 and Lemma 3.4.2, it follows that, for $u \in \mathcal{F}$,

$$\mathcal{E}(u_{m+1} - u_m, u_{m+1} - u_m)$$

$$= \sum_{w \in W_m} \frac{1}{r_w} \mathcal{E}((u_{m+1} - u_m) \circ F_w, (u_{m+1} - u_m) \circ F_w)$$

$$\geq c^{-1} \sum_{w \in W_m} \frac{1}{r_w} \int_K ((u_{m+1} - u_m) \circ F_w)^2 d\mu^w$$

$$= c^{-1} \sum_{w \in W_m} \frac{1}{\mu(K_w) r_w} \int_{K_w} (u_{m+1} - u_m)^2 d\mu$$

$$\geq c^{-1} \frac{1}{R_m} \int_K (u_{m+1} - u_m)^2 d\mu.$$

Hence $cR_m \mathcal{E}(u_{m+1} - u_m, u_{m+1} - u_m) \geq \|u_{m+1} - u_m\|_\mu^2$, where $\|u\|_\mu^2 = \int_K u^2 d\mu$. So

$$\sqrt{c} \sum_{k=m}^{n-1} \sqrt{R_k} \sqrt{\mathcal{E}(u_{k+1} - u_k, u_{k+1} - u_k)} \geq \|u_n - u_m\|_\mu$$

for $n \geq m$. By Lemma 3.2.16, we have $\sum_{k=m}^{n-1} \mathcal{E}(u_{k+1} - u_k, u_{k+1} - u_k) = \mathcal{E}(u_n - u_m, u_n - u_m)$. Hence

$$c\Big(\sum_{k=m}^{n-1} R_k \Big) \mathcal{E}(u_n - u_m, u_n - u_m) \geq \|u_n - u_m\|_\mu^2. \qquad (3.4.1)$$

This shows that $\{u_m\}_{m\geq 0}$ is a Cauchy sequence in $L^2(K,\mu)$. $\qquad\square$

Define $\iota_\mu(u)$ for $u \in \mathcal{F}$ as the limit of $\{u_m\}_{m\geq 0}$ in $L^2(K,\mu)$ as $m \to \infty$. Then $\iota_\mu : \mathcal{F} \to L^2(K,\mu)$ is a linear map. ι_μ will be shown to be an injective map.

Lemma 3.4.4. *Let $\mu \in \widetilde{M}(K)$. If $\sup_{w\in W_*} \sum_{m\geq 0} R_m(\mu^w) < \infty$, then $\iota_\mu : \mathcal{F} \to L^2(K,\mu)$ is injective.*

Proof. By (3.4.1), we have

$$c\Big(\sum_{k\geq m} R_k \Big)\mathcal{E}(u-u_m, u-u_m) \geq ||\iota_\mu(u) - u_m||_\mu^2. \qquad (3.4.2)$$

Note that c is independent of μ. First we will show the following claim.
Claim: There exists $c_1 > 0$ such that, for any $\mu \in M(K)$ and for any $u \in \mathcal{F}$ with $\iota_\mu(u) = 0$,

$$\max\{1, \sum_{m\geq 0} R_m(\mu)\}\mathcal{E}(u,u) \geq c_1 \max_{p\in V_0} |u(p)|^2. \qquad (3.4.3)$$

Proof of Claim. Let $S = \max\{1, \sum_{m\geq 0} R_m(\mu)\}$. If $\iota_\mu(u) = 0$, set $m = 0$ in (3.4.2). Then we have

$$cS\mathcal{E}(u,u) \geq cS\mathcal{E}(u_0,u_0) + \int_K |u_0|^2 d\mu$$
$$\geq c\mathcal{E}(u_0,u_0) + \int_K |u_0|^2 d\mu \geq c\mathcal{E}(u_0,u_0) + (\tilde{u}_0)^2, \qquad (3.4.4)$$

where $\tilde{u}_0 = \int_K u_0 d\mu = \sum_{p\in V_0} u(p) \int_K \psi_p d\mu$. Let $\mathcal{H}_0$ be the collection of all harmonic functions. If $\alpha_p \geq 0$ and $\sum_{p\in V_0} \alpha_p = 1$ for $\alpha = (\alpha_p)_{p\in V_0}$, we define $N_\alpha(v) = \sqrt{c\mathcal{E}(v,v) + (\sum_{p\in V_0} \alpha_p v(p))^2}$ for $v \in \mathcal{H}_0$. Then N_α is a norm on $\mathcal{H}_0$. Since the parameter space α is compact, there exists $c_2 > 0$ such that $N_\alpha(v) \geq c_2 \max_{p\in V_0} |v(p)|$ for any $v \in \mathcal{H}_0$ and for any α. Hence, it is easy to see that (3.4.4) implies the claim. $\qquad\square$

Now we prove Lemma 3.4.4 by using the claim. If $\iota_\mu(u) = 0$, then $\iota_{\mu^w}(u \circ F_w) = 0$ for $w \in W_*$. Hence, applying (3.4.3) to μ^w and $u \circ F_w$,

$$c_3\mathcal{E}(u \circ F_w, u \circ F_w) \geq \max_{p\in F_w(V_0)} |u(p)|^2,$$

where $c_3 = c_1^{-1} \max\{1, \sup_{w\in W_*} \sum_{m\geq 0} R_m(\mu^w)\}$. Note that c_3 is independent of $w \in W_*$. Since $\mathcal{E}(u,u) \geq r_w^{-1}\mathcal{E}(u \circ F_w, u \circ F_w)$, we have

$$c_3 r_w \mathcal{E}(u,u) \geq \max_{p\in F_w(V_0)} |u(p)|^2$$

So if $p = \pi(\omega) \in V_m$, then

$$c_3 r_{\omega_1 \omega_2 \ldots \omega_m} \mathcal{E}(u, u) \geq |u(p)|^2.$$

Using Proposition 3.1.8, we see that $r_{\omega_1 \omega_2 \ldots \omega_m} \to 0$ as $m \to \infty$. Hence $u(p) = 0$ for any $p \in V_*$. Therefore $u = 0$. $\square$

Now we may identify $\mathcal{F}$ as a subset of $L^2(K, \mu)$ through ι_μ.

Lemma 3.4.5. *Let* $\mu \in \widetilde{M}(K)$. *Assume* $\sup_{w \in W_*} \sum_{m \geq 0} R_m(\mu^w) < \infty$. *If* $\mathcal{E}_*(u, v) = \mathcal{E}(u, v) + \int_K uv d\mu$, *then* ι_μ *is a compact operator from* $(\mathcal{F}, \mathcal{E}_*)$ *to* $(L^2(K, \mu), (\cdot, \cdot)_\mu)$.

Proof. By (3.4.2),

$$\sqrt{c \sum_{k \geq m} R_k} \geq \|\iota_\mu u - P_m u\|_\mu / \sqrt{\mathcal{E}_*(u, u)}.$$

Hence $\|\iota_\mu - P_m\|_{(\mathcal{F}, \mathcal{E}_*) \to L^2(K, \mu)} \to 0$ as $m \to \infty$. Since P_m is a finite rank operator (Definition B.1.9), Proposition B.1.10 and Theorem B.1.11 imply that ι_μ is a compact operator. $\square$

Now, we are ready to show that $(\mathcal{E}, \mathcal{F})$ is a local regular Dirichlet form on $L^2(K, \mu)$. See Definitions B.3.1 and B.3.2 for the definitions concerning Dirichlet forms.

Theorem 3.4.6. *Assume* $\sup_{w \in W_*} \sum_{m \geq 0} R_m(\mu^w) < \infty$. *Then* $(\mathcal{E}, \mathcal{F})$ *is a local regular Dirichlet form on* $L^2(K, \mu)$. *The corresponding non-negative self-adjoint operator* H_N *on* $L^2(K, \mu)$ *has compact resolvent.*

Remark. If μ is a self-similar measure with weight $(\mu_i)_{i \in S}$, then the condition $\sup_{w \in W_*} \sum_{m \geq 0} R_m(\mu^w) < \infty$ is equivalent to $r_i \mu_i < 1$ for all $i \in S$.

Recall that H_N is characterized as a non-negative self-adjoint operator on $L^2(K, \mu)$ that satisfies $\mathcal{E} = Q_{H_N}$ and $\mathcal{F} = \text{Dom}(H_N^{1/2})$. Existence of such a self-adjoint operator is ensured by Theorem B.1.6. In other words, for $u \in \mathcal{F}$, $u \in \text{Dom}(H_N)$ if and only if there exists $f \in \mathcal{F}$ such that $\mathcal{E}(u, v) = (f, v)_\mu$ for any $v \in \mathcal{F}$. In this case, $f = H_N u$. The subscript "N" in H_N represents the first letter of "Neumann" because $-H_N$ corresponds to the Laplacian with Neumann boundary condition. We call $-H_N$ the Neumann Laplacian. See 3.7 for details.

As we mentioned at the beginning of this section, the above theorem is an easy corollary of theorems in Chapter 2 if $(D, \mathbf{r})$ is regular or, equivalently, $K = \Omega$. In such a case, the condition $\sup_{w \in W_*} \sum_{m \geq 0} R_m(\mu^w) < \infty$ is automatically satisfied by any $\mu \in \widetilde{M}(K)$ because $r_i < 1$ for any $i \in S$.

Proof. Closedness: Note that $L^2(K,\mu) \subset L^1(K,\mu)$. Set

$$\hat{\mathcal{F}} = \{u : u \in \mathcal{F}, \int_K u d\mu = 0\}.$$

Then $\mathcal{E}$ is an inner-product on $\hat{\mathcal{F}}$ and $(\hat{\mathcal{F}}, \mathcal{E})$ is complete. For $u \in \mathcal{F}$, define $\hat{u} = u - \int_K u d\mu$. If $\{u_n\}_{n\geq 0}$ is a Cauchy sequence in $(\mathcal{F}, \mathcal{E}_*)$, then $\{\hat{u}_n\}_{n\geq 0}$ is a Cauchy sequence in $(\hat{\mathcal{F}}, \mathcal{E})$. Hence $\mathcal{E}(\hat{u}_n - v, \hat{u}_n - v) \to 0$ as $n \to \infty$ for some $v \in \hat{\mathcal{F}}$. Using the fact that $\int_K u^2 d\mu = \int_K \hat{u}^2 d\mu + (\int_K u d\mu)^2$, we see that $\{\int_K u_n d\mu\}_{n\geq 0}$ is a Cauchy sequence. Let c be the limit of $\int_K u_n d\mu$ as $n \to \infty$ and let $u = v + c$. Then it is easy to see that $\{u_n\}_{n\geq 0}$ is convergent to u in $(\mathcal{F}, \mathcal{E}_*)$ as $n \to \infty$. Hence $(\mathcal{F}, \mathcal{E}_*)$ is complete.

Regularity: Set $\mathcal{H}_* = \{u : u$ is an m-harmonic function for some $m \geq 0\}$. Then $\mathcal{H}_* \subset C(K)$. The maximum principle shows that $\{P_m u\}_{m\geq 0}$ converges to u uniformly on K for $u \in C(K)$. Hence $\mathcal{H}_*$ is dense in $C(K)$. Also, $\mathcal{H}_*$ is dense in $(\mathcal{F}, \mathcal{E}_*)$. Therefore $\mathcal{H}_*$ is a core for $(\mathcal{E}, \mathcal{F})$.

Markov property: For $u \in \mathcal{F}$, define $\bar{u}$ in the same way as in (DF3) of Definition 2.1.1. As $(\mathcal{E}, \mathcal{F})$ is a resistance form, $\bar{u} \in \mathcal{F}$ and $\mathcal{E}(\bar{u}, \bar{u}) \leq \mathcal{E}(u, u)$. We need to show that $\iota_\mu \bar{u} = \overline{\iota_\mu u}$. This is obvious for any $u \in \mathcal{F} \cap C(K)$ because $\{P_m u\}_{m\geq 0}$ converges to u uniformly on K as $m \to \infty$. Therefore, $(\mathcal{E}, \mathcal{F} \cap C(K))$ has the Markov property. Since $\mathcal{F} \cap C(K) \supset \mathcal{H}_*$, it follows that $(\mathcal{E}, \mathcal{F})$ is the minimal closed extension of $(\mathcal{E}, \mathcal{F} \cap C(K))$. Therefore, by [43, Theorem 3.1.1], $(\mathcal{E}, \mathcal{F})$ has the Markov property.

Local property: Assume $\iota_\mu(u) = 0$ on a open set $O \subset K$. For any $p \in \Omega \cap O$, there exists $w \in W_*$ such that $p \in K_w \subset O$. Then $\iota_{\mu^w}(u \circ F_w) = 0$. As μ^w satisfies the condition of Lemma 3.4.4, it follows that $u \circ F_w = 0$ in $\mathcal{F}$. Hence $u = 0$ on $\Omega \cap O$. Now if $\operatorname{supp}(\iota_\mu(u)) \cap \operatorname{supp}(\iota_\mu(v)) = \emptyset$ for $u, v \in \mathcal{F}$, then $\operatorname{supp}(u) \cap \operatorname{supp}(v) = \emptyset$. Since $\mathcal{E}_m(u, v) = 0$ for large m, we obtain $\mathcal{E}(u, v) = 0$.

Finally, by Theorem B.1.13, Lemma 3.4.5 shows that H has compact resolvent. $\qquad\square$

As an easy corollary of Theorem 3.4.6, we also have the Dirichlet form which corresponds to Dirichlet boundary conditions.

Corollary 3.4.7. *Define* $\mathcal{F}_0 = \{u : u \in \mathcal{F}, u|_{V_0} = 0\}$. *Suppose that* $\sup_{w \in W_*} \sum_{m\geq 0} R_m(\mu^w) < \infty$. *Then* $(\mathcal{E}, \mathcal{F}_0)$ *is a local regular Dirichlet form on* $L^2(K, \mu)$. *Moreover the corresponding non-negative definite self-adjoint operator* H_D *on* $L^2(K, \mu)$ *has compact resolvent. The operator* $-H_D$ *is called the Dirichlet Laplacian.*

Remark. Strictly speaking, as $\mathcal{F}_0$ is not dense in $C(K)$ with respect to the supremum norm, $(\mathcal{E}, \mathcal{F}_0)$ is not a regular Dirichlet form on $L^2(K, \mu)$. It is, however, a regular Dirichlet form on $L^2(K\backslash V_0, \mu)$.

As with H_N, we see that H_D is characterized by $\mathcal{E}|_{\mathcal{F}_0 \times \mathcal{F}_0} = Q_{H_D}$ and $\mathcal{F}_0 = \mathrm{Dom}(H_D^{1/2})$. Also $u \in \mathrm{Dom}(H_D)$ if and only if there exists $f \in \mathcal{F}_0$ such that $\mathcal{E}(u, v) = (f, v)_\mu$ for any $v \in \mathcal{F}_0$. In this case, $f = H_D u$.

Proposition 3.4.8. H_D *is invertible. Moreover,* $G_D = (H_D)^{-1}$ *is a compact operator on* $L^2(K, \mu)$ *characterized by* $\mathcal{E}(G_D f, v) = (f, v)_\mu$ *for any* $v \in \mathcal{F}_0$ *and for any* $f \in L^2(K, \mu)$.

G_D will be called the Green's operator. See 3.6 for details.

We will give two proofs of this proposition below. For those who are not familiar with the subjects in B.1, the second proof may be more accessible.

Proof. Note that $(\mathcal{F}_0, \mathcal{E})$ is a Hilbert space. First we will give a proof using some general theory. Since $\mathcal{E}$ is positive definite on $\mathcal{F}_0$, 0 is not an eigenvalue of H_D. As H_D has compact resolvent, $G_D = H_D^{-1}$ is a compact operator. As $\mathcal{E}(u, v) = (H_D u, v)_\mu$ for any $u \in \mathrm{Dom}(H_D)$ and for any $v \in \mathcal{F}_0$, we have $\mathcal{E}(G_D f, v) = (f, v)_\mu$ for any $v \in \mathcal{F}_0$ and for any $f \in L^2(K, \mu)$.

Next we give another proof, which is less abstract. By (3.4.2), we have $c\mathcal{E}(u, u) \geq (u, u)_\mu$ for any $u \in \mathcal{F}_0$. Hence, for any $f \in L^2(K, \mu)$, $|(u, f)_\mu| \leq \|u\|_\mu \|f\|_\mu \leq \sqrt{c\mathcal{E}(u, u)}\|f\|_\mu$ for $u \in \mathcal{F}_0$. This means that $u \to (u, f)_\mu$ is a bounded functional on a Hilbert space $(\mathcal{F}_0, \mathcal{E})$. Therefore, there exists $h \in \mathcal{F}_0$ such that $\mathcal{E}(u, h) = (u, f)_\mu$ for any $u \in \mathcal{F}_0$. It is easy to see that h is uniquely determined by f and the correspondence $f \to h$ is linear. Hence we write $h = G_D f$. Since

$$(G_D f, f)_\mu = \mathcal{E}(G_D f, G_D f) \geq c(G_D f, G_D f)_\mu, \qquad (3.4.5)$$

we have $c\|G_D f\|_\mu \leq \|f\|_\mu$. So G_D is a bounded operator from $L^2(K, \mu)$ to itself. It is obvious that $G_D = H_D^{-1}$. Moreover, we also see that $\|f\|_\mu \geq \sqrt{\mathcal{E}(G_D f, G_D f)}$ by (3.4.5). Hence G_D is a bounded operator from $L^2(K, \mu)$ to $(\mathcal{F}_0, \mathcal{E})$. So, if U is a bounded subset of $L^2(K, \mu)$, then $G_D(U)$ is a bounded subset of $(\mathcal{F}_0, \mathcal{E})$. Now, by Lemma 3.4.5, $\iota_\mu : (\mathcal{F}_0, \mathcal{E}) \to L^2(K, \mu)$ is a compact operator. Hence $G_D(U) = \iota_\mu(G_D(U))$ is a relatively compact subset of $L^2(K, \mu)$. Therefore $G_D : L^2(K, \mu) \to L^2(K, \mu)$ is a compact operator. □

3.5 Green's function

We continue to assume that $(D, \mathbf{r})$ is a harmonic structure on a connected p. c. f. self-similar structure $(K, S, \{F_i\}_{i \in S})$, where $S = \{1, 2, \ldots, N\}$. In this section, we will construct the Green's function for $(\mathcal{E}, \mathcal{F})$ derived from the harmonic structure $(D, \mathbf{r})$.

The Green's function will be the integral kernel of Green's operator in the next section. Roughly speaking, the Green's operator is the inverse of the Dirichlet Laplacian, $G_D = (H_D)^{-1}$, given in Proposition 3.4.8. More precisely,

$$(G_D u)(x) = \int_K g(x, y) u(y) d\mu$$

for $u \in L^2(K, \mu)$, where g is the Green's function which we will define in this section. The above relation is justified in 3.7, in particular, Lemma 3.7.10 and Theorem 3.7.14.

Recall that $H_m : \ell(V_m) \to \ell(V_m)$ is given by $H_m = \sum_{w \in W_m} \frac{1}{r_w} {}^t R_w D R_w$, where $R_w : \ell(V_m) \to \ell(V_0)$ is defined by $R_w u = u \circ F_w$. Also, H_m can be expressed as $\begin{pmatrix} T_m & {}^t J_m \\ J_m & X_m \end{pmatrix}$, where $T_m : \ell(V_0) \to \ell(V_0); J_m : \ell(V_0) \to \ell(V_m \backslash V_0)$ and $X_m : \ell(V_m \backslash V_0) \to \ell(V_m \backslash V_0)$.

First we will study a discrete version of Green's function $-X_m^{-1}$. For this purpose, we need the following lemma.

Lemma 3.5.1. *Let $X = (X_{ij})$ be a symmetric $n \times n$ real matrix. Assume that X satisfies the following four conditions.*

(1) $X_{ii} < 0$ *for any* $i = 1, 2, \ldots, n$.

(2) $X_{ij} \geq 0$ *if* $i \neq j$.

(3) $\sum_{i=1}^{n} X_{ij} \leq 0$ *for any* $j = 1, 2, \ldots, n$ *and this sum is negative for some* j.

(4) *If* $i \neq j$, *then there exist* $m \geq 1$ *and* $i_0, \ldots, i_m \in \{1, 2, \ldots, n\}$ *such that* $i_0 = i$, $i_m = j$ *and* $X_{i_k i_{k+1}} > 0$ *for all* $k = 0, \ldots, m - 1$.

Then X is invertible. Moreover, if $G = (G_{ij}) = (-X)^{-1}$, then

$$G_{ii} \geq G_{ij} > 0$$

for any $i, j \in \{1, 2, \ldots, n\}$.

Proof. Define an $n \times n$ diagonal matrix A by $A_{ii} = -X_{ii}$. Also define an $n \times n$ symmetric matrix B by $B_{ij} = X_{ij}$ if $i \neq j$ and $B_{ii} = 0$ for any $i = 1, 2, \ldots, n$. Then $-X A^{-1} = I - B A^{-1}$, where I is the $n \times n$ unit matrix. Let $C = B A^{-1}$. Then $C_{ij} = B_{ij} / A_{jj} \geq 0$. By the assumption (3), we see that $0 < \sum_{i=1}^{n} C_{ij} \leq 1$ for any $j = 1, 2, \ldots, n$ and that this sum is strictly less than 1 for some j, say, $l \in \{1, 2, \ldots, n\}$. Set $|x| = \sum_{i=1}^{n} |x_i|$ for $x = (x_i)_{i=1,2,\ldots,n} \in \mathbb{R}^n$. Then $|\cdot|$ is a norm of

$\mathbb{R}^n$. It follows that

$$|Cx| \leq \sum_{i,j} C_{ij}|x_j| = \sum_{j=1}^{n}(\sum_{i=1}^{n} C_{ij})|x_j| \leq |x|$$

and that $|Cx| < |x|$ if $x_l \neq 0$. For $j \in \{1, 2, \ldots, n\}$, define $e_j \in \mathbb{R}^N$ by $(e_j)_i = \delta_{ij}$ for all $i \in \{1, 2, \ldots, n\}$, where δ is Kronecker's delta. By the assumption (4), we see that there exist m_j and $i_0, i_1, \ldots, i_{m_j} \in \{1, 2, \ldots, n\}$ such that $i_0 = l, i_{m_j} = j$ and $C_{i_k i_{k+1}} > 0$ for any $k = 0, 1, \ldots, m_j - 1$. This implies that $(C^{m_j} e_j)_l > 0$. Hence it follows that $|C^{m_j+1} e_j| < |C^{m_j} e_j| \leq |e_j| = 1$. Let $s = \max_{j \in \{1,2,\ldots,n\}} m_j + 1$. If $x \in \mathbb{R}^n$ and $x \neq 0$, then $|C^s x| \leq \sum_{j=1}^{n} |x_j| |C^s e_j| < |x|$. Hence it follows that $\|C^s\| < 1$, where $\|C^s\|$ is the operator norm of C^s with respect to the norm $|\cdot|$ on $\mathbb{R}^n$. Hence $-XA^{-1} = I - C$ is invertible and $(I - C)^{-1} = I + C + C^2 + \cdots$. This implies that X is invertible and that

$$G_{ij} = (A_{ii})^{-1}(\delta_{ij} + \sum_{m \geq 1} \sum_{(i_0,\ldots,i_m) \in S_m(i,j)} C_{i_0 i_1} C_{i_1 i_2} \cdots C_{i_{m-1} i_m}),$$

where

$$S_m(i,j) = \{(i_0, \ldots, i_m) : i_0 = i, i_m = j \text{ and } \{i_k\}_{k=0,\ldots,m} \subset \{1, 2, \ldots, n\}\}.$$

By condition (4), we see that $G_{ij} > 0$ for any $i, j \in \{1, 2, \ldots, n\}$.

Now define $M_i = \max_{k=1,2,\ldots,n} G_{ik}$. Suppose that $G_{ij} = M_i$ and that $i \neq j$. Then

$$G_{ij} = \sum_{k:k \neq j} G_{ik}(-X_{kj}/X_{jj}).$$

By condition (3), it follows that $\sum_{k:k \neq j} -X_{kj}/X_{jj} \leq 1$. Hence $G_{ik} = G_{ij}$ for any k with $X_{kj} > 0$. This implies that $G_{ik} = M_i$ if there exists $m \geq 1$ and $i_0, \ldots, i_m$ such that $i_0 = k, i_m = j$ and $X_{i_l i_{l+1}} > 0$ for any $l = 0, \ldots, m - 1$. By the condition (4), it follows that $G_{ik} = M_i$ for any $i = 1, 2, \ldots, n$. Hence $G_{ii} \geq G_{ij}$ for any i, j. $\square$

Lemma 3.5.2. X_m *is negative definite. If* $G_m = -X_m^{-1}$, *then* $(G_m)_{pp} \geq (G_m)_{pq} \geq 0$ *for any* $p, q \in V_m \backslash V_0$, *where* $G_m = ((G_m)_{pq})_{p,q \in V_m \backslash V_0}$. *Moreover, for* $p, q \in V_m \backslash V_0$, $(G_m)_{pq} > 0$ *if and only if* $p \underset{m}{\sim} q$.

Recall Definition 3.2.9 for the definition of $\underset{m}{\sim}$.

Proof. Let $(V_m \backslash V_0)/\underset{m}{\sim} = \{V_m^{(1)}, \ldots, V_m^{(k)}\}$, where each $V_m^{(i)} \subset V_m \backslash V_0$ is an equivalence class. Then X_m is decomposed as $X_m^{(i)} : \ell(V_m^{(i)}) \to \ell(V_m^{(i)})$ for $i = 1, 2, \ldots, k$. Applying Lemma 3.5.1 to each $X_m^{(i)}$, we obtain the first half of the lemma. Also, by Lemma 3.5.1, it follows that $(G_m)_{pq} > 0$ if and only if $p \underset{m}{\sim} q$. □

Set $X = X_1$, $X_{pq} = (X_1)_{pq}$, $G = G_1$ and $G_{pq} = (G_1)_{pq}$ for any $p, q \in V_1 \backslash V_0$.

Definition 3.5.3. Define $\Psi(x, y) = \sum_{p,q \in V_1 \backslash V_0} G_{pq} \psi_p(x) \psi_q(y)$ for $x, y \in K$. Also for $w \in W_*$, define

$$\Psi_w(x, y) = \begin{cases} \Psi((F_w)^{-1}(x), (F_w)^{-1}(y)) & \text{if } x, y \in K_w, \\ 0 & \text{otherwise}, \end{cases}$$

for $x, y \in K$. Write $\Psi^x(y) = \Psi(x, y)$ and $\Psi_w^x(y) = \Psi_w(x, y)$.

Ψ_w is a non-negative continuous function on $K \times K$. Ψ_w^x is an $m + 1$-harmonic function if $w \in W_m$.

Lemma 3.5.4. *For any $u \in \mathcal{F}$,*

$$\mathcal{E}(\Psi_w^x, u) = \begin{cases} r_w^{-1}(u_{m+1}(x) - u_m(x)) & \text{if } x \in K_w, \\ 0 & \text{otherwise}. \end{cases}$$

Proof. For $u \in \mathcal{F}$,

$$\mathcal{E}(\Psi^x, u) = \mathcal{E}(\Psi^x, u - u_0) = - \sum_{p,q \in V_1 \backslash V_0} X_{pq} \Psi^x(p)(u(q) - u_0(q))$$

$$= \sum_{q \in V_1 \backslash V_0} (u(q) - u_0(q)) \psi_q(x) = u_1(x) - u_0(x).$$

Therefore, if $x \in K_w$, $w \in W_m$ and $z = F_w^{-1}(x)$, then

$$\mathcal{E}(\Psi_w^x, u) = \sum_{v \in W_m} r_v^{-1} \mathcal{E}(\Psi_w^x \circ F_v, u \circ F_v) = r_w^{-1} \mathcal{E}(\Psi^z, u \circ F_w)$$

$$= r_w^{-1}((u \circ F_w)_1(z) - (u \circ F_w)_0(z)) = r_w^{-1}(u_{m+1}(x) - u_m(x)).$$

□

Set

$$g_m(x, y) = \sum_{k=0}^{m-1} \sum_{w \in W_k} r_w \Psi_w(x, y)$$

and set $g_m^x(y) = g_m(x, y)$. Then g_m^x is an m-harmonic function and $g_m^x(y) = 0$ if $y \in V_0$. By Lemma 3.5.4, we see that

$$\mathcal{E}(g_m^x, u) = u_m(x) - u_0(x) \tag{3.5.1}$$

for any $u \in \mathcal{F}$. Therefore, $g_m(x, y) = \sum_{p,q \in V_m \backslash V_0} (G_m)_{pq} \psi_p^m(x) \psi_q^m(y)$.

Now note that $\Psi_w(x,y) \geq 0$. Hence $\{g_m(x,y)\}_{m\geq 0}$ is a non-decreasing sequence. Therefore the following definition makes sense, allowing ∞ as a possible value of the limit below.

Proposition 3.5.5. *For $x, y \in K$, define*

$$g(x,y) = \lim_{m\to\infty} g_m(x,y) = \sum_{w\in W_*} r_w \Psi_w(x,y).$$

Then g is continuous on $\{(x,y) \in K \times K : x \neq y\}$. Moreover,
(1) Let $x, y \in K\backslash V_0$. Then $g(x,y) > 0$ if and only if x and y belong to the same connected component of $K\backslash V_0$.
(2) If $(D, \mathbf{r})$ is regular, then g is a continuous function on $K \times K$.

g is called the Green's function associated with the harmonic structure $(D, \mathbf{r})$.

We need the following lemma to prove this proposition.

Lemma 3.5.6. *For any $p, q \in V_m\backslash V_0$, the following three conditions are equivalent.*
(1) p and q belong to the same connected component of $K\backslash V_0$,
(2) $p \underset{m}{\sim} q$,
(3) $(G_m)_{pq} > 0$.

Proof. By Lemma 3.5.2, we immediately see that (2) and (3) are equivalent. Also, Lemma 3.2.13 implies that (2) $\Rightarrow$ (1). Now the remaining part is to show that (1) implies (3). Let C be the connected component of $K\backslash V_0$ which contains both p and q. Then since C is arcwise connected by Proposition 1.6.6, there exists a continuous path γ such that $\gamma(0) = p$, $\gamma(1) = q$ and $\gamma([0,1]) \subseteq C$. Let $\{p_1, p_2, \ldots, p_k\} = V_n \cap \gamma([0,1])$, where $p_i = p(t_i)$ and $0 = t_1 < t_2 < \cdots < t_k = 1$. Note that $p_1 = p$ and $p_k = q$. Then there exist $w(0), \ldots, w(k-1) \in W_n$ such that $p_i, p_{i+1} \in K_{w(i)}$ for all $i = 1, 2, \ldots, k-1$. For sufficiently large n, it follows that $K_{w(i)} \subseteq C$ for $i = 1, 2, \ldots, k-1$. By Lemma 3.2.10-(1), there exists an H_n-path between p_i and p_{i+1} for any $i = 1, 2, \ldots, k-1$. Combining those H_n-paths, we have an H_n-path between p and q. Hence $(G_n)_{pq} > 0$. Note that $(G_m)_{pq} = g_m(p,q) = g_n(p,q) = (G_n)_{pq}$ if $m \leq n$ and $p, q \in V_m$. Therefore we see that $(G_m)_{pq} > 0$. This completes the proof. $\square$

Proof of Proposition 3.5.5. Set $K_{m,p} = \cup_{w\in W_m, p\in K_w} K_w$ for $p \in K$. If $x \neq y$, then $K_{m,x} \cap K_{m,y}$ is empty for sufficiently large m. Hence the sum in the above definition is really a finite sum if $x \neq y$. Since Ψ_w is continuous, we can verify that g is continuous on $\{(x,y) \in K \times K : x \neq y\}$.

(1) Choose m so that $K_{m,x} \cap V_0 = K_{m,y} \cap V_0 = \emptyset$. Set $V_{m,x} = K_{m,x} \cap V_m$ and $V_{m,y} = K_{m,y} \cap V_0$. Since $K_{m,x}$ and $K_{m,y}$ are connected, there exist connected components C_1 and C_2 of $K\backslash V_0$ such that $K_{m,x} \in C_1$ and $K_{m,y} \in C_2$. Note that if $\psi_p^m(a) > 0$, then $p \in V_{m,a}$ for $a = x, y$.

First assume that $C_1 = C_2$. We may choose $p \in V_{m,x}$ and $q \in V_{m,y}$ that satisfy $\psi_p^m(x) > 0$ and $\psi_q^m(y) > 0$. Then Lemma 3.5.6 implies $(G_m)_{pq} > 0$. Therefore,

$$g_m(x,y) \geq (G_m)_{pq}\psi_p^m(x)\psi_q^m(y) > 0.$$

Next suppose that $C_1 \neq C_2$. If $\psi_p^m(x) > 0$ and $\psi_q^m(y) > 0$, then $p \in V_{m,x}$ and $q \in V_{m,y}$. So p and q belong to different connected components of $K\backslash V_0$. Hence, by Lemma 3.5.6, $(G_m)_{pq} = 0$. Therefore $g_m(x,y) = 0$ for sufficiently large m. This immediately implies that $g(x,y) = 0$.

(2) As $(D,\mathbf{r})$ is regular, $r = \max_{i \in S} r_i < 1$. There exists $C > 0$ such that $\sup_{x,y \in K} \sum_{w \in W_m} r_w \Psi_w(x,y) \leq Cr^m$. Since $\sum_{w \in W_m} r_w \Psi_w$ is continuous, $g = \sum_{m \geq 0} (\sum_{w \in W_m} r_w \Psi_w)$ is continuous. $\square$

If $(D,\mathbf{r})$ is not regular, it could happen that $g(x,x) = \infty$. In fact, we will show that $g(x,x) < \infty$ if and only if $x \in \Omega$.

Theorem 3.5.7. *The following three conditions are equivalent.*
(1) $g(x,x) < \infty$
(2) $g^x \in \mathcal{F}$, *where* $g^x(y) = g(x,y)$
(3) $x \in \Omega$.

Moreover, if any of the above conditions is satisfied, then $\mathcal{E}(g^x, u) = u(x) - u_0(x)$ *for any* $u \in \mathcal{F}$.

Lemma 3.5.8. *There exist* $c_1, c_2 > 0$ *such that, for any* $\omega \in \Sigma$,

$$c_1 \sum_{m \geq 0} r_{\omega_1\omega_2...\omega_m} \varphi(\pi(\sigma^m\omega))^2 \leq g(\pi(\omega), \pi(\omega))$$

$$\leq c_2 \sum_{m \geq 0} r_{\omega_1\omega_2...\omega_m} \varphi(\pi(\sigma^m\omega))^2,$$

where $\varphi = \sum_{p \in V_1 \backslash V_0} \psi_p$.

Proof. Let $x = \pi(\omega)$. Then $g(x,x) = \sum_{m \geq 0} r_{\omega_1\omega_2...\omega_m} \Psi_{\omega_1\omega_2...\omega_m}(x,x)$. Note that there exist $c_1, c_2 > 0$ such that, for any $u \in \ell(V_1\backslash V_0)$,

$$c_1 \Big(\sum_{p \in V_1 \backslash V_0} |u(p)| \Big)^2 \leq \sum_{p,q \in V_1 \backslash V_0} G_{pq}u(p)u(q) \leq c_2 \Big(\sum_{p \in V_1 \backslash V_0} |u(p)| \Big)^2. \quad (3.5.2)$$

Note that $\Psi_{\omega_1\omega_2\ldots\omega_m}(x,x) = \sum_{p,q\in V_1\backslash V_0} G_{pq}\psi_p(\pi(\sigma^m\omega))\psi_q(\pi(\sigma^m\omega))$ and $\psi_p \geq 0$. Also, by the definition of φ, (3.5.2) implies

$$c_1\varphi(\pi(\sigma^m\omega))^2 \leq \Psi_{\omega_1\omega_2\ldots\omega_m}(x,x) \leq c_2\varphi(\pi(\sigma^m\omega))^2.$$

$\square$

Proof of Theorem 3.5.7. (3) $\Rightarrow$ (2): Set $\mathcal{F}_0 = \{u \in \mathcal{F} : u|_{V_0} = 0\}$. Then $(\mathcal{F}_0, \mathcal{E})$ is a Hilbert space. If $x \in \Omega$ and $p \in V_0$, then $|u(x)|^2 \leq \mathcal{E}(u,u)R(x,p)$ for $u \in \mathcal{F}_0$. This shows that $u \to u(x)$ is a continuous functional $\mathcal{F}_0 \to \mathbb{R}$. Hence there exists $h \in \mathcal{F}_0$ such that $\mathcal{E}(h,u) = u(x)$ for any $u \in \mathcal{F}_0$. Now $\mathcal{E}(h_m, u) = \mathcal{E}(h, u_m) = u_m(x) = \mathcal{E}(g_m^x, u)$ for any $u \in \mathcal{F}_0$. Hence $h_m = g_m^x$. So $h = g^x$ and $g^x \in \mathcal{F}_0$.

(2) $\Rightarrow$ (1): This is obvious by the following equality.

$$\mathcal{E}(g^x, g^x) = \lim_{m\to\infty} \mathcal{E}(g_m^x, g_m^x) = \lim_{m\to\infty} g_m(x,x) = g(x,x).$$

(1) $\Rightarrow$ (3): If $g(x,x) < \infty$, then, by Lemma 3.5.8, $\sum_{m\geq 0} r_{\omega_1\omega_2\ldots\omega_m}\varphi(x_m)^2 < \infty$, where $\pi(\omega) = x$ and $x_m = \pi(\sigma^m\omega)$. Hereafter we will use Ψ^m and $r(m)$ to denote $\Psi_{\omega_1\omega_2\ldots\omega_m}$ and $r_{\omega_1\omega_2\ldots\omega_m}$ respectively.

First, if $\lim_{m\to\infty}\varphi(x_m) = 0$, then $\min_{\tau\in\mathcal{P}}\delta(\sigma^m\omega,\tau) \to 0$ as $m \to \infty$, where δ is a metric on Σ. Then, by Proposition 3.1.8, we see that $\sum_{m\geq 0} r(m) < \infty$. So, Lemma 3.3.9 implies that $x \in \Omega$.

Next assume that $\limsup_{m\to\infty}\varphi(x_m) > 0$. Choose $\{m_i\}_{i\geq 0}$ so that $m_i < m_{i+1}$ and $\varphi(x_{m_i}) \geq \limsup_{m\to\infty}\varphi(x_m)/2$. We may assume that $\sigma^{m_i}\omega \to \tau \in \Sigma\backslash\mathcal{P}$ as $i \to \infty$. Set $q = \pi(\tau)$. Then we can choose k so that $K_{k,q} = \cup_{w\in W_k:q\in K_w}K_w$ does not intersect V_0. Also we may assume that $x_{m_i} \in K_{k,q}$ for all $i \geq 1$. Choose $q_* \in K_{k,q} \cap V_k$ and define $p_i = F_{\omega_1\omega_2\ldots\omega_{m_i}}(q_*)$.

Claim 1: There exists $c_1 > 0$ such that

$$\Psi^m(p_i, p_i) \leq c_1\Psi^m(x,x) \quad \text{and} \quad \Psi^m(p_i, x) \leq c_1\Psi^m(x,x)$$

for any i and for any $m \in [0, m_i - 1] \cup [m_i + k, \infty)$.

Proof of Claim 1: If $m \geq m_i + k$, then $\Psi^m(p_i, y) = 0$ for any $y \in K$. So the claim is obvious in this case. Assume that $m \leq m_i - 1$. If $u(y) = \Psi^m(p, F_{\omega_1\omega_2\ldots\omega_{m_i}}(y))$, u is a non-negative harmonic function for any $p \in K$. Set $X = K_{k,q}$. Then the Harnack inequality (Proposition 3.2.7) implies that there exists $c > 0$ such that $u(y_1) \leq cu(y_2)$ for any $y_1, y_2 \in X$. Hence, we see that $\Psi^m(p, x_1) \leq c\Psi^m(p, x_2)$ for any $x_1, x_2 \in F_{\omega_1\omega_2\ldots\omega_m}(K_{k,q})$. Therefore

$$\Psi^m(p_i, p_i) \leq c\Psi^m(p_i, x) \leq c^2\Psi^m(x,x).$$

This immediately implies the claim.

Claim 2: Let $\mathbb{N}_1 = \cup_{i \geq 1}\{m_i, \ldots, m_i + k - 1\}$. Then $\sum_{m \in \mathbb{N}_1} r(m) < \infty$.

Proof of Claim 2: Since $\liminf_{i \to \infty} \varphi(x_{m_i}) > 0$ and $\sum_{m \geq 0} r(m)\varphi(x_m)^2 < \infty$, it follows that $\sum_{i \geq 1} r(m_i) < \infty$. So

$$\sum_{m \in \mathbb{N}_1} r(m) \leq \sum_{i \geq 1} (1 + R + R^2 + \cdots + R^{k-1}) r(m_i) < \infty,$$

where $R = \max_{s \in S} r_s$.

Claim 3: $\lim_{i \to \infty} g(p_i, p_i) = \lim_{i \to \infty} g(p_i, x) = g(x, x)$.

Proof of Claim 3: Define

$$a_m = r(m) \times \begin{cases} \sup_{x,y \in K} \Psi(x, y) & \text{if } m \in \mathbb{N}_1, \\ c_1 \Psi^m(x, x) & \text{otherwise.} \end{cases}$$

Then Claim 2 along with the fact that $g(x, x) < \infty$ implies $\sum_{m \geq 0} a_m < \infty$. Also, by Claim 1, it is easy to see that $\Psi^m(p_i, p_i) \leq a_m$ and $\Psi^m(p_i, x) \leq a_m$ for all m. As $\Psi^m(x, x) = \lim_{i \to \infty} \Psi^m(p_i, p_i) = \lim_{i \to \infty} \Psi^m(p_i, x)$, the dominated convergence theorem proves the claim.

Claim 4: There exists $c_3 > 0$ such that

$$R(z, y) \leq 2\mathcal{E}(g^z - g^y, g^z - g^y) + c_3 \max_{p \in V_0} |\psi_p(z) - \psi_p(y)|^2.$$

for any $z, y \in \Omega$.

Proof of Claim 4: Since $u(z) - u(y) = \mathcal{E}(g^z - g^y, u) + (u_0(x) - u_0(y))$ for any $u \in \mathcal{F}$, we have

$$|u(z) - u(y)|^2 \leq 2\mathcal{E}(g^z - g^y, u)^2 + 2|u_0(z) - u_0(y)|^2$$
$$\leq 2\mathcal{E}(g^z - g^y, g^z - g^y)\mathcal{E}(u, u) + 2\Big(\sum_{p \in V_0} |u(p) - u(p_*)||\psi_p(z) - \psi_p(y)|\Big)^2$$
$$\leq 2\mathcal{E}(g^z - g^y, g^z - g^y)\mathcal{E}(u, u) + 2\Big(M \max_{p \in V_0} |u(p) - u(p_*)||\psi_p(z) - \psi_p(y)|\Big)^2,$$

where $p_* \in V_0$ and $M = \#(V_0)$. Also recall that $\mathcal{E}(u_0, u_0) = \mathcal{E}_0(u, u) = \mathcal{E}_D(u|_{V_0}, u|_{V_0})$, where $D \in \mathcal{LA}(V_0)$. So there exists $c > 0$ such that

$$\Big(\max_{p \in V_0} |u(p) - u(p_*)|\Big)^2 \leq c\mathcal{E}(u_0, u_0) \leq c\mathcal{E}(u, u)$$

for any $u \in \mathcal{F}$. Hence we have

$$\frac{|u(z) - u(y)|^2}{\mathcal{E}(u, u)} \leq 2\mathcal{E}(g^z - g^y, g^z - g^y) + c_3 \max_{p \in V_0} |\psi_p(z) - \psi_p(y)|^2.$$

This immediately implies the claim.

Claim 5: $\{p_i\}_{i \geq 1}$ is a Cauchy sequence in (Ω, R).

Proof of Claim 5: Note that $\{p_i\}_{i\geq 1}$ is a Cauchy sequence in (K, d), where d is a metric on K that is compatible with the original topology of K. Hence $\max_{p\in V_0} |\psi_p(p_i) - \psi_p(p_j)| \to 0$ as $i, j \to \infty$. Also, by Claim 3, as $i \to \infty$,

$$\mathcal{E}(g^{p_i} - g^x, g^{p_i} - g^x) = g(p_i, p_i) - 2g(p_i, x) + g(x, x) \to 0.$$

As $\mathcal{E}(u - v, u - v) \leq (\sqrt{\mathcal{E}(u - h, u - h)} + \sqrt{\mathcal{E}(h - v, h - v)})^2$, it follows that $\mathcal{E}(g^{p_i} - g^{p_j}, g^{p_i} - g^{p_j}) \to 0$ as $i, j \to \infty$. Therefore, by Claim 4, we have $R(p_i, p_j) \to 0$ as $i, j \to \infty$. So $\{p_i\}_{i\geq 1}$ is a Cauchy sequence in (Ω, R).

Now as $\lim_{i\to\infty} d(x, p_i) = 0$, the limit point of $\{p_i\}$ in (Ω, R) is identified with x. (Recall that Ω is identified as $\theta(\Omega) \subset K$. See Proposition 3.3.2.) Hence $x \in \Omega$. □

Remark. The idea of using the Harnack inequality is originally due to Dr. A. Teplyaev.

3.6 Green's operator

As in the previous sections, $(D, \mathbf{r})$ is a harmonic structure on a self-similar structure $(K, S, \{F_i\}_{i\in S})$, where $S = \{1, 2, \ldots, N\}$. In the last section, we defined the Green's function g. In this section we define the (extended) Green's operator G_μ, which is an integral operator whose kernel is the Green's function.

Let $\mu \in \widetilde{M}(K)$.

Theorem 3.6.1. *Define $R_m^t(\mu) = \max_{w\in W_m} r_w \mu(K_w)^{1/t}$ for $1 \leq t \leq \infty$. (If $t = \infty$, $R_m^t = \max_{w\in W_m} r_w$.) Assume $\sum_{m\geq 0} R_m^t(\mu) < \infty$. If s is the constant (including ∞) that satisfies $1/t + 1/s = 1$, then for $f \in L^s(K, \mu)$,*

$$(G_\mu f)(x) = \int_K g(x, y) f(y) \mu(dy) \tag{3.6.1}$$

is well-defined for all $x \in K$ and $G_\mu f \in C(K) \cap \mathcal{F}_0$. Moreover, $G_\mu : L^s(K, \mu) \to C(K)$ is a compact operator. Also

$$\mathcal{E}(u, G_\mu f) = \int_K (u - u_0) f d\mu \tag{3.6.2}$$

for any $u \in \mathcal{F}$. The operator G_μ is called the (extended) Green's operator.

Remark. If μ is a self-similar measure with weight $(\mu_1, \mu_2, \ldots, \mu_N)$, then the condition $\sum_{m\geq 0} R_m^t(\mu) < \infty$ is equivalent to $r_i(\mu_i)^{1/t} < 1$ for all $i \in S$.

Remark. If $(D, \mathbf{r})$ is regular, then g is a continuous function on $K \times K$. Therefore the above theorem is almost obvious in this case. In particular, we see that $\sum_{m \geq 0} R_m^\infty < \infty$. Hence $G_\mu : L^1(K, \mu) \to C(K)$.

Comparing the above theorem with Proposition 3.4.8, we see that $G_\mu = G_D = H_D^{-1}$ if $\sup_{w \in W_*} \sum_{m \geq 0} R_m(\mu^w) < \infty$. More precisely, (3.6.2) implies that $G_\mu f = G_D f$ for $f \in L^s(K, \mu) \cap L^2(K, \mu)$. Hence, if $t \geq 2$, then the (extended) Green's operator G_μ is really a natural extension of the Green's operator G_D and $G_D(L^2(K, \mu)) \subset C(K) \cap \mathcal{F}_0$. Hence $\mathrm{Dom}(H_D) \subset C(K) \cap \mathcal{F}_0$. Also, if $1 \leq t \leq 2$, then $G_D|_{L^s(K,\mu)} = G_\mu$ and $G_D(L^s(K, \mu)) \subset C(K) \cap \mathcal{F}_0$.

To prove Theorem 3.6.1, we need several lemmas.

Lemma 3.6.2. *If $\sum_{m \geq 0} R_m^t(\mu) < \infty$, then $g^x \in L^t(K, \mu)$ for any $x \in K$. Moreover, $x \to g^x$ is a continuous map from K to $L^t(K, \mu)$.*

For $u \in L^a(K, \mu)$, we set $\|u\|_{\mu,a} = (\int_K |u|^a d\mu)^{1/a}$.

Proof. For $x = \pi(\omega_1 \omega_2 \ldots)$, $g^x(y) = \sum_{m \geq 0} r_{\omega_1 \omega_2 \ldots \omega_m} \Psi_{\omega_1 \omega_2 \ldots \omega_m}(x, y)$. Hence

$$\|g^x\|_{\mu,t} \leq \sum_{m \geq 0} r_{\omega_1 \omega_2 \ldots \omega_m} \|\Psi^x_{\omega_1 \omega_2 \ldots \omega_m}\|_{\mu,t}$$
$$\leq \sum_{m \geq 0} r_{\omega_1 \omega_2 \ldots \omega_m} \mu(K_{\omega_1 \omega_2 \ldots \omega_m})^{1/t} \leq \sum_{m \geq 0} R_m^t(\mu) < \infty.$$

Hence $g^x \in L^t(K, \mu)$. Note that

$$\|g^x - g^y\|_{\mu,t} \leq \|g_m^x - g_m^y\|_{\mu,t} + \|g_m^x - g^x\|_{\mu,t} + \|g_m^y - g^y\|_{\mu,t}.$$

Since $\|g_m^x - g^x\|_{\mu,t} \leq \sum_{k \geq m+1} R_k^t(\mu)$ and g_m is continuous on $K \times K$, we see that $\|g^x - g^y\|_{\mu,t} \to 0$ if $d(x, y) \to 0$. $\qquad \square$

Lemma 3.6.3.

$$\sum_{q \in V_m} (H_m)_{pq} g^q(y) = \begin{cases} -\psi_p^m(y) & \text{if } p \in V_m \backslash V_0, \\ -\psi_p^m(y) + \psi_p(y) & \text{if } p \in V_0. \end{cases}$$

Proof. Note that $\mathcal{E}(\psi_p^m, u) = \mathcal{E}_m(\psi_p^m, u) = -(H_m u)(p)$ for any $u \in \mathcal{F}$. Since $\mathcal{E}(g^p, u) = u(p) - u_0(p)$, it follows that

$$\mathcal{E}\left(\sum_{p \in V_m} (H_m)_{pq} g^q, u \right) = (H_m u)(p) - (H_m u_0)(p) = \mathcal{E}(-\psi_p^m, u) + \mathcal{E}(\psi_p^m, u_0).$$

Since u_0 is harmonic, Lemma 3.2.16 implies $\mathcal{E}(\psi_p^m - \psi_p, u_0) = 0$ for $p \in V_0$.

Also $(H_m u_0)(p) = 0$ for $p \in V_m \backslash V_0$. Hence we have

$$\mathcal{E}(\sum_{p \in V_m} (H_m)_{pq} g^q, u) = \begin{cases} \mathcal{E}(-\psi_p^m, u) & \text{if } p \in V_m \backslash V_0, \\ \mathcal{E}(-\psi_p^m + \psi_p, u) & \text{if } p \in V_0 \end{cases}$$

for any $u \in \mathcal{F}$. This implies the desired equality. $\quad\square$

Lemma 3.6.4.

$$(H_m(G_\mu f))(p) = \begin{cases} -\int_K \psi_p^m f d\mu & \text{if } p \in V_m \backslash V_0, \\ -\int_K (\psi_p^m - \psi_p) f d\mu & \text{if } p \in V_0. \end{cases}$$

Proof. Note that

$$(H_m(G_\mu f))(p) = \sum_{q \in V_m} (H_m)_{pq} \int_K g^q f d\mu = \int_K (\sum_{q \in V_m} (H_m)_{pq} g^q) f d\mu.$$

The result follows by Lemma 3.6.3. $\quad\square$

Proof of Theorem 3.6.1. By Hölder's inequality,

$$|(G_\mu f)(x) - (G_\mu f)(y)| \leq ||g^x - g^y||_{\mu,t} ||f||_{\mu,s}.$$

Hence Lemma 3.6.2 implies that $G_\mu f \in C(K)$. Let $\{f_n\}_{n \geq 1}$ be a bounded sequence in $L^s(K, \mu)$. Then $\{G_\mu f_n\}_{n \geq 1}$ is uniformly bounded and equicontinuous. So, by Ascoli–Arzelà's theorem, $\{G_\mu f_n\}_{n \geq 1}$ contains a subsequence that is convergent in $C(K)$. Hence G_μ is a compact operator from $L^s(K, \mu)$ to $C(K)$.

By Lemma 3.6.4,

$$\mathcal{E}_m(u, G_\mu f) = -\sum_{p \in V_m} u(p)(H_m(G_\mu f))(p) = \int_K (u_m - u_0) f d\mu. \quad (3.6.3)$$

Set $u = G_\mu f$. Then

$$\mathcal{E}_m(G_\mu f, G_\mu f) = \int_K (G_\mu f)_m f d\mu \leq ||f||_{\mu,1} \sup_{x \in K} |(G_\mu f)(x)|.$$

(Note that $(G_\mu f)_0 = 0$.) Hence $\mathcal{E}(G_\mu f, G_\mu f) < \infty$ and $G_\mu f \in \mathcal{F}_0$. For any $u \in \mathcal{F}$, letting $m \to \infty$ in (3.6.3), we obtain (3.6.2). $\quad\square$

Next we study the n-th power of the Green's operator G_μ for $n \in \mathbb{N}$. The integral kernel $g^{(n)}$ for G_μ^n is given by the following inductive formula. Set $g^{(1)}(x, y) = g(x, y)$ and

$$g^{(n+1)}(x, y) = \int_K g(x, z) g^{(n)}(z, y) \mu(dz)$$

for $n \in \mathbb{Z}$.

Theorem 3.6.5. *Assume that* $\sup_{w \in W_*} \sum_{m \geq 0} R_m(\mu^w) < \infty$. *Then, for any* n, *there exist* $c_n > 0$ *and* $\Psi_w^{(n)}(x, y)$ *such that* $\Psi_w^{(n)}$ *is continuous on* $K \times K$,

$$g^{(n)}(x, y) = \sum_{w \in W_*} r_w{}^n \mu(K_w)^{n-1} \Psi_w^{(n)}(x, y), \qquad (3.6.4)$$

and

$$0 \leq \Psi_w^{(n)}(x, y) \leq c_n \chi_{K_w}(x) \chi_{K_w}(y) \qquad (3.6.5)$$

for any $w \in W_*$. *Moreover, if* μ *is a self-similar measure on* K, *then there exists a non-negative continuous function* $\Psi^{(n)}$ *on* $K \times K$ *such that*

$$\Psi_w^{(n)}(x, y) = \begin{cases} \Psi^{(n)}(F_w^{-1}(x), F_w^{-1}(y)) & \textit{if } x, y \in K_w, \\ 0 & \textit{otherwise.} \end{cases} \qquad (3.6.6)$$

From the above theorem, it is easy to see that $g^{(n)}$ is continuous on $\{(x, y) : x, y \in K, x \neq y\}$ because the sum in (3.6.4) is, in fact, a finite sum if $x \neq y$. Note that $g^{(1)} = g$ is independent of μ but this is no longer true if $n \geq 2$.

Corollary 3.6.6. *Assume that* $\sup_{w \in W_*} \sum_{m \geq 0} R_m(\mu^w) < \infty$. *If there exists* $t > 1$ *such that* $\sum_{m \geq 0} R_m^t(\mu) < \infty$, *then, for any* $n \geq (1 - 1/t)^{-1}$, $g^{(n)}$ *is continuous on* $K \times K$ *and* G_μ^n *is extended to a compact operator from* $L^1(K, \mu)$ *to* $C(K)$. *In particular, suppose that* μ *is a self-similar measure with weight* $(\mu_i)_{i \in S}$ *which satisfies* $\mu_i r_i < 1$ *for any* $i \in S$. *Then* $g^{(n)}$ *is a continuous function on* $K \times K$ *for any* $n > \max_{i \in S} \log \mu_i / \log \mu_i r_i$.

Remark. If μ is a self-similar measure with weight $(\mu_i)_{i \in S}$, the quantity $\max_{i \in S} \log \mu_i / \log \mu_i r_i$ coincides with $d_S \overline{\delta}(\nu, \mu)/2$, where d_S is the unique positive number that satisfies $\sum_{i \in S} (\mu_i r_i)^{d_S/2} = 1$, ν is a self-similar measure with weight $((\mu_i r_i)^{d_S/2})_{i \in S}$ and $\overline{\delta}(\nu, \mu)$ is the distortion between ν and μ defined in Proposition 4.5.2. In 4.1, d_S is called the spectral exponent and it determines the asymptotic behavior of the eigenvalue counting function of the Laplacian. See Theorem 4.1.5 for details.

Proof of Corollary 3.6.6. Set

$$h_m^{(n)}(x, y) = \sum_{w \in W_m} r_w{}^n \mu(K_w)^{n-1} \Psi_w^{(n)}(x, y).$$

Then $h_m^{(n)}$ is continuous on $K \times K$ and $|h_m^{(n)}(x, y)| \leq c_n (R_m^{t'}(\mu))^n$, where $t' = (1 - 1/n)^{-1}$. As $1 \geq 1/n + 1/t$, it follows that $t' < t$ and hence $\sum_{m \geq 0} R_m^{t'}(\mu) < \infty$. So $\sum_{m \geq 0} c_n (R_m^{t'}(\mu))^n < c_n (\sum_{m \geq 0} R_m^{t'}(\mu))^n < \infty$.

Therefore $\sum_{m\geq 0} h_m^{(n)}$ is uniformly convergent on $K \times K$. Hence $g^{(n)}$ is continuous on $K \times K$. The rest of the statement is straightforward. □

Proof of Theorem 3.6.5. The proof is by induction on n. For $n = 1$, we have $g(x,y) = \sum_{w\in W_*} r_w \Psi_w(x,y)$ and $0 \leq \Psi_w(x,y) \leq c_1 \chi_{K_w}(x)\chi_{K_w}(y)$, where $c_1 = \max_{x,y\in K} \Psi(x,y)$.

Next assume that the claim of the theorem is true for n. Set $\Psi_{\alpha,\beta}^{(n+1)} = \int_K \Psi_\alpha(x,z)\Psi_\beta^{(n)}(z,y)\mu(dz)$ for $\alpha, \beta \in W_*$. Note that

$$\Psi_\alpha(x,z)\Psi_\beta^{(n)}(z,y) \leq c_1 c_n \chi_{K_\alpha}(x)\chi_{K_\alpha \cap K_\beta}(z)\chi_{K_\beta}(y),$$

Hence we have

$$0 \leq \Psi_{\alpha,\beta}^{(n+1)}(x,y) \leq c_1 c_n \mu(K_\alpha \cap K_\beta)\chi_{K_\alpha}(x)\chi_{K_\beta}(y). \qquad (3.6.7)$$

Now, let $h_{w,m}(x,y) = \sum_{\gamma\in W_m} r_w r_{w\gamma}{}^n \mu(K_{w\gamma})^{n-1}\Psi_{w,w\gamma}^{(n+1)}(x,y)$. By (3.6.7), $0 \leq \Psi_{w,w\gamma}^{(n+1)}(x,y) \leq c_1 c_n \mu(K_{w\gamma})\chi_{K_w}(x)\chi_{K_{w\gamma}}(y)$. Therefore,

$$|h_{w,m}(x,y)| \leq c_1 c_n r_w{}^{n+1}\mu(K_w)^n \max_{\gamma\in W_m} r_\gamma{}^n(\mu^w(K_\gamma))^n$$
$$= c_1 c_n r_w{}^{n+1}\mu(K_w)^n(R_m(\mu^w))^n.$$

Since $\sum_{m\geq 0}(R_m(\mu^w))^n \leq M^n$, where $M = \sup_{w\in W_*}\sum_{m\geq 0} R_m(\mu^w)$, we see that $\sum_{m\geq 0} h_{w,m}$ is uniformly convergent on $K \times K$. Hence, if $h_w(x,y) = \sum_{m\geq 0} h_{w,m}(x,y)$, h_w is continuous on $K \times K$. Also

$$0 \leq h_w(x,y) \leq c_1 c_n M^n r_w{}^{n+1}\mu(K_w)^n\chi_{K_w}(x)\chi_{K_w}(y). \qquad (3.6.8)$$

Next define $f_{w,m}(x,y) = \sum_{\gamma\in W_m} r_{w\gamma} r_w{}^n \mu(K_w)^{n-1}\Psi_{w\gamma,w}^{(n+1)}(x,y)$. Similar arguments to those given above for $h_{w,m}$ will show that $f_w = \sum_{m\geq 1} f_{w,m}$ is continuous on $K \times K$ and that

$$0 \leq f_w(x,y) \leq c_1 c_n M r_w{}^{n+1}\mu(K_w)^n\chi_{K_w}(x)\chi_{K_w}(y). \qquad (3.6.9)$$

So define $\Psi_w^{(n+1)}$ by $r_w{}^{n+1}\mu(K_w)^n\Psi_w^{(n+1)}(x,y) = h_w(x,y)+f_w(x,y)$. Then, by (3.6.8) and (3.6.9), it follows that $\Psi_w^{(n+1)}$ is continuous on $K \times K$ and that $0 \leq \Psi_w^{(n+1)}(x,y) \leq c_{n+1}\chi_{K_w}(x)\chi_{K_w}(y)$, where $c_{n+1} = c_n c_1(M^n + M)$. Now by definition,

$$g^{(n+1)}(x,y) = \sum_{\alpha,\beta\in W_*} r_\alpha r_\beta{}^n \mu(K_\beta)^{n-1}\Psi_{\alpha,\beta}^{(n+1)}(x,y)$$
$$= \sum_{w\in W_*} (h_w(x,y) + f_w(x,y)) = \sum_{w\in W_*} r_w{}^{n+1}\mu(K_w)^n\Psi_w^{(n+1)}(x,y).$$

Finally assume that μ is a self-similar measure. Note that (3.6.6) is true for $n = 1$. So suppose (3.6.6) is true for n. Then

$$\Psi^{(n+1)}_{w\alpha,w\beta}(x,y) = \begin{cases} \mu_w \Psi^{(n+1)}_{\alpha,\beta}(F_w^{-1}(x), F_w^{-1}(y)) & \text{if } x, y \in K_w, \\ 0 & \text{otherwise.} \end{cases} \tag{3.6.10}$$

Recall that $r_w{}^{n+1} \mu(K_w)^n \Psi_w^{(n+1)}(x,y) = h_w(x,y) + f_w(x,y)$. By (3.6.10), routine calculations show that (3.6.6) is true for $n+1$. $\qquad\square$

Obviously, $g^{(n)}(x,y)$ is non-negative by definition. Moreover, Theorem 3.6.5 implies that $g^{(n)}(x,y) > 0$ if x and y belong to the same connected component of $K \backslash V_0$. More precisely, we have the following result.

Corollary 3.6.7. *Let*

$$g_m^{(n)}(x,y) = \sum_{k=0}^m \sum_{w \in W_k} r_w{}^n \mu(K_w)^{n-1} \Psi_w^{(n)}(x,y)$$

for any $m, n \in \mathbb{N}$. If x and y belong to the same connected component of $K \backslash V_0$, then there exists $m \geq 0$ such that $g_m^{(n)}(x,y) > 0$ for any $n \geq 1$.

Proof. By Proposition 3.5.5-(1), there exists $m \geq 0$ such that $g_m^{(1)}(x,y) = g_m(x,y) > 0$ and $g_m^{(1)}(x,x) > 0$. Suppose that $g_m^{(n)}(x,y) > 0$. By (3.6.4),

$$\int_K g_m(x,z) g_m^{(n)}(z,y) \mu(dz) \leq g_m^{(n+1)}(x,y). \tag{3.6.11}$$

Note that g_m and $g_m^{(n)}$ are continuous on $K \times K$. Since $g_m(x,x) > 0$ and $g_m^{(n)}(x,y) > 0$, (3.6.11) implies that $g_m^{(n+1)}(x,y) > 0$. Now the corollary follows by induction. $\qquad\square$

3.7 Laplacians

As in the previous sections, $(D, \mathbf{r})$ is a harmonic structure on a self-similar structure $(K, S, \{F_i\}_{i \in S})$, where $S = \{1, 2, \dots, N\}$.

In this section, we define the Laplacian Δ_μ associated with $(D, \mathbf{r})$ and μ, where $\mu \in \widetilde{M}(K)$. If $\sup_{w \in W_m} \sum_{m \geq 0} R_m(\mu^w) < \infty$, we already introduced in 3.4 non-negative definite self-adjoint operators H_N and H_D associated with Dirichlet forms $(\mathcal{E}, \mathcal{F})$ and $(\mathcal{E}, \mathcal{F}_0)$, respectively, on $L^2(K, \mu)$. (See Theorem 3.4.6 and Corollary 3.4.7.) We might think of $-H_N$ and $-H_D$ as Laplacians for the corresponding boundary conditions. However, in this section, we will give a more direct way of defining the Laplacian. Our Laplacian, Δ_μ, corresponds to the classical second derivative and its domain $\text{Dom}(\Delta_\mu)$ corresponds to the C^2-class of functions. See Example 3.7.2.

Definition 3.7.1. Let $\mu \in \widetilde{M}(K)$. Define

$$\mathcal{D}_\mu = \{u \in C(K) : \text{there exists } f \in C(K) \text{ such that}$$
$$\lim_{m \to \infty} \max_{p \in V_m \setminus V_0} |\mu_{m,p}^{-1}(H_m u)(p) - f(p)| = 0\},$$

where $\mu_{m,p} = \int_K \psi_p^m d\mu$. Then for $u \in \mathcal{D}_\mu$, we write $f = \Delta_\mu u$, where f is the function appearing in the definition of $\mathcal{D}_\mu$. Δ_μ is called the Laplacian associated with $(D, \mathbf{r})$ and μ.

Obviously, $\Delta_\mu : \mathcal{D}_\mu \to C(K)$ is a linear map and $\mathcal{D}_\mu = \mathrm{Dom}(\Delta_\mu)$. If u is a harmonic function on K, then it immediately follows that $u \in \mathcal{D}_\mu$ and $\Delta_\mu u = 0$.

Example 3.7.2 (Interval). Recall Example 3.1.4. Set $\mathbf{r} = (1/2, 1/2)$. Then $(D, \mathbf{r})$ is a harmonic structure. For this harmonic structure, if $p = i/2^m \in V_m$, then

$$(H_m u)(p) = \frac{1}{2^{-m}} \begin{cases} u(p + 2^{-m}) + u(p - 2^{-m}) - 2u(p) & \text{if } p \neq 0, 1, \\ u(2^{-m}) - u(0) & \text{if } p = 0, \\ u(1 - 2^{-m}) - u(1) & \text{if } p = 1. \end{cases}$$
$$(3.7.1)$$

Obviously, a harmonic function is a linear function. Now let μ be the self-similar measure with weight $(1/2, 1/2)$. Then μ is, in fact, a restriction of Lebesgue measure on $\mathbb{R}$ to $[0, 1]$. For $p \in V_m \setminus V_0$, since ψ_p^m is a piecewise linear function, we see that $\mu_{m,p} = 2^{-m}$. Hence

$$\mu_{m,p}^{-1}(H_m u)(p) = \frac{1}{4^{-m}}(u(p + 2^{-m}) + u(p - 2^{-m}) - 2u(p)).$$

Routine calculus arguments show that $\mathcal{D}_\mu = C^2([0, 1])$ and $\Delta_\mu u = d^2 u/dx^2$ for $u \in \mathcal{D}_\mu$.

Example 3.7.3 (Sierpinski gasket). Let $(D, \mathbf{r})$ be the harmonic structure on the Sierpinski gasket given in Example 3.1.5. Define a linear operator $L_m : \ell(V_m) \to \ell(V_m)$ by

$$(L_m f)(p) = \sum_{q \in V_{m,p}} (f(q) - f(p)),$$

where, for $p \in V_m$,

$$V_{m,p} = \{q \in V_m : q \neq p, \text{ there exists } w \in W_m \text{ such that } p, q \in F_w(V_0)\}.$$

Then $H_m = (5/3)^m L_m$. Note that

$$\#(V_{m,p}) = \begin{cases} 4 & \text{if } p \in V_m \backslash V_0, \\ 2 & \text{if } p \in V_0. \end{cases}$$

Now let μ be the self-similar measure on the Sierpinski gasket with weight $(1/3, 1/3, 1/3)$. Then it follows that $\int_K \psi_p d\mu = 1/3$ for any $p \in V_0$. Using self-similarity and symmetry, we obtain,

$$\int_K \psi_p^m d\mu = \begin{cases} 2/3^{m+1} & \text{if } p \in V_m \backslash V_0, \\ 1/3^{m+1} & \text{if } p \in V_0 \backslash. \end{cases}$$

Hence

$$\mu_{m,p}^{-1}(H_m u)(p) = \frac{3}{2} 5^m (L_m u)(p)$$

for any $p \in V_m \backslash V_0$. The associated Laplacian Δ_μ is called the standard Laplacian on the Sierpinski gasket.

Lemma 3.7.4. *For $v \in C(K)$ and $u \in \mathcal{D}_\mu$,*

$$\sum_{p \in V_m \backslash V_0} v(p)(H_m u)(p) \to \int_K v \Delta_\mu u \, d\mu$$

as $m \to \infty$.

Proof. If $f_m(x) = \sum_{p \in V_m \backslash V_0} v(p)\mu_{m,p}^{-1}(H_m u)(p)\psi_p^m(x)$, then $f_m \to v\Delta_\mu u$ as $m \to \infty$ uniformly on any compact subset of $K \backslash V_0$. Also $\{f_m\}_{m \geq 0}$ is uniformly bounded. Hence, $\int_K f_m d\mu \to \int_K v\Delta_\mu u \, d\mu$ as $m \to \infty$. $\square$

Lemma 3.7.5. *For $f \in \mathcal{D}_\mu$ and $p \in V_0$,*

$$\lim_{m \to \infty} -(H_m f)(p) = -(Df)(p) + \int_K \psi_p \Delta_\mu f \, d\mu.$$

Proof. Recall that $H_m = \begin{pmatrix} T_m & {}^t J_m \\ J_m & X_m \end{pmatrix}$, where $T_m : \ell(V_0) \to \ell(V_0)$, $J_m : \ell(V_0) \to \ell(V_m \backslash V_0)$ and $X_m : \ell(V_m \backslash V_0) \to \ell(V_m \backslash V_0)$. Note that $D = T_m - {}^t J_m X_m^{-1} J_m$, hence we have

$${}^t J_m X_m^{-1}(J_m f_0 + X_m f_1) = -Df_0 + T_m f_0 + {}^t J_m f_1,$$

where $f_0 = f|_{V_0}$ and $f_1 = f|_{V_1 \backslash V_0}$. This implies

$$-Df_0 + (H_m f)|_{V_0} = {}^t J_m X_m^{-1}(H_m f)|_{V_m \backslash V_0}.$$

Since $({}^{t}J_{m}X_{m}^{-1})_{pq} = -\psi_{p}(q)$ for $p \in V_{0}$ and $q \in V_{m}\backslash V_{0}$, we have

$$-(Df)(p) + (H_{m}f)(p) = -\sum_{q \in V_{1}\backslash V_{0}} \psi_{p}(q)(H_{m}f)(q).$$

Letting $m \to \infty$, Lemma 3.7.4 implies the lemma. $\qquad \square$

Definition 3.7.6. For $f \in \mathcal{D}_{\mu}$ and $p \in V_{0}$, we define

$$(df)_{p} = \lim_{m \to \infty} -(H_{m}f)(p).$$

$(df)_{p}$ is called the Neumann derivative of f at $p \in V_{0}$.

For the case of Example 3.7.2, (3.7.1) implies that $(df)_{0} = -f'(0)$ and $(df)_{1} = f'(1)$.

Lemma 3.7.7. *For $v \in C(K)$ and $u \in \mathcal{D}_{\mu}$,*

$$\lim_{m \to \infty} \mathcal{E}_{m}(v, u) = \sum_{p \in V_{0}} v(p)(du)_{p} - \int_{K} v\Delta_{\mu}u\,d\mu.$$

Proof. Note that

$$\mathcal{E}_{m}(v, u) = \sum_{p \in V_{0}} -v(p)(H_{m}u)(p) - \sum_{p \in V_{m}\backslash V_{0}} v(p)(H_{m}u)(p).$$

Apply Lemma 3.7.4 to complete the proof. $\qquad \square$

Using the above lemma, we can immediately obtain the following theorem.

Theorem 3.7.8 (Gauss–Green's formula). *$\mathcal{D}_{\mu} \subseteq \mathcal{F}$. For any $v \in \mathcal{F}$ and $u \in \mathcal{D}_{\mu}$,*

$$\mathcal{E}(v, u) = \sum_{p \in V_{0}} v(p)(du)_{p} - \int_{K} v\Delta_{\mu}u\,d\mu. \qquad (3.7.2)$$

If both u and v belong to $\mathcal{D}_{\mu}$, then

$$\sum_{p \in V_{0}} (v(p)(du)_{p} - u(p)(dv)_{p}) = \int_{K} (v\Delta_{\mu}u - u\Delta_{\mu}v)d\mu.$$

(3.7.2) is a counterpart of the Gauss–Green's formula in ordinary calculus on $\mathbb{R}^{n}$ and Riemannian manifolds. For example, let U be a bounded domain in $\mathbb{R}^{2}$ and assume that ∂U is a smooth curve. For $u, v \in C^{2}(U)$, set

$$\mathcal{E}(v, u) = \int_{U} \left(\frac{\partial v}{\partial x}\frac{\partial u}{\partial x} + \frac{\partial v}{\partial y}\frac{\partial u}{\partial y}\right)dxdy.$$

Then the Gauss–Green's formula says

$$\mathcal{E}(v,u) = \int_{\partial U} v\frac{\partial u}{\partial n}ds - \int_U v\Delta u\, dxdy.$$

where $\Delta u = \frac{\partial^2 u}{\partial x^2} + \frac{\partial^2 u}{\partial y^2}$.

Next we will study relations between the Laplacian Δ_μ and the abstract non-negative definite self-adjoint operators H_N and H_D.

Theorem 3.7.9. *Assume that $\mu \in \widetilde{M}(K)$ and $\sup_{w\in W_*}\sum_{m\geq 0} R_m(\mu^w) < \infty$. If $\mathcal{D}_{D,\mu} = \{u \in \mathcal{D}_\mu : u|_{V_0} = 0\}$, then $\mathcal{D}_{D,\mu} = \mathrm{Dom}(H_D) \cap \mathcal{D}_\mu$, $H_D|_{\mathcal{D}_{D,\mu}} = -\Delta_\mu|_{\mathcal{D}_{D,\mu}}$ and H_D is the Friedrichs extension of $-\Delta_\mu$ on $\mathcal{D}_{D,\mu}$.*

Also define $\mathcal{D}_{N,\mu} = \{u \in \mathcal{D}_\mu : (du)|_p = 0 \text{ for all } p \in V_0\}$. Then $\mathcal{D}_{N,\mu} = \mathrm{Dom}(H_N)\cap\mathcal{D}_\mu$, $H_N|_{\mathcal{D}_{N,\mu}} = -\Delta_\mu|_{\mathcal{D}_{N,\mu}}$ and H_N is the Friedrichs extension of $-\Delta_\mu$ on $\mathcal{D}_{N,\mu}$.

This theorem shows that $-H_N$ is the Neumann Laplacian associated with $(\mathcal{E},\mathcal{F})$ and μ and that $-H_D$ is the Dirichlet Laplacian associated with $(\mathcal{E},\mathcal{F})$ and μ.

We write $\mathcal{D}_D$ and $\mathcal{D}_N$ instead of $\mathcal{D}_{D,\mu}$ and $\mathcal{D}_{N,\mu}$ respectively if no confusion can occur.

Lemma 3.7.10. *For any $f \in C(K)$, $G_\mu f \in \mathcal{D}_D$ and $\Delta_\mu(G_\mu f) = -f$.*

Proof. By Lemma 3.6.4, for $p \in V_m\backslash V_0$,

$$|\mu_{m,p}^{-1}H_m(G_\mu f)(p) + f(p)| \leq \mu_{m,p}^{-1}\int_K \psi_p^m(y)|f(y) - f(p)|\mu(dy) \leq \epsilon_m,$$

where $\epsilon_m = \max_{p\in V_m}(\sup_{y\in K_{m,p}}|f(y) - f(p)|)$. As f is uniformly continuous on K, $\epsilon_m \to 0$ as $m \to \infty$. $\square$

Proof of Theorem 3.7.9, Part I. Recall that, for $u \in \mathcal{F}_0$, $u \in \mathrm{Dom}(H_D)$ and $H_D u = f$ if and only if $\mathcal{E}(v,u) = (v,f)_\mu$ for any $v \in \mathcal{F}_0$. By (3.7.2), if $u \in \mathcal{D}_D$, $\mathcal{E}(v,u) = (v, -\Delta_\mu u)_\mu$ for any $v \in \mathcal{F}_0$. Hence $\mathcal{D}_D \subseteq \mathrm{Dom}(H_D)$ and $H_D|_{\mathcal{D}_D} = -\Delta_\mu|_{\mathcal{D}_D}$. Since $G_\mu|_{C(K)} = H_D^{-1}|_{C(K)}$, if $u \in \mathcal{D}_D$ and $-\Delta_\mu u = f$, then $G_\mu f = u$. Hence $\mathcal{D}_D \subseteq G_\mu(C(K))$. On the other hand, by Lemma 3.7.10, we have $\mathcal{D}_D \supseteq G_\mu(C(K))$. Hence $\mathcal{D}_D = G_\mu(C(K))$.

Next set $H_D u = f$ for $u \in \mathrm{Dom}(H_D)$. Then, there exists $\{f_n\}_{n\geq 0} \subset C(K)$ such that $f_n \to f$ as $n \to \infty$ in $L^2(K,\mu)$. If $u_n = G_\mu f_n$, then $u_n \in \mathcal{D}_D$. Recall that $G_D = H_D^{-1}$. By (3.4.5) (and the discussions following it), $\mathcal{E}_*(G_D h, G_D h) \leq c(h,h)_\mu$ for any $h \in L^2(K,\mu)$, where $\mathcal{E}_* = \mathcal{E} + (\cdot,\cdot)_\mu$. As $G_D|_{C(K)} = G_\mu|_{C(K)}$, we have $\mathcal{E}_*(u_n - u, u_n - u) \leq c(f_n - f, f_n - f)_\mu \to 0$ as $n \to \infty$. Hence the completion of $\mathcal{D}_D$ with respect to $\mathcal{E}_*$ contains $\mathrm{Dom}(H_D)$. Now obviously, the completion of $\mathrm{Dom}(H_D)$ with respect to $\mathcal{E}_*$ equals $\mathcal{F}_0$

because the Friedrichs extension of H_D is H_D and $\mathcal{F}_0 = \text{Dom}(H_D{}^{1/2})$. Therefore the completion of $\mathcal{D}_D$ with respect to $\mathcal{E}_*$ equals $\mathcal{F}_0$. This means that H_D is the Friedrichs extension of $-\Delta_\mu|_{\mathcal{D}_D}$.   □

To prove the rest of Theorem 3.7.9, we need several preliminaries. For $f \in L^1(K, \mu)$, define $Pf = f - \int_K f d\mu$. Set $\widetilde{L}^p(K, \mu) = P(L^p(K, \mu))$, $\widetilde{\mathcal{F}} = P(\mathcal{F})$, $\widetilde{\mathcal{D}}_N = P(\mathcal{D}_N)$ and $\widetilde{C}(K) = P(C(K))$. Note that $(\widetilde{\mathcal{F}}, \mathcal{E})$ is a Hilbert space, hence $(\mathcal{E}, \widetilde{\mathcal{F}})$ is a closed form on $\widetilde{L}^2(K, \mu)$. Let $\widetilde{H}_N$ be the non-negative self-adjoint operator associated with $(\mathcal{E}, \widetilde{\mathcal{F}})$ on $\widetilde{L}^2(K, \mu)$. Then, for $u \in \widetilde{\mathcal{F}}$ and $f \in \widetilde{L}^2(K, \mu)$, $u \in \text{Dom}(\widetilde{H}_N)$ and $\widetilde{H}_N u = f$ if and only if $\mathcal{E}(v, u) = (v, f)$ for any $v \in \widetilde{\mathcal{F}}$. So it is easy to see that $\text{Dom}(\widetilde{H}_N) = \text{Dom}(H_N) \cap \widetilde{\mathcal{F}}$ and $\widetilde{H}_N = H_N|_{\text{Dom}(H_N) \cap \widetilde{\mathcal{F}}}$. Note that $\widetilde{H}_N$ is injective because $\mathcal{E}$ is positive definite on $\widetilde{\mathcal{F}}$.

The following lemma is obvious from Lemma 3.6.4.

Lemma 3.7.11. *For* $f \in L^\infty(K, \mu)$, $(d(G_\mu f))_p = -\int_K \psi_p f d\mu$ *for any* $p \in V_0$.

Lemma 3.7.12. *For any* $f \in \widetilde{L}^2(K, \mu)$, *there exists a unique* $u \in \widetilde{\mathcal{F}}$ *such that* $\mathcal{E}(v, u) = (v_0, f)_\mu$ *for any* $v \in \widetilde{\mathcal{F}}$, *where* $v_0 = \sum_{p \in V_0} v(p) \psi_p$. *Moreover, define* $Q : \widetilde{L}^2(K, \mu) \to \widetilde{L}^2(K, \mu)$ *by* $Qf = u$. *Then* Qf *is harmonic and* $(d(Qf))_p = \int_K \psi_p f d\mu$ *for any* $p \in V_0$. *Also* $Q : \widetilde{L}^2(K, \mu) \to \widetilde{L}^2(K, \mu)$ *is a compact operator.*

Proof. Let $a_p = \int_K \psi_p f d\mu$ for any $p \in V_0$. Then $(v_0, f)_\mu = \sum_{p \in V_0} a_p v(p)$. Since $f \in \widetilde{L}^2(K, \mu)$, $\sum_{p \in V_0} a_p = 0$. Choose any $q \in \Omega$. Then

$$|\sum_{p \in V_0} a_p v(p)| = |\sum_{p \in V_0} a_p (v(p) - v(q))| \le \sum_{p \in V_0} |a_p||v(p) - v(q)|$$
$$\le \sqrt{\mathcal{E}(v, v)} \sum_{p \in V_0} |a_p| \sqrt{R(p, q)}.$$

Therefore, $v \to (v_0, f)_\mu$ is a continuous functional on $(\widetilde{\mathcal{F}}, \mathcal{E})$. Hence there exists $u \in \widetilde{\mathcal{F}}$ such that $\mathcal{E}(v, u) = (v_0, f)$ for any $v \in \widetilde{\mathcal{F}}$. Let $u = Qf$. Then, for $p \in V_m \backslash V_0$,

$$-(H_m(Qf))(p) = \mathcal{E}(\psi_p^m, Qf) = \mathcal{E}(P(\psi_p^m), Qf) = ((P(\psi_p^m))_0, f)_\mu = 0,$$

because $(P(\psi_p^m))_0$ is a constant. Hence Qf is harmonic. Since $\Delta_\mu(Qf) = 0$, (3.7.2) implies $\mathcal{E}(v, Qf) = \sum_{p \in V_0} v(p)(d(Qf))_p$. Therefore $(d(Qf))_p = a_p$. Since Qf is harmonic,

$$\mathcal{E}_0(Qf, Qf) = \mathcal{E}(Qf, Qf) = (Qf, f)_\mu \le \|Qf\|_\mu \|f\|_\mu. \tag{3.7.3}$$

Let $\widetilde{\mathcal{H}}_0 = \{Pu : u$ is a harmonic funtion on $K\}$. As $\dim \widetilde{\mathcal{H}}_0 < \infty$, there exists $c > 0$ such that $c(u,u)_\mu \leq \mathcal{E}_0(u,u)$ for any $u \in \widetilde{\mathcal{H}}_0$. Hence, by (3.7.3), we have $c\|Qf\|_\mu \leq \|f\|_\mu$. Therefore $Q : \widetilde{L}^2(K,\mu) \to \widetilde{L}^2(K,\mu)$ is a bounded operator. Also, as $\dim \widetilde{\mathcal{H}}_0 < \infty$, Q is a finite rank operator and hence it is a compact operator by Proposition B.1.10. $\qquad\square$

Proof of Theorem 3.7.9, Part 2. For $f \in \widetilde{L}^2(K,\mu)$, define $\widetilde{G}_N f = P(G_D f + Qf)$. Then for any $u \in \widetilde{\mathcal{F}}$,

$$\mathcal{E}(u, \widetilde{G}_N f) = (u - u_0, f)_\mu + (u_0, f)_\mu = (u, f)_\mu.$$

Hence $\widetilde{G}_N f \in \mathrm{Dom}(\widetilde{H}_N)$ and $\widetilde{H}_N(\widetilde{G}_N f) = f$. As G_D and Q are compact operators, we see that $\widetilde{G}_N : \widetilde{L}^2(K,\mu) \to \widetilde{L}^2(K,\mu)$ is a compact operator. Since $\widetilde{H}_N$ is injective, it follows that $\widetilde{H}_N$ is invertible and $\widetilde{G}_N = \widetilde{H}_N^{-1}$. Also we obtain

$$\mathcal{E}(\widetilde{G}_N f, \widetilde{G}_N f) = (\widetilde{G}_N f, f)_\mu \leq \|\widetilde{G}_N f\|_\mu \|f\|_\mu \leq c\|f\|_\mu^2. \qquad (3.7.4)$$

Next for, $u \in \widetilde{\mathcal{D}}_N$, (3.7.2) implies that $\mathcal{E}(v, u) = (v, -\Delta_\mu u)_\mu$ for any $v \in \widetilde{\mathcal{F}}$. Also, by (3.7.2), $0 = \mathcal{E}(1, u) = -\int_K \Delta_\mu u d\mu$. This implies $\Delta_\mu u \in \widetilde{C}(K)$. Therefore, we have $u \in \mathrm{Dom}(\widetilde{H}_N)$ and $\widetilde{H}_N u = -\Delta_\mu u$. Thus we obtain that $\widetilde{\mathcal{D}}_N \subseteq \mathrm{Dom}(\widetilde{H}_N)$ and that $\widetilde{H}_N|_{\widetilde{\mathcal{D}}_N} = -\Delta_\mu|_{\widetilde{\mathcal{D}}_N}$.

On the other hand, for $f \in \widetilde{C}(K)$, $\widetilde{G}_N f = P(G_\mu f + Qf)$. Hence, it follows from Lemma 3.7.11 and Lemma 3.7.12 that $\widetilde{G}_N f \in \widetilde{\mathcal{D}}_N$. Therefore $\widetilde{G}_N(\widetilde{C}(K)) = \widetilde{\mathcal{D}}_N$.

Now for $u \in \mathrm{Dom}(\widetilde{H}_N)$, let $f = \widetilde{H}_N u$. Then there exists $\{f_n\}_{n\geq 1} \subset \widetilde{C}(K)$ such that $f_n \to f$ as $n \to \infty$ in $\widetilde{L}^2(K,\mu)$. If $u_n = \widetilde{G}_N f_n$, then, by (3.7.4), $\mathcal{E}(u_n - u, u_n - u) \leq c\|f_n - f\|_\mu^2$. This implies that $u_n \to u$ as $n \to \infty$ in $(\mathcal{F}, \mathcal{E}_*)$. Since $u_n \in \widetilde{\mathcal{D}}_N$, we have shown that the completion of $\widetilde{\mathcal{D}}_N$ with respect to $\mathcal{E}_*$ contains $\mathrm{Dom}(\widetilde{H}_N)$. Since $\widetilde{H}_N$ is itself a non-negative self-adjoint operator, the Friedrichs extension of $\widetilde{H}_N$ on $\mathrm{Dom}(\widetilde{H}_N)$ is $\widetilde{H}_N$ and $\widetilde{\mathcal{F}} = \mathrm{Dom}(\widetilde{H}_N^{1/2})$. Hence, the completion of $\mathrm{Dom}(\widetilde{H}_N)$ with respect to $\mathcal{E}_*$ equals to $\widetilde{\mathcal{F}}$. Therefore the completion of $\widetilde{\mathcal{D}}_N$ with respect to $\mathcal{E}_*$ is $\widetilde{\mathcal{F}}$.

Finally, note that $\mathcal{F} = \widetilde{\mathcal{F}} \oplus \{\text{constants}\}$ and $\mathcal{D}_N = \widetilde{\mathcal{D}}_N \oplus \{\text{constants}\}$. Also we obtained $\widetilde{H}_N = H_N|_{\mathrm{Dom}(H_N)\cap\widetilde{\mathcal{F}}}$. Since $\widetilde{\mathcal{D}}_N \subset \mathrm{Dom}(\widetilde{H}_N)$ and $\widetilde{H}_N|_{\widetilde{\mathcal{D}}_N} = -\Delta_\mu|_{\widetilde{\mathcal{D}}_N}$, we see that $\mathcal{D}_N \subset \mathrm{Dom}(H_N)$ and $H_N|_{\mathcal{D}_N} = -\Delta_\mu|_{\mathcal{D}_N}$. Moreover, since the completion of $\widetilde{\mathcal{D}}_N$ with respect to $\mathcal{E}_*$ is $\widetilde{\mathcal{F}}$, it follows that the completion of $\mathcal{D}_N$ with respect to $\mathcal{E}_*$ equals $\mathcal{F}$. Therefore H_N is the Friedrichs extension of $-\Delta_\mu|_{\mathcal{D}_N}$. $\qquad\square$

By the arguments in the above proof, we also see the following fact.

Corollary 3.7.13. *For $b = D$ and N, if $f \in \mathrm{Dom}(H_b)$ and $H_b f \in C(K)$, then $f \in \mathrm{Dom}(H_b) \cap \mathcal{D}_\mu = \mathcal{D}_{b,\mu}$.*

To end this section, we study the Dirichlet problem for Poisson's equation.

Theorem 3.7.14. *For any $f \in C(K)$ and $\rho \in \ell(V_0)$, there exists a unique $u \in \mathcal{D}_\mu$ such that*

$$\begin{cases} \Delta_\mu u & = f, \\ u|_{V_0} & = \rho. \end{cases} \tag{3.7.5}$$

Moreover, u is given by $u = \sum_{p \in V_0} \rho(p) \psi_p - G_\mu f$.

Proof. If $u = \sum_{p \in V_0} \rho(p) \psi_p - G_\mu f$, then, by Lemma 3.7.10, u satisfies (3.7.5).

Assume that both u_1 and u_2 satisfy (3.7.5); then $v = u_1 - u_2 \in \mathcal{D}_D$ and $\Delta_\mu v = 0$. Since $-G_\mu = (\Delta_\mu|_{\mathcal{D}_D})^{-1}$, it follows that $v = 0$. Therefore $u_1 = u_2$. □

Corollary 3.7.15. *The collection of harmonic functions on K coincides with $\{u : u \in \mathcal{D}_\mu, \Delta_\mu u = 0\}$.*

Combining Lemma 3.6.4, Corollary 3.7.13 and Theorem 3.7.14, we obtain the following theorem.

Theorem 3.7.16. *Let $\varphi \in C(K)$. Then the following five conditions are equivalent.*
(1) *$u \in \mathcal{D}_\mu$ and $\Delta_\mu u = \varphi$.*
(2) *$u + G_\mu \varphi$ is a harmonic function.*
(3) *For any $x \in K$,*

$$u(x) = \sum_{p \in V_0} u(p) \psi_p(x) - \int_K g(x, y) \varphi(y) \mu(dy).$$

(4) *$u \in C(K)$ and, for any $m \geq 1$ and any $p \in V_m \backslash V_0$,*

$$(H_m u)(p) = \int_K \psi_p^m \varphi d\mu.$$

(5) *$u \in \mathcal{F}$ and, for any $v \in \mathcal{D}_{\mu,D}$,*

$$\mathcal{E}(v, u) = -\int_K v \varphi d\mu.$$

3.8 Nested fractals

In 3.1, we introduced the notion of harmonic structures. In the following sections, we then constructed Dirichlet forms, Green's functions and Laplacians under the assumption of the existence of a harmonic structure. In this section, we will show that there exists a harmonic structure on a class of highly symmetric p. c. f. self-similar structures based on the idea of Lindstrøm [116], where the notion of nested fractals was introduced. Recall 3.1, where we observed that the existence problem for harmonic structures is equivalent to the eigenvalue problem (or fixed point problem) of the renormalization operator $\mathcal{R}_r$. In [116], Lindstrøm showed the existence of a fixed point of a probabilistic counterpart of the renormalization map $\mathcal{R}_r$ for nested fractals. In this section, we will extend the notion of nested fractal and introduce strongly symmetric p. c. f. self-similar structures. In Theorem 3.8.10, we will show existence of harmonic structures for strongly symmetric p. c. f. self-similar structures.

Throughout this section, $\mathcal{L} = (K, S, \{F_i\}_{i \in S})$ be a connected p. c. f. self-similar structure, where $S = \{1, 2, \ldots, N\}$.

First we introduce the notion of symmetry of a p. c. f. self-similar structure.

Definition 3.8.1. A homeomorphism $g : K \to K$ is called a symmetry of $\mathcal{L}$ if and only if, for any $m \geq 0$, there exists an injective map $g^{(m)} : W_m \to W_m$ such that $g(F_w(V_0)) = F_{g^{(m)}(w)}(V_0)$ for any $w \in W_m$.

In this section we also assume that $K \subset \mathbb{R}^d$ and that F_i is the restriction of a similitude with $\mathrm{Lip}(F_i) = c_i$ on K for any $i \in S$. Recall that a map $f : \mathbb{R}^d \to \mathbb{R}^d$ is called a similitude with $\mathrm{Lip}(f) = c$ if there exist $U \in O(d)$ and $a \in \mathbb{R}^d$ such that $f(x) = cUx + a$ for any $x \in \mathbb{R}$. See Definition 1.1.2 and the following remark.

Let $\#(V_0) = M$ and let $V_0 = \{p_i\}_{i=1,\ldots,M}$. Then, without loss of generality, we may assume that $\sum_{i=1}^{M} p_i = 0$.

Definition 3.8.2. For $x, y \in \mathbb{R}^d$, $x \neq y$, let

$$H_{xy} = \{z \in \mathbb{R}^d : |x - z| = |y - z|\}.$$

(H_{xy} is the hyperplane bisecting the line segment $[x, y]$.) Also let $g_{xy} : \mathbb{R}^d \to \mathbb{R}^d$ be reflection in H_{xy}.

Remark. Let g be an affine transformation from $\mathbb{R}^d$ to itself. Suppose that $g|_K$ is a symmetry of $\mathcal{L}$. Then $g(V_0) = V_0$. Since $\sum_{i=1}^{M} p_i = 0$, it follows that $g(0) = 0$. Hence g is a linear transformation.

Definition 3.8.3. (1) Let $m_* = \#\{|x - y| : x, y \in V_0, x \neq y\}$. Define $l_0 = \min\{|x - y| : x, y \in V_0, x \neq y\}$. Also we define $\{l_i\}_{i=0,\dots,m_*-1}$ inductively by $l_{i+1} = \min\{|x - y| : x, y \in V_0, |x - y| > l_i\}$.
(2) Let $x_i \in V_m$ for $i = 1, 2, \dots, n$. Then $(x_i)_{i=1,2,\dots,n}$ is called an m-walk (between x_1 and x_n) if and only if there exist $w^1, \dots, w^{n-1} \in W_m$ such that $x_i, x_{i+1} \in F_{w^i}(V_0)$ for all $i = 1, 2, \dots, n - 1$.
(3) A 0-walk $(x_i)_{i=1,2,\dots,n}$ is called a strict 0-walk (between x_1 and x_n) if and only if $|x_i - x_{i+1}| = l_0$ for any $i = 1, 2, \dots, n - 1$.
(4) $\mathcal{G}_s = \{g : g \in O(d), g|_K \text{ is a symmetry of } \mathcal{L}\}$.

$\mathcal{G}_s$ is a group with respect to the natural composition.

Recall the remark after Definition 3.8.2. Then $g \in \mathcal{G}_s$ is equivalent to the condition that $g : \mathbb{R}^d \to \mathbb{R}^d$ is an affine transformation which preserves the Euclidean metric and $g|_K$ is a symmetry of $\mathcal{L}$.

Definition 3.8.4. (1) We say that $\mathcal{L} = (K, S, \{F_i\}_{i \in S})$ is strongly symmetric if it satisfies the following four conditions:
(SS1) For any $x \neq y \in V_0$, there exists a strict 0-walk between x and y.
(SS2) If $x, y, z \in V_0$ and $|x - y| = |x - z|$, then there exists $g \in \mathcal{G}_s$ such that $g(x) = x$ and $g(y) = z$.
(SS3) For any $i = 0, \dots, m_* - 2$, there exist x, y and $z \in V_0$ such that $|x - y| = l_i$, $|x - z| = l_{i+1}$ and $g_{yz}|_K$ is a symmetry of $\mathcal{L}$.
(SS4) V_0 is $\mathcal{G}_s$-transitive, that is, for any $x \neq y \in V_0$, there exists $g \in \mathcal{G}_s$ such that $g(x) = y$.
(2) We say that $\mathcal{L} = (K, S, \{F_i\}_{i \in S})$ is an affine nested fractal if $g_{xy}|_K$ is a symmetry of $\mathcal{L}$ for any $x \neq y \in V_0$.

It is easy to see that if $\mathcal{L}$ is strongly symmetric, then it is weakly symmetric, which has been defined in 1.6.

The following proposition is obvious by (SS4).

Proposition 3.8.5. *Let* $V_0 = \{p_i\}_{i=1,2,\dots,M}$*. If* $\mathcal{L}$ *is strongly symmetric, then* $|p_1| = \cdots = |p_M|$*.*

It follows that an affine nested fractal is strongly symmetric. In fact, the conditions (SS2), (SS3) and (SS4) are obvious. We can verify (SS1) as well by virtue of the following lemma.

Lemma 3.8.6. *Let* $\mathcal{L} = (K, S, \{F_i\}_{i \in S})$ *be an affine nested fractal. Then there exists a strict 0-walk between x and y for any $x \neq y \in V_0$.*

Proof. For $p \in V_0$, set

$$B_p = \{q \in V_0 : \text{there exists a strict 0-walk between } p \text{ and } q\}.$$

Define $A = \{(x,y) : x, y \in V_0, B_x \neq B_y\}$. (Note that there exists no strict 0-walk between x and y if $(x,y) \in A$.) Assume that $A \neq \emptyset$. Set $\alpha = \min\{|x-y| : (x,y) \in A\}$ and choose $(x,y) \in A$ satisfying $|x-y| = \alpha$. (Note that $\alpha > l_0$.) Let $z \in B_x$ with $|z-x| = l_0$. Then $(z,y) \in A$ and therefore $|y-z| \geq |y-x| = \alpha$. Since $|z-x| = l_0 < |x-y| = \alpha$, it follows that x and z belong to the same side of the hyperplane H_{yz}. Hence, if $e = (y-z)/|y-z|$, then $(x,e) < 0$. Let $\gamma = -(x,e)$ and let $x_0 = g_{yz}(x)$. Then $|y-x_0| = l_0$ and $x_0 \in B_y$. Note that $x_0 = x + 2\gamma e$. Using Proposition 3.8.5, we obtain

$$(y-x, e) = -(x, e) + (y, e) = \gamma + (y - (y+z)/2, e) = \gamma + |y-z|/2.$$

Therefore,

$$|y-x| = \alpha \geq (y-x, e) = \gamma + |y-z|/2 \geq \gamma + \alpha/2. \qquad (3.8.1)$$

This implies that $|y-x| \geq 2\gamma = |x-x_0| \geq \alpha = |y-x|$. Therefore equalities hold in (3.8.1). Thus we obtain that $|y-x| = (y-x, e)$ and that $|y-z| = |y-x|$. So $e = (y-z)/|y-z| = (y-x)/|y-x| = (y-x)/|y-z|$. Hence $x = z$. This contradiction immediately implies that $A = \emptyset$. $\qquad \square$

So we obtain the following proposition.

Proposition 3.8.7. *If* $\mathcal{L} = (K, S, \{F_i\}_{i \in S})$ *is an affine nested fractal, then it is strongly symmetric.*

The converse of the above proposition is not true. See Example 3.8.13 below for an example.

Originally Lindstrøm introduced the notion of nested fractals in [116]. Afterwards, the notion of affine nested fractals was introduced as an extension of nested fractals in [37]. The definition of (affine) nested fractals in [116] and [37] contained the following conditions:

Connectivity: For any $i, j \in S$, there exist $i_1, \ldots, i_m$ such that $i_1 = i$, $i_m = j$ and $F_{i_k}(V_0) \cap F_{i_{k+1}}(V_0) \neq \emptyset$ for any $k = 1, \ldots, m-1$.

Nesting: For any m and any $w, w' \in W_m$ with $w \neq w'$,

$$K_w \cap K_{w'} = F_w(V_0) \cap F_{w'}(V_0).$$

Since we assume that K is connected, the connectivity condition is satisfied by Corollary 1.6.5. Also the nesting condition is immediately verified by Proposition 1.3.5-(2).

Since the boundary V_0 of an affine nested fractal is highly symmetric, the geometry of V_0 seems quite restricted. In fact, it is conjectured in [6,

§5] that V_0 is a regular planar polygon, a d-dimensional tetrahedron or a d-dimensional simplex.

Hereafter in this section, we always assume that $\mathcal{L} = (K, S, \{F_i\}_{i \in S})$ is strongly symmetric. We also impose the following assumption on $\mathcal{L}$:

$$\#(F_i(V_0) \cap F_j(V_0)) \leq 1 \quad \text{whenever } i \neq j. \tag{3.8.2}$$

This may be regarded as a technical assumption in order to avoid non-essential difficulties. The forthcoming results in this section may hold without (3.8.2). In [37], it is claimed that (3.8.2) holds for any affine nested fractals. However, this does not seem quite clear at this moment. See also [6, §5] for some comments.

Definition 3.8.8. (1) We define the collection of symmetric Laplacians on V_0, $\mathcal{L}_s(V_0)$, by

$$\mathcal{L}_s(V_0) = \{D \in \mathcal{LA}(V_0) : D_{xy} = D_{x'y'} \text{ if } |x - y| = |x' - y'|\}.$$

(2) Let $\mathbf{r} = (r_1, r_2, \dots, r_N) \in (0, \infty)^N$. We say that $\mathbf{r}$ is $\mathcal{G}_s$-invariant if $r_i = r_j$ whenever there exists $g \in \mathcal{G}_s$ such that $g(F_i(V_0)) = F_j(V_0)$.

(3) For a symmetry g, $D \in \mathcal{LA}(V_0)$ is said to be g-invariant if $D_{xy} = D_{g(x)g(y)}$ for any $x, y \in V_0$. Also, if $D \in \mathcal{LA}(V_0)$ is g-invariant for any $g \in \mathcal{G}_s$, then D is said to be $\mathcal{G}_s$-invariant.

Obviously, if $\mathbf{r} = (1, \dots, 1)$ then $\mathbf{r}$ is $\mathcal{G}_s$-invariant.

Proposition 3.8.9. *The collection of all $\mathcal{G}_s$-invariant Laplacians equals $\mathcal{L}_s(V_0)$.*

Proof. Since $\mathcal{G}_s \subset O(d)$, D is $\mathcal{G}_s$-invariant for any $D \in \mathcal{L}_s(V_0)$. Conversely, suppose that $D \in \mathcal{LA}(V_0)$ is $\mathcal{G}_s$-invariant. Let $x, y, x', y' \in V_0$ and assume that $|x - y| = |x' - y'|$. By (SS4), there exists $g_1 \in \mathcal{G}_s$ such that $g_1(x) = x'$. Let $y'' = g_1(y)$. Then $|x' - y'| = |x' - y''|$. By (SS2), there exists $g_2 \in \mathcal{G}_s$ such that $g_2(x') = x'$ and $g_2(y'') = y'$. Set $g = g_2 \circ g_1$. Then $g(x) = x'$ and $g(y) = y'$. Hence $D_{xy} = D_{x'y'}$. So $D \in \mathcal{L}_s(V_0)$. $\qquad\square$

Theorem 3.8.10. *Let $\mathcal{L} = (K, S, \{F_i\}_{i \in S})$ be strongly symmetric. If $\mathbf{r} = (r_1, r_2, \dots, r_N) \in (0, \infty)^N$ is $\mathcal{G}_s$-invariant, then there exist $D \in \mathcal{L}_s(V_0)$ and $\lambda > 0$ such that $(D, \lambda\mathbf{r})$ is a harmonic structure on $\mathcal{L}$. Moreover, the resistance form $(\mathcal{E}, \mathcal{F})$ constructed from the harmonic structure $(D, \lambda\mathbf{r})$ is $\mathcal{G}_s$-invariant, i.e., $u \circ g \in \mathcal{F}$ and $\mathcal{E}(u, u) = \mathcal{E}(u \circ g, u \circ g)$ for any $g \in \mathcal{G}_s$ and $u \in \mathcal{F}$.*

Remark. In 3.1, we have seen that $(D, \lambda\mathbf{r})$ is a harmonic structure if and only if $\mathcal{R}_{\mathbf{r}}(D) = \lambda D$ (or equivalently $\mathcal{R}_{\lambda\mathbf{r}}(D) = D$), where $\mathcal{R}_{\mathbf{r}} : \mathcal{LA}(V_0) \to$

$\mathcal{LA}(V_0)$ is the renormalization map on $\mathcal{LA}(V_0)$ defined by $\mathcal{R}_{\mathbf{r}}(D) = [H_1]_{V_0}$. It is easy to show that such a λ is uniquely determined. (i.e., if there exist $D, D' \in \mathcal{LA}(V_0)$ and $\lambda, \lambda' > 0$ such that $\mathcal{R}_{\mathbf{r}}(D) = \lambda D$ and $\mathcal{R}_{\mathbf{r}}(D') = \lambda' D'$, then $\lambda = \lambda'$.) Note that $\mathcal{R}_{\mathbf{r}}(\mathcal{L}_s(V_0)) \subset \mathcal{L}_s(V_0)$, which will be shown in Lemma 3.8.22. Using Sabot's arguments in [160], we can also show that the fixed point of $\mathcal{R}_{\lambda\mathbf{r}}|_{\mathcal{L}_s(V_0)}$ is unique if $(K, S, \{F_i\}_{i \in S})$ is an affine nested fractal. See also Metz [129] for another proof of the uniqueness of the fixed point of $\mathcal{R}_{\lambda\mathbf{r}}|_{\mathcal{L}_s(V_0)}$.

Remark. The harmonic structure $(D, \lambda\mathbf{r})$ obtained in the above theorem is not always regular. For example, in Example 3.2.3, the self-similar structure $(K, S, \{F_i\}_{i \in S})$ is an affine nested fractal and the harmonic structure $(D, \lambda\mathbf{r})$ is $\mathcal{G}_s$-invariant. In this example, $(D, \lambda\mathbf{r})$ is not regular if $t \geq \sqrt{5}$.

Example 3.8.11 (Pentakun). Let $S = \{1, 2, \dots, 5\}$. For $k \in S$, let $p_k = e^{2k\pi\sqrt{-1}/5} \in \mathbb{C}$ and define a contraction $F_i : \mathbb{C} \to \mathbb{C}$ by

$$F_i(z) = \frac{3 - \sqrt{5}}{2}(z - p_i) + p_i,$$

where we naturally identify $\mathbb{C}$ with $\mathbb{R}^2$. The pentakun† is the self-similar set with respect to $\{F_i\}_{i \in S}$. See Figure 3.2. The self-similar structure that corresponds to the pentakun is post critically finite. In fact

$$\mathcal{C} = \bigcup_{k=1}^{5} \{[k-2]_5[k+1]_5, [k+2]_5[k-1]_5\},$$

$\mathcal{P} = \{\dot{k}\}_{k \in S}$ and $p_k = \pi(\dot{k})$ for $k \in S$, where $[i]_5 \in S$ is defined by $[i]_5 \equiv i$ mod 5. It follows that $V_0 = \{p_k\}_{k \in S}$ is the collection of vertices of a regular pentagon.

The pentakun is an affine nested fractal. In fact, $\mathcal{G}_s$ coincides with

$$\{g \in O(2) : g(V_0) = V_0\},$$

which is generated by the reflections $\{g_{xy} : x, y \in V_0, x \neq y\}$. Therefore, $D \in \mathcal{LA}(V_0)$ belongs to $\mathcal{L}_s(V_0)$ if and only if

$$D_{p_i p_j} = \begin{cases} a & \text{if } |i - j| = \pm 1 \mod 5 \\ b & \text{if } |i - j| = \pm 2 \mod 5, \end{cases}$$

where $a, b \geq 0$ with $a + b > 0$. In this case, $\mathbf{r} = (r_k)_{k=1,2,\dots,5}$ is $\mathcal{G}_s$-invariant if and only if $r_1 = \cdots = r_5$. Let $\mathbf{r} = (1, \dots, 1)$. By Theorem 3.8.10, there exist $\lambda > 0$ and $D \in \mathcal{L}_s(V_0)$ such that $\mathcal{R}_{\mathbf{r}}(D) = \lambda D$, or equivalently

---

† In the same way, we can also define hexakun, heptakun, octakun and so on. 'kun' is 'Mr' in Japanese.

Fig. 3.2. Pentakun

$(D, \lambda \mathbf{r})$ is a harmonic structure. In this case, it is also easy to check uniqueness of such λ and D directly. (Use a symbolic calculus program, such as Mathematica or Maple V.)

Example 3.8.12 (Snowflake). Let $p_k = e^{2k\pi\sqrt{-1}/6} \in \mathbb{C}$ for $k = 1, \ldots, 6$ and let $p_7 = 0$. We naturally identify $\mathbb{C}$ with $\mathbb{R}^2$. Define $F_i : \mathbb{C} \to \mathbb{C}$ by $F_k(z) = (z - p_k)/3 + p_k$ for $k = 1, 2, \ldots, 7$. The snowflake is the self-similar set with respect to the contractions $\{F_k\}_{k=1,2,\ldots,7}$. See Figure 3.3. The self-similar structure associated with the snowflake is post critically finite. In fact, $p_k = \pi(\dot{k})$ for $k = 1, 2, \ldots, 7$,

$$C = \bigcup_{k=1,2,\ldots,6} \{7\dot{k}, k[k \dot{+} 2]_6, k[k \dot{+} 3]_6, k[k \dot{+} 4]_6\},$$

$\mathcal{P} = \{\dot{k}\}_{k=1,2,\ldots,6}$ and $V_0 = \{p_k\}_{k=1,2,\ldots,6}$, where $[i]_6 \in \{1, 2, \ldots, 6\}$ is defined by $[i]_6 \equiv i \mod 6$.

The snowflake is an affine nested fractal. In fact, $\mathcal{G}_s$ coincides with

$$\{g \in O(2) : g(V_0) = V_0\},$$

which is generated by the reflections $\{g_{xy} : x, y \in V_0, x \neq y\}$. In this case, we see that $\#\{|x - y| : x, y \in V_0, x \neq y\} = 3$ and hence $\mathcal{L}_s(V_0)$ is a 3-dimensional manifold. Also $\mathbf{r} = (r_k)_{k=1,2,\ldots,7}$ is $\mathcal{G}_s$-invariant if and only if $r_1 = \cdots = r_6$. Let $\mathbf{r} = (1, \ldots, 1, t)$ for $t > 0$. Then, by Theorem 3.8.10, for any $t > 0$, there exist $\lambda > 0$ and $D \in \mathcal{L}_s(V_0)$ such that $\mathcal{R}_{\mathbf{r}}(D) = \lambda D$, or equivalently, $(D, \lambda \mathbf{r})$ is a harmonic structure.

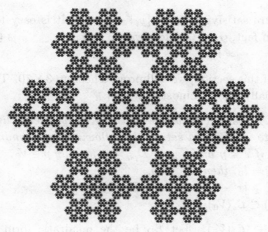

Fig. 3.3. Snowflake

The next example is strongly symmetric but it is not an affine nested fractal.

Example 3.8.13. Let $(K, S, \{F_i\}_{i \in S})$ be a connected p. c. f. self-similar set, where $K \subset \mathbb{R}^3 = \{(x, y, z) : x, y, z \in \mathbb{R}\}$ and $F_i : \mathbb{R}^3 \to \mathbb{R}^3$. We assume that V_0 is the set of vertices of a cube, that is,

$$V_0 = \{((-1)^i, (-1)^j, (-1)^k) : i, j, k \in \{0, 1\}\}.$$

This case is not included in the class of affine nested fractals, because $g_{pq}(V_0) \neq V_0$ if $p = (-1, -1, -1)$ and $q = (1, 1, 1)$. Let $\mathcal{G} = \{g \in O(3) : g(V_0) = V_0\}$. Then it follows that $\#(\mathcal{G}) = 48$. We define three subgroups $\mathcal{G}^0, \mathcal{G}^1$ and $\mathcal{G}^2$ of $\mathcal{G}$. First, let $\mathcal{G}^0$ be the group generated by the three reflections in the xy, yz and xz-planes. Then $\#(\mathcal{G}^0) = 8$. Next, let $\mathcal{G}^1$ be the group generated by the four rotations by $2\pi/3$ around the lines $x = y = z$, $x = y = -z$, $x = -y = -z$ and $x = -y = z$. Then $\#(\mathcal{G}^1) = 12$. Also let $\mathcal{G}^2$ be the group generated by $\mathcal{G}^0$ and $\mathcal{G}^1$. Then $\#(\mathcal{G}^2) = 24$. Note that $\mathcal{G}^2$ is a proper subgroup of $\mathcal{G}$ For example, $g_{xy} \notin \mathcal{G}^2$ if $x = (1, 1, 1)$ and $y = (1, -1, -1)$. Also the rotation by $\pi/2$ around the x-axis does not belong to $\mathcal{G}^2$. Now if $\mathcal{G}_s$ includes $\mathcal{G}^2$, then $(K, S, \{F_i\}_{i \in S})$ is strongly symmetric.

Let us present a concrete example. Let $\{p_i\}_{i=1,2,\ldots,8}$ be the vertices of a cube. Also let $p_9 = 0$. For any $i = 1, 2, \ldots, 9$, define $F_i(a) = (a - p_i)/3 + p_i$ for any $a \in \mathbb{R}^3$. If K is the self-similar set with respect to $\{F_i\}_{i=1,2,\ldots,9}$, then $(K, S, \{F_i\}_{i \in S})$, where $S = \{1, 2, \ldots, 9\}$, is a connected p. c. f. self-similar structure satisfying $V_0 = \{p_i\}_{i=1,2,\ldots,8}$. It is easy to see that $\mathcal{G}_s$ contains $\mathcal{G}^2$. In fact, $\mathcal{G}_s = \mathcal{G}$. Moreover, $\mathbf{r} = (r_i)_{i=1,2,\ldots,9}$ is $\mathcal{G}_s$-invariant if and only if $r_1 = \cdots = r_8$.

In the rest of this section, we will prove Theorem 3.8.10. The idea of the proof is essentially due to Lindstrøm [116].

Lemma 3.8.14. *Let $\mathcal{L}_*(V_0)$ be the collection of symmetric linear operators D from $\ell(V_0)$ to itself which satisfy the following three conditions:*
(V1) $D_{xy} \geq 0$ if $x \neq y$ and $\sum_{q \in V_0} D_{pq} = 0$ for any $p \in V_0$.
(V2) If $|x - y| = l_0$, then $D_{xy} > 0$.
(V3) If $|x - y| \geq |x' - y'| > 0$, then $D_{xy} \leq D_{x'y'}$.
Then $\mathcal{L}_*(V_0) \subset \mathcal{L}_s(V_0)$.

Proof. Let $D \in \mathcal{L}_*(V_0)$. Let $\mathcal{E}_D$ be the quadratic form on $\ell(V_0)$ defined by $\mathcal{E}_D(u, v) = -(u, Dv)$. Then obviously $\mathcal{E}_D$ satisfies (DF1) because $\mathcal{E}_D(u, u) = \sum_{p,q \in V_0} D_{pq}(u(p) - u(q))^2/2$. (Recall Definitions 2.1.1 and 2.1.2.) Assume that $Du = 0$ for $u \in \ell(V_0)$. Then $u(p) = u(q)$ if $D_{pq} > 0$. Now by (SS1), for any $x, y \in V_0$, there exists a strict 0-walk between x and y, say, $(x_i)_{i=1,2,\ldots,n}$. Then, by (V2), $D_{x_i x_{i+1}} > 0$ for any $i = 1, 2, \ldots, n-1$. Therefore $u(x) = u(y)$. This shows (DF2). Obviously D satisfies (L3) and hence $\mathcal{E}_D$ satisfies (DF3). By Proposition 2.1.3, it follows that $D \in \mathcal{LA}(V_0)$. Moreover, by (V3), $D_{xy} = D_{x'y'}$ if $|x-y| = |x'-y'|$. Hence $D \in \mathcal{L}_s(V_0)$. $\square$

For $(D, \mathbf{r}) \in \mathcal{LA}(V_0) \times (0, \infty)^N$, we defined $H_1 \in \mathcal{LA}(V_1)$ in Definition 3.1.1. Recall that the renormalization operator $\mathcal{R}_\mathbf{r}$ is defined by $\mathcal{R}_\mathbf{r}(D) = [H_1]_{V_0}$.

Definition 3.8.15. For any $x, y \in V_1$ and $n \geq 0$, define

$$P_n(x, y) = \{\mathbf{x} = (x_0, x_1, \ldots, x_n, x_{n+1}) : x_0 = x, x_{n+1} = y,$$

$\mathbf{x}$ is a 1-walk between x and y and $x_i \in V_1 \backslash V_0$ for any $i = 1, 2, \ldots, n.\}$

Moreover, let $D \in \mathcal{LA}(V_0)$. Define $h(\mathbf{x})$ for $\mathbf{x} = (x_0, x_1, \ldots, x_n, x_{n+1}) \in P_n(x, y)$ by

$$h(\mathbf{x}) = \frac{h_{x_0 x_1} h_{x_1 x_2} \cdots h_{x_{n-1} x_n} h_{x_n x_{n+1}}}{h_{x_1 x_1} h_{x_2 x_2} \cdots h_{x_n x_n}},$$

where

$$h_{pq} = \begin{cases} (H_1)_{pq} & \text{if } p \neq q, \\ -(H_1)_{pp} & \text{if } p = q. \end{cases}$$

In particular, if $n = 0$, then $P_0(x, y) = \{(x, y)\}$ and $h((x, y)) = h_{xy}$.

The following lemma holds for any connected p. c. f. self-similar structure and any $(D, \mathbf{r}) \in \mathcal{LA}(V_0) \times (0, \infty)^N$.

Lemma 3.8.16. *For any* $x \neq y \in V_0$,

$$\mathcal{R}_{\mathbf{r}}(D)_{xy} = \sum_{n \geq 0} 2^{-n} \sum_{\mathbf{x} \in P_n(x,y)} h(\mathbf{x}).$$

Proof. We decompose H_1 into the following form:

$$H_1 = \begin{pmatrix} T & {}^t J \\ J & X \end{pmatrix},$$

where $T : \ell(V_0) \to \ell(V_0)$, $J : \ell(V_0) \to \ell(V_1 \backslash V_0)$ and $X : \ell(V_1 \backslash V_0) \to \ell(V_1 \backslash V_0)$. Then, by Theorem 2.1.6, it follows that

$$[H_1]_{V_0} = T - {}^t J X^{-1} J.$$

Note that $h_{pp} = -(H_1)_{pp} > 0$ and $h_{pp} \geq h_{pq}$ for any $p, q \in V_1 \backslash V_0$. Define $A : \ell(V_0) \to \ell(V_0)$ by $A_{pp} = h_{pp} = -X_{pp}$ for any $p \in V_1 \backslash V_0$ and $A_{pq} = 0$ if $p \neq q$. Also define $L : \ell(V_1 \backslash V_0) \to \ell(V_1 \backslash V_0)$ by $L_{pq} = h_{pq}/h_{qq}$ for any $p, q \in V_1 \backslash V_0$. Then $X = -2(I - L/2)A$, where I is the identity map. Therefore, $(-X)^{-1} = A^{-1}(I - L/2)^{-1}/2$.

Recall the proof of Lemma 3.5.1. Let C be the counterpart of the C defined there. Define an equivalence relation $\sim$ on $V_1 \backslash V_0$ by $p \sim q$ if and only if there exists $\mathbf{x} \in \cup_{n \geq 0} P_n(x, y)$ such that $h(\mathbf{x}) > 0$. Let $(V_0 \backslash V_1)/\!\!\sim \; = \{U_1, \dots, U_m\}$, where $U_i \subseteq V_0 \backslash V_1$ is an equivalence class. Then X and C are decomposed as $X_i : \ell(U_i) \to \ell(U_i)$ and $C_i : \ell(U_i) \to \ell(U_i)$, respectively, for $i = 1, 2, \dots, m$. Then, by the argument in the proof of Lemma 3.5.1, it follows that $\|(C_i)^{n_i}\| < 1$ for some $n_i \in \mathbb{N}$. This implies that $\|C^n\| < 1$ for some n. Note that $\|C\| \leq 1$. Since $L = (C + I)$, we obtain $\|(L/2)^n\| < 1$. Hence,

$$(-X)^{-1} = A^{-1}(I - L/2)^{-1}/2 = A^{-1} \sum_{k \geq 0} 2^{-(k+1)} L^k,$$

where $L^0 = I$. Hence, for $p, q \in V_1 \backslash V_0$,

$$(-X)_{pq}^{-1} = \frac{\delta_{pq}}{2h_{pp}} + \sum_{k \geq 1} 2^{-(k+1)} \sum_{(x_i)_{i=1,2,\dots,k+1} \in P_{k-1}(p,q)} \frac{h_{x_1 x_2} \cdots h_{x_k x_{k+1}}}{h_{x_1 x_1} \cdots h_{x_{k+1} x_{k+1}}}.$$

This immediately implies the desired expression of $\mathcal{R}_{\mathbf{r}}(D)_{xy}$. □

Remark. Let $(K, S, \{F_i\}_{i \in S})$ be strongly symmetric and let $D \in \mathcal{L}_*(V_0)$. Then we can show that X itself satisfies the assumptions of Lemma 3.5.1. (This fact is not used afterwards.)

Lemma 3.8.17. *Let $D \in \mathcal{L}_*(V_0)$ and let $x, y, z \in F_s(V_0)$ for some $s \in S$. If $|x - y| \leq |x - z|$, then $h_{xy} \geq h_{xz}$.*

Proof. First suppose that $x = y$. Then

$$h_{xy} = h_{xx} = -(H_1)_{xx} = \sum_{p \in V_1 \backslash \{x\}} (H_1)_{px} \geq (H_1)_{xz} = h_{xz}.$$

Next suppose that $x \neq y$. Then, by (3.8.2), there exist $x_0, y_0, z_0 \in V_0$ such that $F_s(x_0) = x, F_s(y_0) = y, F_s(z_0) = z, h_{xy} = (r_s)^{-1} D_{x_0 y_0}$ and $h_{xz} = (r_s)^{-1} D_{x_0 z_0}$. Since $D_{x_0 y_0} \geq D_{x_0 z_0}$, it follows that $h_{xy} \geq h_{xz}$. $\qquad \square$

Lemma 3.8.18. *Let $w \in W_m$ and let $x \in K \backslash V_m$. Then $x \in K_w \backslash F_w(V_0)$ if and only if $\overline{J} \cap V_m = F_w(V_0)$, where J is the connected component of $K \backslash V_m$ containing x.*

Proof. Suppose that $x \in K_w \backslash F_w(V_0)$ and let J be the connected component of $K \backslash V_m$ containing x. Then, by Proposition 1.6.9, it follows that $J = K_w \backslash F_w(V_0)$. Hence, $\overline{J} \cap V_m = F_w(V_0)$.

Conversely, if $x \notin K_w \backslash F_w(V_0)$, there exists $w' \in W_m$ such that $w \neq w'$ and $x \in K_{w'} \backslash F_{w'}(V_0)$. Then $\overline{J} \cap V_m = F_{w'}(V_0)$. Hence $F_w(V_0) = F_{w'}(V_0)$. This is impossible because $\#(F_w(V_0) \cap F_{w'}(V_0)) \leq 1$ by the assumption (3.8.2). $\qquad \square$

Proposition 3.8.19. *Let g be a symmetry of $\mathcal{L}$. If $F_{w'}(V_0) = g(F_w(V_0))$ for $w, w' \in W_m$, then $K_{w'} = g(K_w)$.*

Proof. Let $x \in K_w \backslash F_w(V_0)$ and let J be the connected component of $K \backslash V_m$ containing x. Then $g(J)$ is the connected component of $K \backslash V_m$ containing $g(x)$. Since $\overline{g(J)} \cap V_m = F_{w'}(V_0)$, Lemma 3.8.18 implies that $g(x) \in K_{w'}$. Hence $g(K_w) \subseteq K_{w'}$. Note that g^{-1} is also a symmetry and $g^{-1}(F_{w'}(V_0)) = F_w(V_0)$. Using entirely the same arguments, we see that $g(K_w) \supseteq K_{w'}$. $\qquad \square$

Originally F_i was a map from K to itself. From now on, we think of F_i as a similitude on $\mathbb{R}^d$.

Proposition 3.8.20. *For any $w \in W_*$ and any $g \in \mathcal{G}_s$, set*

$$U_{g,w} = F_{w'}^{-1} \circ g \circ F_w,$$

where $w' \in W_*$ is the unique word that satisfies $F_{w'}(V_0) = g(F_w(V_0))$. Then $U_{g,w} \in \mathcal{G}_s$.

Note that $U_{g,\emptyset} = g$.

Proof. By Proposition 3.8.19, we see that $g(K_w) = K_{w'}$. Therefore, it follows that, for any $v \in W_*$, there exists $v' \in W_*$ such that $g(F_{wv}(V_0)) = F_{w'v'}(V_0)$. This implies that $U_{g,w}(F_v(V_0)) = F_{v'}(V_0)$. Hence $U_{g,w}$ is a symmetry. Note that $U_{g,w}$ is a linear transformation. Also, from the definition, $U_{g,w}$ preserves distance. Hence $U_{g,w} \in O(d)$. □

Let $A(\cdot, \cdot) \in \mathcal{DF}(V_m)$. We say that $A(\cdot, \cdot)$ is $\mathcal{G}_s$-invariant if and only if $A(u \circ g, v \circ g) = A(u, v)$ for any $u, v \in \ell(V_m)$ and any $g \in \mathcal{G}_s$.

Corollary 3.8.21. *Let $D \in \mathcal{L}_s(V_0)$. If $\mathbf{r}$ is $\mathcal{G}_s$-invariant, then $\mathcal{E}^{(m)}(\cdot, \cdot)$ is $\mathcal{G}_s$-invariant, where $\mathcal{E}^{(m)} \in \mathcal{DF}(V_m)$ is defined in Definition 3.1.1.*

Proof. We use induction on m. First $\mathcal{E}^{(0)} = \mathcal{E}_D$ is $\mathcal{G}_s$-invariant. Now assume that $\mathcal{E}^{(m)}$ is $\mathcal{G}_s$-invariant. Let $g \in \mathcal{G}_s$. By (3.1.1),

$$\mathcal{E}^{(m+1)}(u \circ g, v \circ g) = \sum_{i=1}^{N} \frac{1}{r_i} \mathcal{E}^{(m)}(u \circ g \circ F_i, v \circ g \circ F_i) \qquad (3.8.3)$$

for $u, v \in \ell(V_{m+1})$. Now let $g(K_i) = K_j$. Then $r_i = r_j$ because $\mathbf{r}$ is $\mathcal{G}_s$-invariant. Also the induction hypothesis along with Proposition 3.8.20 implies

$$\mathcal{E}^{(m)}(u \circ g \circ F_i, v \circ g \circ F_i) = \mathcal{E}^{(m)}(u \circ F_j \circ U_{g,i}, v \circ F_j \circ U_{g,i})$$
$$= \mathcal{E}^{(m)}(u \circ F_j, v \circ F_j).$$

Again, by (3.1.1), it follows that $\mathcal{E}^{(m+1)}(u \circ g, u \circ g) = \mathcal{E}^{(m+1)}(u, u)$. □

Lemma 3.8.22. *If $\mathbf{r}$ is $\mathcal{G}_s$-invariant, then $\mathcal{R}_{\mathbf{r}}(\mathcal{L}_s(V_0)) \subseteq \mathcal{L}_s(V_0)$.*

Proof. Let $D \in \mathcal{L}_s(V_0)$. By Corollary 3.8.21, H_1 is $\mathcal{G}_s$-invariant. Hence $(H_1)_{pq} = (H_1)_{g(p)g(q)}$ for any $p, q \in V_1$ and any $g \in \mathcal{G}_s$. Now, as we have seen in the proof of Proposition 3.8.9, if $x, x', y, y' \in V_0$ and $|x - y| = |x' - y'|$, then there exists $g \in \mathcal{G}_s$ such that $g(x) = x'$ and $g(y) = y'$. Define $g(\mathbf{x}) = (g(x_i))_{i=0,1,\dots,n+1}$ for $\mathbf{x} = (x_i)_{i=0,1,\dots,n+1} \in P_n(x, y)$. Then g is a bijective mapping between $P_n(x, y)$ and $P_n(x', y')$. Also it follows that $h(\mathbf{x}) = h(g(\mathbf{x}))$ because H_1 is $\mathcal{G}_s$-invariant. Hence, by Lemma 3.8.16, $\mathcal{R}_{\mathbf{r}}(D)_{xy} = \mathcal{R}_{\mathbf{r}}(D)_{x'y'}$. □

Lemma 3.8.23. *Let $p, q \in V_0$ and assume that $g_{pq} \in \mathcal{G}_s$. If $x, y \in F_w(V_0)$ for some $w \in W_*$, $|x-q| < |x-p|$ and $|y-q| > |y-p|$, then $g_{pq}(K_w) = K_w$.*

Proof. Let $J = K_w \backslash F_w(V_0)$. Then, as $K \backslash V_0$ is connected, J is a connected component of $K \backslash V_m$. If $g_{pq}(K_w) = K_{w'}$, then $g_{pq}(J) = K_{w'} \backslash F_{w'}(V_0)$ and $g_{pq}(J)$ is also a connected component of $K \backslash V_m$. By Proposition 1.6.6, there exists a continuous path between x and y contained in $J \cup \{x, y\}$. Since

x and y belong to different sides of the hyperplane H_{pq}, this continuous path between x and y intersects H_{xy}. Hence $J \cap H_{pq} \neq \emptyset$. By the fact that $g_{pq}(J \cap H_{pq}) = J \cap H_{pq}$, we see that $J \cap g_{pq}(J) \neq \emptyset$. Hence $J = g_{pq}(J)$ and therefore $w = w'$. $\qquad \square$

Lemma 3.8.24. *Let $\alpha_i \in \mathbb{R}$ for $i = 1, 2, \dots, n$. Then,*

$$\sum_{J:J \subseteq \{1,2,\dots,n\}} (-1)^{\#(J)} \prod_{j \in J} \alpha_j = \prod_{i=1}^{n} (1 - \alpha_i).$$

Lemma 3.8.25. *If $\mathbf{r}$ is $\mathcal{G}_s$-invariant, then $\mathcal{R}_{\mathbf{r}}(\mathcal{L}_*(V_0)) \subseteq \mathcal{L}_*(V_0)$.*

Proof. Let $D \in \mathcal{L}_*(V_0)$. Since $D \in \mathcal{L}_s(V_0)$, Lemma 3.8.22 implies that $\mathcal{R}_{\mathbf{r}}(D) \in \mathcal{L}_s(V_0)$. Therefore we need to show (V3). Now by (SS3), there exist $x, y, z \in V_0$ such that $|x - y| = l_k$, $|x - z| = l_{k+1}$ and $g_{yz} \in \mathcal{G}_s$. Let $V = \{p \in \mathbb{R}^d : |p - y| \leq |p - z|\}$ and define $T : \mathbb{R}^d \to \mathbb{R}^d$ by

$$T(p) = \begin{cases} p & \text{if } p \in V, \\ g_{yz}(p) & \text{otherwise.} \end{cases}$$

Note that $x \in V$ and $T(x) = x$. Also for $\mathbf{x} = (x_i)_{i=0,\dots,n+1} \in P_n(x,z) \cup P_n(x,z)$, define $T(\mathbf{x}) = (T(x_i))_{i=0,\dots,n+1}$. Then, by Lemma 3.8.23, $T(\mathbf{x}) \in P_n(x,y)$.

Now let $\mathbf{x} = (x_i)_{i=0,\dots,n+1} \in \operatorname{Im} T$ and let us consider $T^{-1}(\mathbf{x})$. Define

$$I(\mathbf{x}) = \{i \in \{1,\dots,n+1\} : \text{there exists } s \in S \text{ such that}$$
$$x_{i-1}, g_{yz}(x_i) \in F_s(V_0) \text{ and } x_i \notin H_{yz}\}.$$

For $J \subseteq I(\mathbf{x})$, we define $\tau(\mathbf{x}, J) = (y_i)_{i=0,\dots,n+1}$ by $y_i = (g_{zy})^{s(i)}(x_i)$, where $s(i) = \#\{j \in J : j \leq i\}$. Then $T(\tau(\mathbf{x}, J)) = \mathbf{x}$. Moreover,

$$(T)^{-1}(\mathbf{x}) \cap P_n(x,z) = \{\tau(\mathbf{x}, J) : J \subseteq I(\mathbf{x}), \#(J) \text{ is odd}\}$$

and

$$(T)^{-1}(\mathbf{x}) \cap P_n(x,y) = \{\tau(\mathbf{x}, J) : J \subseteq I(\mathbf{x}), \#(J) \text{ is even}\}.$$

Next, consider $h(\tau(\mathbf{x}, J))$. Let $\tau(\mathbf{x}, J) = (y_i)_{i=0,\dots,n+1}$. Note that $h_{x_i x_i} = h_{y_i y_i}$ because $y_i = g_{yz}(x_i)$ or $y_i = x_i$ and H_1 is g_{yz}-invariant. Also, if $i \notin J$, then $(y_{i-1}, y_i) = (x_{i-1}, x_i)$ or $(g_{yz}(x_{i-1}), g_{yz}(x_i))$. Hence $h_{x_{i-1} x_i} = h_{y_{i-1} y_i}$. Therefore, we obtain that

$$h(\tau(\mathbf{x}, J)) = h(\mathbf{x}) \prod_{j \in J} \alpha_j(\mathbf{x}),$$

where $\alpha_j(\mathbf{x}) = h_{x_{j-1}g_{yz}(x_j)}/h_{x_{j-1}x_j}$. Since $|x_{j-1} - x_j| \leq |x_{j-1} - g_{yz}(x_j)|$, Lemma 3.8.17 implies that $\alpha_j(\mathbf{x}) \leq 1$. So, using Lemma 3.8.24, we obtain

$$\sum_{\mathbf{x} \in P_n(x;y)} h(\mathbf{x}) - \sum_{\mathbf{x} \in P_n(x,z)} h(\mathbf{x})$$

$$\geq \sum_{\mathbf{x} \in \text{Im } \mathcal{T}} \sum_{\substack{J:J \subseteq I(\mathbf{x}) \\ \#(J) \text{ is even.}}} h(\tau(\mathbf{x}, J)) - \sum_{\mathbf{x} \in \text{Im } \mathcal{T}} \sum_{\substack{J:J \subseteq I(\mathbf{x}) \\ \#(J) \text{ is odd.}}} h(\tau(\mathbf{x}, J))$$

$$= \sum_{\mathbf{x} \in \text{Im } \mathcal{T}} h(\mathbf{x}) \sum_{J:J \subseteq I(\mathbf{x})} (-1)^{\#(J)} \prod_{j \in J} \alpha_j(\mathbf{x})$$

$$= \sum_{\mathbf{x} \in \text{Im } \mathcal{T}} h(\mathbf{x}) \prod_{j \in I(\mathbf{x})} (1 - \alpha_j(\mathbf{x})) \geq 0.$$

We also see that $h_{xy} \geq h_{xz}$. (If $h_{xz} > 0$, then $x, z \in F_s(V_0)$ for some $s \in V_0$. Lemma 3.8.23 implies $y \in F_s(V_0)$. By Lemma 3.8.17, we obtain $h_{xy} \geq h_{xz}$.) Hence, by Lemma 3.8.16, we verify that $\mathcal{R}_{\mathbf{r}}(D)_{xy} \geq \mathcal{R}_{\mathbf{r}}(D)_{xz}$. This immediately shows (V3). □

Proof of Theorem 3.8.10. Choose $x, y \in V_0$ so that $|x - y| = l_0$. Let $\widetilde{\mathcal{L}}_*(V_0) = \{D \in \mathcal{L}_*(V_0) : D_{xy} = 1\}$. Define $\widetilde{\mathcal{R}}(D) = \mathcal{R}_{\mathbf{r}}(D)/\mathcal{R}_{\mathbf{r}}(D)_{xy}$. Then we see that $\widetilde{\mathcal{R}} : \widetilde{\mathcal{L}}_*(V_0) \to \widetilde{\mathcal{L}}_*(V_0)$ and that $\widetilde{\mathcal{R}}$ is continuous. Now $\widetilde{\mathcal{L}}_*(V_0)$ is homeomorphic to a compact ball in $\mathbb{R}^{m-1}$, where $m = \#\{|p-q| : p, q \in V_0, p \neq q\}$. By Brouwer's fixed point theorem, it follows that $\widetilde{\mathcal{R}}$ has a fixed point $D \in \widetilde{\mathcal{L}}_*(V_0)$. Let $\lambda = \mathcal{R}_{\mathbf{r}}(D)_{xy}$. Then $\mathcal{R}_{\mathbf{r}}(D) = \lambda D$. Hence $(D, \lambda \mathbf{r})$ is a harmonic structure.

By Corollary 3.8.21, $\mathcal{E}^{(k)} \in \mathcal{DF}(V_m)$ induced by the harmonic structure $(D, \lambda \mathbf{r})$ is $\mathcal{G}_s$-invariant for any $k \geq 0$. Letting $k \to \infty$, we verify that $(\mathcal{E}, \mathcal{F})$ is $\mathcal{G}_s$-invariant. □

Notes and references

The use of Dirichlet forms to study analysis on fractals was first developed in Fukushima & Shima [44], where they studied the asymptotic behavior of the eigenvalues of the standard Laplacian on the Sierpinski gasket. Then, in [107] and [83], such an approach using Dirichlet forms was extended to more general classes of (finitely ramified) self-similar sets. See also Fukushima [41], Kusuoka [108], Kumagai [100] and Metz [125, 128].

Although the main source of this chapter is [83], we add many results for the case of non-regular harmonic structures.

3.1 The notion of harmonic structures was introduced in [83]. The counterpart of the renormalization map $\mathcal{R}_{\mathbf{r}}$ on $\mathcal{DF}(V_0)$ had already been introduced in Hattori, Hattori & Watanabe [70], Kusuoka [107] and Lindstrøm [116]. In particular, Kusuoka constructed a self-similar Dirichlet form from a fixed point of such a renormalization map in [107]. Also, Lindstrøm constructed a diffusion process (called Brownian motion) on a nested fractal from decimation invariant random walks in [116].

The fixed point problem of $\mathcal{R}_\mathbf{r}$ has been studied by many authors, for example, Lindstrøm [116], Metz [126, 127, 130, 129] and Sabot [160]. All of those works focus on (affine) nested fractals. The existence of the fixed point was shown by Lindstrøm in [116]. Moreover, Sabot proved the uniqueness of the fixed point in [160]. See Theorem 3.8.10 and the remark after it. Also, Metz obtained a kind of stability of the fixed point in [129].

3.2 Theorems 3.2.8, 3.2.11 and 3.2.14 have not been explicitly stated elsewhere before.

3.3 Self-similarity of Dirichlet forms, Proposition 3.3.1, was first observed in Fukushima [41] for nested fractals. Theorem 3.3.4 is an extension of [85, Theorem 3.1]. Theorem 3.3.8 is a new result.

3.4 The key ideas of the proof of Lemmas 3.4.3, 3.4.4 and 3.4.5 are due to Kumagai [100]. Those lemmas play an essential role in proving Theorem 3.4.6 when the harmonic structure is not regular.

3.5 The explicit definition of the Green's function (as we see in Proposition 3.5.5) was given in [83]. Theorem 3.5.7 is a new result. In [94], one can find a figure of the graph of the Green's function for the standard harmonic structure on the Sierpinski gasket.

3.7 The explicit definition of the Laplacian in Definition 3.7.1 was given in [83]. Theorem 3.7.9 has been known to specialists for sometime. In [44], Fukushima & Shima showed this fact for the standard Laplacian on the Sierpinski gasket with Dirichlet boundary condition.

Exercises

Exercise 3.1. Let $\mathcal{L}$ be the harmonic structure associated with the Sierpinski gasket. (See Example 3.1.5.) Set

$$
D = \begin{pmatrix} -2 & 1 & 1 \\ 1 & -(1+h) & h \\ 1 & h & -(1+h) \end{pmatrix}
$$

and $\mathbf{r} = (s, st, st)$, where h, s and t are positive real numbers. Show that if we fix $h > 0$, there exist unique s and t such that $(D, \mathbf{r})$ becomes a harmonic structure on the Sierpinski gasket. Also prove that $(D, \mathbf{r})$ is a regular harmonic structure.

Hint: let $R = 1/h$. Then calculate the effective resistances for D and H_1 by using the Δ–Y transform (Lemma 2.1.15). Then apply Corollary 2.1.13. You will find that the condition for $(D, \mathbf{r})$ being a harmonic structure is

$$s\left(1 + \frac{(R+t)^2}{2(tR+t+R)}\right) = s\left(t + \frac{t(R+t)}{tR+t+R}\right) = 1.$$

Exercise 3.2 (modified Sierpinski gasket). Let $\{p_i\}_{i=1,2,3}$ be the vertices of a regular triangle in the complex plane $\mathbb{C}$. Set $p_4 = (p_2 + p_3)/2$, $p_5 = (p_3 + p_1)/2$ and $p_6 = (p_1 + p_2)/2$. Choose real numbers α and β so that $2\alpha + \beta = 1$ and $\alpha > \beta > 0$. We define $F_i(z) = \alpha(z - p_i) + p_i$ for $i = 1, 2, 3$ and $F_i(z) = \beta(z - p_i) + p_i$ for $i = 4, 5, 6$. Let K be the self-similar set with respect to $\{F_i\}_{i \in S}$, where $S = \{1, 2, \dots, 6\}$.
(1) Prove that the self-similar structure $\mathcal{L}$ associated with K is post critically finite with $V_0 = \{p_1, p_2, p_3\}$.
(2) Define $D \in \mathcal{LA}(V_0)$ by

$$D = \begin{pmatrix} -2 & 1 & 1 \\ 1 & -2 & 1 \\ 1 & 1 & -2 \end{pmatrix}.$$

Let $\mathbf{r} = (r, r, r, rs, rs, rs)$ where $r, s > 0$. Show that if we fix s, then there exists an unique r such that $(D, \mathbf{r})$ is a harmonic structure on $\mathcal{L}$. Is this harmonic structure regular?

Hint: use the Δ–Y transform and calculate effective resistances as in Exercise 3.1.

Exercise 3.3. Calculate A_i for the harmonic structures on Hata's tree-like set (Example 3.1.6) and the modified Sierpinski gasket (Exercise 3.2).

Exercise 3.4 (Neumann problem for Poisson's equation).
Prove that, for $f \in C(K)$ and $\eta \in \ell(V_0)$, there exists $u \in \mathcal{D}_\mu$ such that

$$\begin{cases} \Delta_\mu u &= f, \\ (du)_p &= \eta(p) \quad \text{for all } p \in V_0 \end{cases}$$

if and only if $\int_K f d\mu = \sum_{p \in V_0} \eta(p)$.

Exercise 3.5. Let $V \subset V_0$ and assume that $V \neq \emptyset$. Define $\mathcal{F}_V = \{u \in \mathcal{F} : u|_V = 0\}$.
(1) Prove that $(\mathcal{F}_V, \mathcal{E})$ is a Hilbert space. Also show that $(\mathcal{E}, \mathcal{F}_V)$ is a local regular Dirichlet form on $L^2(K, \mu)$.
(2) Show that, for any $f \in L^2(K, \mu)$, there exists $u \in \mathcal{F}_V$ such that $\mathcal{E}(v, u) = (v, f)_\mu$ for any $v \in \mathcal{F}_V$. Denote u by $G_V f$. Then prove that

G_V is a bounded operator from $L^2(K,\mu)$ to $L^2(K,\mu)$ and $G_V = H_V^{-1}$, where H_V is the non-negative definite self-adjoint operator associated with the Dirichlet form $(\mathcal{E}, \mathcal{F}_V)$ on $L^2(K,\mu)$.

(3) Set $\mathcal{D}_V = \{u \in \mathcal{D}_\mu : u|_V = 0, (du)_p = 0 \quad \text{for any } p \in V_0 \backslash V\}$. Then prove that $\mathcal{D}_V \subset \text{Dom}(H_V)$ and $H_V|_{\mathcal{D}_V} = -\Delta_\mu|_{\mathcal{D}_V}$. Also show that H_V is the Friedrichs extension of $-\Delta_\mu|_{\mathcal{D}_V}$.

4

Eigenvalues and Eigenfunctions of Laplacians

In this chapter, we will study eigenvalues and eigenfunctions for the Laplacian Δ_μ associated with $(D, \mathbf{r})$ and μ. In particular, we will be interested in the asymptotic behavior of the eigenvalue counting function and present a Weyl-type result (Theorem 4.1.5) in 4.1.

It turns out that the nature of eigenvalues and eigenfunctions of Δ_μ is quite different from that of Laplacians on a bounded domain of $\mathbb{R}^n$. For example, we will find localized eigenfunctions in certain cases. More precisely, in 4.3, we will define the notion of pre-localized eigenfunctions, which are the eigenfunction of Δ_μ satisfying both Neumann and Dirichlet boundary conditions. It is known that such an eigenfunction does not exists for the ordinary Laplacian on a bounded domain of $\mathbb{R}^n$. Proposition 4.3.3 shows that if there exists a pre-localized eigenfunction, then, for any open set $O \subseteq K$, there exists a pre-localized eigenfunction whose support is contained in O.

One important consequence of the existence of pre-localized eigenfunctions is the discontinuity of the integrated density of states. See Theorem 4.3.4 and the remark after it.

We will give a sufficient condition for the existence of pre-localized eigenfunctions in 4.4. In particular, we will see that there exists a pre-localized eigenfunction for the Laplacian on an affine nested fractal associated with the harmonic structure appearing in Theorem 3.8.10. See Corollary 4.4.11.

Throughout this chapter, $\mathcal{L} = (K, S, \{F_i\}_{i \in S})$ is a connected p. c. f. self-similar structure with $S = \{1, 2, \dots, N\}$ and $(D, \mathbf{r})$ is a harmonic structure on $\mathcal{L}$, where $\mathbf{r} = (r_i)_{i \in S}$. Also $\mu \in \widetilde{M}(K)$ and $\sup_{w \in W_*} \sum_{m \geq 0} R_m(\mu^w) < \infty$. Note that if μ is a self-similar measure with weight $(\mu_i)_{i \in S}$, then the above condition on μ is equivalent to $\mu_i r_i < 1$ for any $i \in S$. We use $(\mathcal{E}, \mathcal{F})$ to denote the resistance form associated with $(D, \mathbf{r})$. Also d is a metric on

131

K that is compatible with the original topology of K. (Note that (K, d) is compact.)

4.1 Eigenvalues and eigenfunctions

In this section, we first define eigenvalues and eigenfunctions of the Laplacians Δ_μ. Then we will define the eigenvalue counting function and present a theorem which gives the asymptotic behavior of the eigenvalue counting function.

Definition 4.1.1. Let $E_N(\lambda) = \{\varphi \in \mathcal{D}_N : \Delta_\mu \varphi = -\lambda\varphi\}$. If $\dim E_N(\lambda) \geq 1$, then λ is called a Neumann eigenvalue (N-eigenvalue for short) of Δ_μ and any non-trivial $\varphi \in E_N(\lambda)$ is called a Neumann eigenfunction (N-eigenfunction for short) of Δ_μ belonging to the Neumann eigenvalue λ. Also define $E_D(\lambda) = \{\varphi \in \mathcal{D}_D : \Delta_\mu \varphi = -\lambda\varphi\}$. If $\dim E_D(\lambda) \geq 1$, then λ is called a Dirichlet eigenvalue (D-eigenvalue for short) of Δ_μ and any non-trivial $\varphi \in E_D(\lambda)$ is called a Dirichlet eigenfunction (D-eigenfunction for short) of Δ_μ belonging to the Dirichlet eigenvalue λ.

Of course, eigenvalues and eigenfunctions of Δ_μ are expected to coincide with those of the non-negative self-adjoint operator H_N (or H_D, depending on the boundary condition) associated with the local regular Dirichlet form $(\mathcal{E}, \mathcal{F})$ (or $(\mathcal{E}, \mathcal{F}_0)$) on $L^2(K, \mu)$.

Proposition 4.1.2. *Suppose that $\sum_{m \geq 0} R_m^t(\mu) < \infty$ for some $t > 0$. Then the following three conditions are equivalent.*
(N1) *$\varphi \in \mathrm{Dom}(H_N)$ and $H_N\varphi = \lambda\varphi$,*
(N2) *$\varphi \in \mathcal{F}$ and $\mathcal{E}(\varphi, u) = \lambda(\varphi, u)_\mu$ for any $u \in \mathcal{F}$,*
(N3) *$\varphi \in E_N(\lambda)$.*
Also, the following three conditions are equivalent.
(D1) *$\varphi \in \mathrm{Dom}(H_D)$ and $H_D\varphi = \lambda\varphi$,*
(D2) *$\varphi \in \mathcal{F}_0$ and $\mathcal{E}(\varphi, u) = \lambda(\varphi, u)_\mu$ for any $u \in \mathcal{F}_0$,*
(D3) *$\varphi \in E_D(\lambda)$.*

Remark. Let μ be a self-similar measure. Then $\sum_{m \geq 0} R_m^t(\mu) < \infty$ for some $t > 0$ if and only if $r_i\mu_i < 1$ for all $i \in S$. See the remark after Corollary 3.6.6.

Proof. By Corollary 3.6.6, $g^{(n)}$ is continuous on $K \times K$ for $n \geq (1 - 1/t)^{-1}$. Recall that $G_\mu|_{C(K,d)} = G_D|_{C(K,d)}$. Hence,

$$(G_D{}^n u)(x) = (G_\mu{}^n u)(x) = \int_K g^{(n)}(x, y)u(y)\mu(dy) \tag{4.1.1}$$

for $u \in C(K, d)$. Since $G_D{}^n$ is a bounded operator from $L^2(K, \mu)$ to itself and $g^{(n)}$ is continuous, it follows that (4.1.1) is true for all $u \in L^2(K, \mu)$. This implies that $G_D{}^n u \in C(K, d)$ for any $u \in L^2(K, \mu)$.

Now the equivalence between (D1) and (D2) is obvious from the characterization of a self-adjoint operator associated with a closed form in Lemma B.1.4. Also, by Theorem 3.7.9, (D3) implies (D1). Now suppose $\varphi \in \mathrm{Dom}(H_D)$ and $H_D \varphi = \lambda \varphi$. Then, as $\lambda^n G_D{}^n \varphi = \varphi$, we see that $\varphi \in C(K, d)$. Hence $\lambda G_D \varphi = \lambda G_\mu \varphi = \varphi$. So applying Lemma 3.7.10, we see that $\varphi \in \mathcal{D}_D$. Therefore $\varphi \in E_D(\lambda)$.

Next we consider the Neumann case. It follows immediately that (N1) and (N2) are equivalent by Lemma B.1.4. Assume that $\varphi \in \mathrm{Dom}(H_N)$ and $H_N \varphi = \lambda \varphi$. Let $\varphi_0 = \sum_{p \in V_0} \varphi(p) \psi_p$. Then we see that $\varphi - \varphi_0 - \lambda G_D \varphi \in \mathcal{F}_0$ and that $\mathcal{E}(\varphi - \varphi_0 - \lambda G_D \varphi, u) = 0$ for any $u \in \mathcal{F}_0$. Hence $\varphi = \lambda G_D \varphi + \varphi_0$. This implies that $\varphi = \lambda^n G_D{}^n \varphi + \sum_{k=0}^{n-1} \lambda^k G_D{}^k \varphi_0$. Now, the rest of the discussion is exactly the same as in the Dirichlet case. $\qquad \square$

By Theorem 3.4.6 and Corollary 3.4.7, the self-adjoint operators H_N and H_D have compact resolvents. Therefore, N-eigenvalues (and also D-eigenvalues) are non-negative, of finite multiplicity and the only accumulation point is ∞. In other words, for $b = D$ and N, there exist $\{\lambda_i^b\}_{i \geq 1}$ and $\varphi_i^b \in E_b(\lambda_i^b)$ such that

$$0 \leq \lambda_1^b \leq \cdots \leq \lambda_m^b \leq \lambda_{m+1}^b \leq \cdots$$

and $\{\varphi_i^b\}_{i \geq 1}$ is a complete orthonormal system for $L^2(K, \mu)$. So we can define eigenvalue counting functions as follows.

Notation. Hereafter, the symbol b always represents a boundary condition: b is either N or D, where N represents the Neumann boundary condition and D represents the Dirichlet boundary condition.

Definition 4.1.3 (Eigenvalue counting function). Define $\rho_b(x, \mu)$ for $b = D$ and N by $\rho_b(x, \mu) = \sum_{\lambda \leq x} \dim E_b(\lambda)$. $\rho_b(x, \mu)$ is called the b-eigenvalue counting function.

Obviously, $\rho_b(x, \mu) = \max\{k : \lambda_k^b \leq x\}$ and $\rho_b(x, \mu) \to \infty$ as $x \to \infty$. For ordinary Laplacians on bounded domains in $\mathbb{R}^n$, we have Weyl's famous theorem on the asymptotic behavior of eigenvalue counting functions.

Theorem 4.1.4 (Weyl's theorem). *Let Ω be a bounded domain in $\mathbb{R}^n$. Let λ_i be the i-th eigenvalue of the Dirichlet eigenvalue problem of $-\Delta$ on*

Ω, *that is,*

$$\begin{cases} \Delta u = -\lambda u \\ u|_{\partial\Omega} = 0, \end{cases}$$

where $\Delta = \sum_{i=1}^{n} \partial^2/\partial x_i{}^2$. *Set* $\rho(x) = \#\{i : \lambda_i \leq x\}$. *Then as* $x \to \infty$,

$$\rho(x) = (2\pi)^{-n}\mathcal{B}_n|\Omega|_n x^{n/2} + o(x^{n/2}),$$

where $|\cdot|_n$ *is n-dimensional Lebesgue measure and* $\mathcal{B}_n = |\{x : |x| \leq 1\}|_n$.

Remark. Weyl [182, 183] proved the above result under some extra conditions on the domain D. It is now known that the above result is true for any bounded domain. See Lapidus [111].

We are interested in an analogue of Weyl's theorem for our Laplacian Δ_μ, or more specifically, the asymptotic behavior of the eigenvalue counting functions $\rho_b(x, \mu)$ as $x \to \infty$. If μ is a self-similar measure, we have the following result.

Theorem 4.1.5. *Let μ be a self-similar measure with weight* $(\mu_i)_{i=1,2,\ldots,N}$. *Assume that $\mu_i r_i < 1$ for all $i \in \{1, 2, \ldots, N\}$. Let d_S be the unique real number d that satisfies*

$$\sum_{i=1}^{N} \gamma_i{}^d = 1, \tag{4.1.2}$$

where $\gamma_i = \sqrt{r_i \mu_i}$. *Then*

$$0 < \liminf_{x\to\infty} \rho_b(x, \mu)/x^{d_S/2} \leq \limsup_{x\to\infty} \rho_b(x, \mu)/x^{d_S/2} < \infty$$

for $b = D$ and N. d_S is called the spectral exponent of $(\mathcal{E}, \mathcal{F}, \mu)$. Moreover
(1) *Non-lattice case: If $\sum_{i=1}^{N} \mathbb{Z} \log \gamma_i$ is a dense subgroup of $\mathbb{R}$, then the limit $\lim_{x\to\infty} \rho_b(x, \mu)/x^{d_S/2}$ exists and is independent of the boundary conditions.*
(2) *Lattice case: If $\sum_{i=1}^{N} \mathbb{Z} \log \gamma_i$ is a discrete subgroup of $\mathbb{R}$, let $T > 0$ be its generator. Then, as $x \to \infty$,*

$$\rho_b(x, \mu) = (G(\log x/2) + o(1))x^{d_S/2},$$

where G is a right-continuous, T-periodic function satisfying $0 < \inf G(x) \leq \sup G(x) < \infty$ and $o(1)$ is a term which vanishes as $x \to \infty$. Moreover, the periodic function G is independent of the boundary conditions.

Remark. In the proof, we will get more concrete expressions for the limit for the lattice case and the periodic function G for the non-lattice case. See (4.1.4) and (4.1.5).

Remark. If we use the full version of the renewal theorem (Theorem B.4.3), we get more detailed information about the error term $o(1)x^{d_S/2}$ in the lattice case. For example, if the γ_i are all equal, then

$$\rho_b(x,\mu) = G(\log x/2)x^{d_S/2} + O(1) \qquad (4.1.3)$$

as $x \to \infty$, where $O(1)$ is a term which is bounded as $x \to \infty$.

Remark. According to [41], $G((\log x)/2)x^{d_S/2}$ is the integrated density of states of K associated with the Laplacian Δ_μ.

Example 4.1.6 (Sierpinski gasket). For the standard Laplacian on the Sierpinski gasket in Example 3.7.3, $r_i = 3/5$ and $\mu_i = 1/3$ for all $i = 1, 2, 3$. Let d_S be the spectral exponent. Then, by (4.1.2), we have $d_S = \log 9/\log 5$. In this case, d_S is also called the spectral dimension of the Sierpinski gasket. Note that d_S differs from the value of the Hausdorff dimension of the Sierpinski gasket, which is $\log 3/\log 2$.

Obviously we are in the lattice case. Taking the above remark into account, we have $\rho_b(x,\mu) = G(\log x/2)x^{-d_S/2} + O(1)$ as $x \to \infty$, where G is a $\log 5/2$-periodic positive function. Fukushima and Shima [44] showed that

$$0 < \liminf_{x\to\infty} \rho_b(x,\mu)/x^{d_S/2} < \limsup_{x\to\infty} \rho_b(x,\mu)/x^{d_S/2} < \infty$$

by using the eigenvalue decimation method. See Notes and References of this chapter about the eigenvalue decimation method. Hence we see that G is a non-trivial (i.e., non-constant) function.

The rest of this section is devoted to proving Theorem 4.1.5. First we give a comparison theorem for eigenvalues.

Theorem 4.1.7. *Let $\mathcal{H}$ be a separable Hilbert space. Let $\mathcal{E}_i$ be a closed form on $\mathcal{H}$ for $i = 1, 2$. Assume that $\mathcal{E}_1$ is an extension of $\mathcal{E}_2$. Set $\mathcal{F}_i = \mathrm{Dom}(\mathcal{E}_i)$ for $i = 1, 2$. Also let H_i be the non-negative self-adjoint operator associated with $\mathcal{E}_i$ for $i = 1, 2$. Assume that H_i has compact resolvent for $i = 1, 2$. Let $\{\lambda_n^i\}_{n\geq 1}$ be the set of eigenvalues of H_i appearing in Theorem B.1.13. Then $\lambda_n^1 \leq \lambda_n^2$ for all $n \geq 1$. Moreover, if $\dim \mathcal{F}_1/\mathcal{F}_2 < \infty$, then $\lambda_n^2 \leq \lambda_{n+M}^1$ for all $n \geq 1$, where $M = \dim \mathcal{F}_1/\mathcal{F}_2$.*

Proof. For any finite dimensional subspace L of $\mathcal{F}_i$, set

$$\lambda^i(L) = \sup\{\mathcal{E}_i(f,f) : f \in L, \|f\| = 1\},$$

where $\|\cdot\|$ is the norm of $\mathcal{H}$. Since $\mathcal{E}_1$ is an extension of $\mathcal{E}_2$, we see that

$\lambda^1(L) = \lambda^2(L)$ if $L \subseteq \mathcal{F}_2$. So the variational formula (Theorem B.1.14) implies $\lambda_n^1 \leq \lambda_n^2$ for all $n \geq 1$.

Next, assume that $\dim \mathcal{F}_1/\mathcal{F}_2 < \infty$. Then, for any subspace $L \subset \mathcal{F}_1$, there is a natural injective map from $L/(L \cap \mathcal{F}_2) \to \mathcal{F}_1/(\mathcal{F}_1 \cap \mathcal{F}_2) = \mathcal{F}_1/\mathcal{F}_2$. This shows $\dim L \cap \mathcal{F}_2 \geq \dim L - \dim \mathcal{F}_1/\mathcal{F}_2$. Hence, if

$$\mathcal{L}_{k,l} = \{L : L \text{ is a subspace of } \mathcal{F}_1, \dim L = k, \dim L \cap \mathcal{F}_2 = l\},$$

then

$$\{L : L \text{ is a subspace of } \mathcal{F}_1, \dim L = n + M\} = \cup_{i=0}^{M} \mathcal{L}_{n+M,n+i}.$$

For $L \in \mathcal{L}_{n+M,n+i}$, the variational formula shows that

$$\lambda^1(L) \geq \lambda^1(L \cap \mathcal{F}_2) = \lambda^2(L \cap \mathcal{F}_2) \geq \lambda_{n+i}^2 \geq \lambda_n^2.$$

Again, by the variational formula, we have $\lambda_{n+M}^1 \geq \lambda_n^2$. □

Immediately, from the above theorem, we have the following comparison of eigenvalue counting functions.

Corollary 4.1.8. *Under the same assumptions as in Theorem 4.1.7, define $\rho_i(x) = \#\{n : \lambda_n^i \leq x\}$. If $\dim \mathcal{F}_1/\mathcal{F}_2 = M < \infty$, then $\rho_2(x) \leq \rho_1(x) \leq \rho_2(x) + M$ for any $x \geq 0$.*

For ease of notation, we write $\rho_b(x)$ instead of $\rho_b(x, \mu)$ hereafter.

Lemma 4.1.9. *Set $M_1 = \#(V_1 \backslash V_0)$ and $M_0 = \#(V_0)$. Then, for any $x \in \mathbb{R}$,*

$$\rho_D(x) - M_1 \leq \sum_{i=1}^{N} \rho_D(r_i \mu_i x) \leq \rho_D(x) \leq \rho_N(x) \leq \rho_D(x) + M_0.$$

Proof. First applying Corollary 4.1.8 to $(\mathcal{E}, \mathcal{F})$ and $(\mathcal{E}, \mathcal{F}_0)$, we immediately obtain $\rho_D(x) \leq \rho_N(x) \leq \rho_D(x) + M_0$.

Next, define $\mathcal{F}_1 = \{u : u \in \mathcal{F}, u|_{V_1} = 0\}$. Then it follows that $(\mathcal{E}, \mathcal{F}_1)$ is a closed form on $L^2(K, \mu)$. Note that $\dim \mathcal{F}_0/\mathcal{F}_1 = \#(V_0 \backslash V_1) = M_1$. Hence, if ρ_1 is the eigenvalue counting function associated with $(\mathcal{E}, \mathcal{F}_1)$, then Theorem 4.1.7 implies $\rho_D(x) - M_1 \leq \rho_1(x) \leq \rho_D(x)$. Let H_1 be the non-negative self-adjoint operator associated with $(\mathcal{E}, \mathcal{F}_1)$. Then $H_1 u = \lambda u$ for $u \in \mathrm{Dom}(H_1)$ if and only if $u \in \mathcal{F}_1$ and $\mathcal{E}(v, u) = \lambda(v, u)_\mu$ for all $v \in \mathcal{F}_1$. By the self-similarity of both $\mathcal{E}$ and μ, the last equation is equivalent to $\sum_{i=1}^{N} r_i^{-1} \mathcal{E}(v \circ F_i, u \circ F_i) = \lambda \sum_{i=1}^{N} \mu_i (v \circ F_i, u \circ F_i)_\mu$. Hence, it follows that $H_1 u = \lambda u$ if and only if $H_D(u \circ F_i) = \mu_i r_i \lambda(u \circ F_i)$ for any i. Thus $\rho_1(x) = \sum_{i=1}^{N} \rho_D(\mu_i r_i x)$. This completes the proof. □

Proof of Theorem 4.1.5. Let $R(x) = \rho_D(x) - \sum_{i=1}^{N} \rho_D(\mu_i r_i x)$. Then

$$\rho_D(x) = \sum_{i=1}^{N} \rho_D(\mu_i r_i x) + R(x).$$

Defining $f(t) = e^{-d_S t}\rho_D(e^{2t})$, $u(t) = e^{-d_S t}R(e^{2t})$ and $p_i = \gamma_i{}^{d_S}$, the above equation becomes

$$f(t) = \sum_{i=1}^{N} p_i f(t - (-\log\gamma_i)) + u(t).$$

Note that $\sum_{i=1}^{N} p_i = 1$ by (4.1.2). This is the renewal equation (B.4.1). Recalling Proposition 3.4.8, we see that the first eigenvalue λ_1 of H_D is positive. Since $\rho_D(x) = 0$ for $x < \lambda_1$, it follows that there exists $t_* \in \mathbb{R}$ such that $f(t) = R(t) = 0$ for all $t \le t_*$. Also, Lemma 4.1.9 implies that $|u(t)| \le M_1 e^{-d_S t}$ for any t. Hence all the assumptions of the renewal theorem (Theorem B.4.2) are satisfied. So, for the non-lattice case, we have

$$f(t) \to \Big(\sum_{i=1}^{N}(-\log\gamma_i)\gamma_i{}^{d_S}\Big)^{-1}\int_{\mathbb{R}} e^{-d_S t}R(e^{2t})dt \qquad (4.1.4)$$

as $t \to \infty$. For the lattice case, we have $f(t) - G(t) \to 0$ as $t \to \infty$, where T is the positive generator of the discrete additive group $\sum_{i=1}^{N}\mathbb{Z}\log\gamma_i$ and

$$G(t) = \Big(\sum_{i=1}^{N}(-\log\gamma_i)\gamma_i{}^{d_S}\Big)^{-1}T\sum_{j\in\mathbb{Z}} e^{-d_S(t+jT)}R(e^{2(t+jT)}). \qquad (4.1.5)$$

This implies the theorem for the Dirichlet case. The Neumann case immediately follows as well by Lemma 4.1.9. $\qquad\square$

4.2 Relation between dimensions

In Example 4.1.6, it was observed that the spectral dimension of the Sierpinski gasket is different from the Hausdorff dimension with respect to the restriction of the Euclidean metric. (Note that so far the terminology "spectral dimension" is merely a synonym of spectral exponent. Recalling Theorem 4.1.5, we find that the spectral exponent d_S depends on $(D, \mathbf{r})$ and μ. We will define the spectral dimension of the resistance metric (K, R) later.) In any case, the spectral dimension (exponent) is defined through the eigenvalue distribution of the Laplacian. Hence we may think of the spectral dimension as a dimension from the analytical point of view. On the other hand, the Hausdorff dimension is a dimension from the geometrical point of view. In this section, we study the relation between those two dimensions.

Theorem 4.2.1. *Assume that $(D, \mathbf{r})$ is regular. Let R be the effective resistance metric on K associated with $(D, \mathbf{r})$. Let $\mathcal{H}^d$ be the d-dimensional Hausdorff measure of (K, R). If d_H is the unique positive number that satisfies*

$$\sum_{i=1}^{N} (r_i)^{d_H} = 1, \tag{4.2.1}$$

then $0 < \mathcal{H}^{d_H}(K) < \infty$. In particular, $\dim_H(K, R) = d_H$. Moreover, let ν be the normalized d_H-dimensional Hausdorff measure: $\nu = \mathcal{H}^{d_H}/\mathcal{H}^{d_H}(K)$ and let μ_* be the self-similar measure with weight $((r_i)^{d_H})_{i \in S}$. Then there exist positive numbers c_1 and c_2 such that

$$c_1 \mu_*(A) \leq \nu(A) \leq c_2 \mu_*(A) \tag{4.2.2}$$

for any $A \in \mathcal{B}(K)$. Also, if

$$d_S = \frac{2d_H}{d_H + 1}, \tag{4.2.3}$$

then there exist positive numbers c_3 and c_4 such that

$$c_3 x^{d_S/2} \leq \rho_D(x, \nu) \leq c_4 x^{d_S/2}$$

for sufficiently large x.

It is natural to think of ν as the intrinsic probability measure on (K, R). If we define the spectral dimension of (K, R), $d_S(K, R)$, by

$$d_S(K, R) = 2 \lim_{n \to \infty} \frac{\log \rho_D(x, \nu)}{\log x},$$

then $d_S(K, R) = d_S$ and (4.2.3) gives the relation between the Hausdorff dimension and the spectral dimension of (K, R).

In [93], the quantity d_H given by (4.2.1) is called the similarity dimension of the harmonic structure $(D, \mathbf{r})$. Also, if we denote the solution of (4.1.2) by $d_S(\mu)$, then

$$d_S(K, R) = d_S(\mu_*) = \max\{d_S(\mu) : \mu \text{ is a self-similar measure}\}. \tag{4.2.4}$$

On the contrary, if $(D, \mathbf{r})$ is not regular, then the supremum of the spectral exponent $d_S(\mu)$ is infinite.

In the rest of this section, we will prove Theorem 4.2.1.

Definition 4.2.2. Let Λ be a partition. $f \in \mathcal{F}$ is called a Λ-harmonic function if and only if $f \circ F_w$ is harmonic for any $w \in \Lambda$. For $p \in V(\Lambda)$, ψ_p^{Λ} is the Λ-harmonic function with $\psi_p^{\Lambda}|_{V(\Lambda)} = \chi_p^{V(\Lambda)}$.

Recall Lemma 1.3.10 for the definition of $V(\Lambda) = V(\Lambda, \mathcal{L})$.

Lemma 4.2.3. *Let Λ be a partition. Set $\Lambda_w = \{v \in \Lambda : K_w \cap K_v \neq \emptyset\}$ for any $w \in \Lambda$. Then*

$$\#(\Lambda_w) \leq \#(\mathcal{C}_{\mathcal{L}})\#(V_0)$$

for any $w \in \Lambda$.

Proof. It is enough to show that $\#(\pi^{-1}(p)) \leq \#(\mathcal{C}_{\mathcal{L}})$ for any $p \in K$ because $\#(\Lambda_w) \leq \sum_{p \in F_w(V_0)} \#(\pi^{-1}(p))$. For ease of notation, we use $\mathcal{C}$ instead of $\mathcal{C}_{\mathcal{L}}$ hereafter. Let $k = \#(\mathcal{C})$ and assume that $\pi^{-1}(p) \supseteq \{\omega^1, \ldots, \omega^{k+1}\}$. Let

$$m = \min\{n : \omega_n^i \neq \omega_n^j \text{ for some } i, j \in \{1, \ldots, k+1\}\}.$$

If $v = \omega_1^1 \ldots \omega_{m-1}^1$ and $q = F_v^{-1}(p)$, then $q \in \mathcal{C}_{\mathcal{L}}$ and $\pi^{-1}(q) \subset \mathcal{C}$. Hence $\#(\pi^{-1}(q)) \leq k$. Since $\pi^{-1}(q) \supseteq \{\sigma^{m-1}(\omega^1), \ldots, \sigma^{m-1}(\omega^{k+1})\}$, it follows that $\omega^i = \omega^j$ for some $i < j$. Hence $\#(\pi^{-1}(p)) \leq k$. $\square$

Let $\Lambda(a) = \Lambda(\mathbf{r}, a)$, where $\Lambda(\mathbf{r}, a)$ is defined in Definition 1.5.6. Note that $\Lambda(a)$ is a partition for small $a > 0$.

Lemma 4.2.4. *There exist positive constants c and M such that, for sufficiently small $a > 0$ and for any $x \in K$,*

$$\#\{w : w \in \Lambda(a), d(x, K_w) \leq ca\} \leq M, \qquad (4.2.5)$$

where $d(x, K_w) = \min_{y \in K_w} R(x, y)$.

Proof. For $w \in \Lambda(a)$, define $u = \sum_{p \in F_w(V_0)} \psi_p^{\Lambda(a)}$. Suppose that $v \in \Lambda(a)$ and $K_w \cap K_v = \emptyset$. Then $u|_{K_w} \equiv 1$ and $u|_{K_v} \equiv 0$. Hence Theorem 2.3.4 implies that $R(x, y) \geq \mathcal{E}(u, u)^{-1}$ if $x \in K_w$ and $y \in K_v$. Recalling (3.3.1) we have

$$\mathcal{E}(u, u) = \sum_{h \in \Lambda(a)} \frac{1}{r_h} \mathcal{E}(u \circ F_h, u \circ F_h) = \sum_{h \in \Lambda_w(a)} \frac{1}{r_h} \mathcal{E}(u \circ F_h, u \circ F_h),$$

$$(4.2.6)$$

where $\Lambda_w(a) = \{h \in \Lambda(a) : K_w \cap K_h \neq \emptyset\}$. Set $\underline{r} = \min\{r_i : i = 1, 2, \ldots, N\}$. Then it follows that $r_h \geq a\underline{r}$ for any $h \in \Lambda(a)$. Also define

$$c_1 = \max\{\mathcal{E}(\sum_{p \in V} \psi_p, \sum_{p \in V} \psi_p) : V \text{ is a non-empty subset of } V_0.\}$$

As $u \circ F_h = \sum_{p \in V} \psi_p$ for some $V \subseteq V_0$, we obtain $\mathcal{E}(u \circ F_h, u \circ F_h) \leq c_1$. Thus, applying those estimates along with Lemma 4.2.3, it follows from (4.2.6) that

$$\mathcal{E}(u, u) \leq \#(\mathcal{C})\#(V_0)(a\underline{r})^{-1}c_1.$$

Using the above estimate, if $w, v \in \Lambda(a)$, $K_w \cap K_v = \emptyset$, $x \in K_w$ and $y \in K_v$, then we have

$$R(x, y) \geq \mathcal{E}(u, u)^{-1} \geq c_2 a,$$

where $c_2 = (\#(\mathcal{C})\#(V_0)c_1)^{-1}\underline{r}$. Thus we see that for any $x \in K$, if $d(x, K_v) \leq c_2 a/2$ for $v \in \Lambda(a)$, then there exists $w \in \Lambda(a)$ such that $v \in \Lambda_w(a)$ and $x \in K_w$. Hence, if $c = c_2/2$, Lemma 4.2.3 implies

$$\#\{v \in \Lambda(a) : d(x, K_v) \leq ca\} \leq \#\{w \in \Lambda(a) : x \in K_w\}\#(\mathcal{C})\#(V_0).$$

As is shown in the proof of Lemma 4.2.3, $\#(\pi^{-1}(x)) \leq \#(\mathcal{C})$ for any $x \in K$. Hence $\#\{w \in \Lambda(a) : x \in K_w\} \leq \#(\mathcal{C})$. This completes the proof of this lemma with $M = \#(\mathcal{C})^2\#(V_0)$.   □

Proof of Theorem 4.2.1. We will apply Theorem 1.5.7 for the first part of the theorem on the Hausdorff measure. By Lemma 3.3.5, there exists $C > 0$ such that

$$\mathrm{diam}(K_w) \leq Cr_w$$

for any $w \in W_*$. This is the condition (1.5.2). Also, by Lemma 4.2.4, we can verify the other condition (1.5.3). Hence Theorem 1.5.7 implies the first part of Theorem 4.2.1.

Next we show the second part of the theorem on the asymptotic behavior of $\rho_D(x, \nu)$. Let $\lambda_i(\mu_*)$ and $\lambda_i(\nu)$ be the i-th eigenvalue of $-\Delta_{\mu_*}$ and $-\Delta_\nu$ with Dirichlet boundary conditions, respectively. By (4.2.2), $c_1\|u\|_{\mu_*,2} \leq \|u\|_{\nu,2} \leq c_2\|u\|_{\mu_*,2}$. Therefore, the variational formula, Theorem B.1.14, implies that

$$(c_1)^2\lambda_i(\nu) \leq \lambda_i(\mu_*) \leq (c_2)^2\lambda_i(\nu)$$

for $i \geq 1$. Hence we obtain

$$\rho_D((c_1)^2 x, \mu_*) \leq \rho_D(x, \nu) \leq \rho_D((c_2)^2 x, \mu_*) \tag{4.2.7}$$

for any $x \geq 0$. By Theorem 4.1.5, it follows that

$$0 < \liminf_{x \to \infty} \rho_D(x, \mu_*)/x^{d_S/2} \leq \limsup_{x \to \infty} \rho_D(x, \mu_*)/x^{d_S/2} < \infty,$$

where d_S is the unique positive number that satisfies (4.1.2). By (4.2.7), the above inequality holds if we replace $\rho_D(x, \mu_*)$ by $\rho_D(x, \nu)$. Since $(\mu_*)_i = (r_i)^{d_H}$, we see that $d_S = 2d_H/(d_H + 1)$. $\qquad\square$

4.3 Localized eigenfunctions

In [44], it was observed that there exists an eigenfunction of the standard Laplacian on the Sierpinski gasket whose support is confined in a small subdomain of the Sierpinski gasket. Such an eigenfunction is called a localized eigenfunction, which never appears in the case of the ordinary Laplacian on a connected domain of $\mathbb{R}^n$. So the existence of a localized eigenfunction is one of the unique features of the Laplacian on fractals. In this and the following sections, we study localized eigenfunctions of the Laplacian Δ_μ.

Hereafter we assume that μ is a self-similar measure on K with weight $(\mu_i)_{i \in S}$.

First we give a definition of localized eigenfunctions. It is, however, difficult to say that the support of an eigenfunction is "small". So we define the notion of a pre-localized eigenfunction instead. We will see that a pre-localized eigenfunction produces genuinely "localized" eigenfunctions in Proposition 4.3.3 below.

Definition 4.3.1. $u \in \mathcal{F}$ is called a pre-localized eigenfunction of Δ_μ belonging to the eigenvalue λ if $u \in E_D(\lambda) \cap E_N(\lambda)$ and $u \neq 0$.

In other words, a pre-localized eigenfunction is both a Neumann and a Dirichlet eigenfunction of Δ_μ:

$$\Delta_\mu u = -\lambda u, \quad u|_{V_0} \equiv 0 \quad \text{and} \quad (du)_p = 0 \text{ for all } p \in V_0.$$

Proposition 4.3.2. *Define $P_w : \mathcal{F}_0 \to \mathcal{F}$ by*

$$(P_w u)(x) = \begin{cases} u(F_w^{-1}(x)) & \text{if } x \in K_w, \\ 0 & \text{otherwise.} \end{cases}$$

Let u be a pre-localized eigenfunction belonging to the eigenvalue λ. Set $u_w = P_w(u)$. Then u_w is a pre-localized eigenfunction belonging to the eigenvalue $\lambda/\mu_w r_w$.

Proof. Let $w \in W_m$. The self-similarity of $(\mathcal{E}, \mathcal{F})$ (Proposition 3.3.1) along with Proposition 4.1.2 implies

$$\mathcal{E}(u_w, v) = \sum_{\tau \in W_m} \frac{1}{r_\tau} \mathcal{E}(u_w \circ F_\tau, v \circ F_\tau)$$

$$= \frac{1}{r_w} \mathcal{E}(u, v \circ F_w) = \frac{\lambda}{r_w}(u, v \circ F_w)_\mu$$

for any $v \in \mathcal{F}$. On the other hand,

$$(u_w, v)_\mu = \sum_{\tau \in W_m} \mu_\tau (u_w \circ F_\tau, v_w \circ F_\tau)_\mu = \mu_w (u, v \circ F_w)_\mu.$$

Therefore $\mathcal{E}(u_w, v) = \lambda(\mu_w r_w)^{-1}(u_w, v)$ for any $v \in \mathcal{F}$. Hence, by Proposition 4.1.2, $u_w \in E_N((\mu_w r_w)^{-1}\lambda)$. Also, obviously $u_w|_{V_0} \equiv 0$. This completes the proof. $\qquad\square$

By the above proposition, once we find one pre-localized eigenfunction, then there exists an infinite sequence of pre-localized eigenfunctions, which are actually "localized" eigenfunctions.

Proposition 4.3.3. *There exists a pre-localized eigenfunction of $-\Delta_\mu$ if and only if for any non-empty open subset $O \subset K$, there exists a pre-localized eigenfunction u such that* $\mathrm{supp}(u) \subset O$.

Proof. For any non-empty open subset $O \subset K$, there exists $w \in W_*$ such that $K_w \subset O$. If there exists a pre-localized eigenfunction u, then u_w is a pre-localized eigenfunction with $\mathrm{supp}(u_w) \subset K_w \subset O$. The converse direction is obvious. $\qquad\square$

A pre-localized eigenfunction causes several notable phenomena. First we consider the lattice case in Theorem 4.1.5.

Theorem 4.3.4. *For the lattice case, the following four conditions are equivalent:*
(1) There exists a pre-localized eigenfunction of $-\Delta_\mu$.
(2) G is discontinuous.
(3) For any $M \in \mathbb{N}$, there exists a (Neumann or Dirichlet) eigenvalue of $-\Delta_\mu$ whose multiplicity is greater than M.
(4) There exists a (Neumann or Dirichlet) eigenvalue of $-\Delta_\mu$ whose multiplicity is greater than $\#(V_0)$.

Remark. Recall that $G((\log x)/2)x^{d_s/2}$ is called the integrated density of states. (See the third remark after Theorem 4.1.5.) So if there exists a pre-localized eigenfunction, then the integrated density of states is discontinuous.

By the above theorem, the T-periodic function G in Theorem 4.1.5 turns out to be non-constant if there exists a pre-localized eigenfunction. So, we have the following corollary.

Corollary 4.3.5. *For the lattice case, if there does exist a pre-localized eigenfunction, then $\rho_b(x, \mu)/x^{d_s/2}$ does not converge as $x \to \infty$ for $b = D, N$.*

So far, this corollary is the only known result which can be used to show that the T-periodic function G is not constant. We will give a proof of Theorem 4.3.4 after stating another consequence of the existence of a pre-localized eigenfunction.

The second theorem concerns approximation by polynomials.

Notation. Define spaces of multi-harmonic functions Π_m for $m \geq 1$ inductively by

$$\begin{cases} \Pi_1 & = \{u \in \mathcal{D}_\mu : \Delta_\mu u = 0\} \\ \Pi_{m+1} & = \{u \in \mathcal{D}_\mu : \Delta_\mu u \in \Pi_m\} \quad \text{for } m \geq 1. \end{cases}$$

Also set $\Pi_\infty = \cup_{m \geq 1} \Pi_m$.

Obviously, Π_1 is the collection of harmonic functions. In the case of the ordinary Laplacian on $[0,1]$, Π_∞ corresponds to the collection of all polynomials. So we think of $u \in \Pi_\infty$ as a polymonial on K with respect to the Laplacian Δ_μ. By Weierstrass's theorem (see Yosida [186, §0.2] for instance), the collection of all polynomials on $[0,1]$ is known to be dense in $L^2([0,1])$. On the contrary, the following theorem tells us that a pre-localized eigenfunction cannot be approximated by polynomials in $L^2(K, \mu)$. In other words, Π_∞ is not dense in $L^2(K, \mu)$ if there exists a pre-localized eigenfunction.

Theorem 4.3.6. *Let E^{loc} be the L^2-closure of the linear space spanned by all pre-localized eigenfunctions. Also let $\overline{\Pi}_\infty$ be the L^2-closure of Π_∞. Then E^{loc} is the orthogonal complement of $\overline{\Pi}_\infty$ in $L^2(K, \mu)$, so that $L^2(K, \mu) = \overline{\Pi}_\infty \oplus E^{loc}$.*

The rest of this section is devoted to proving these theorems.

First we prove Theorem 4.3.4. Recall the notation for the lattice case in 4.1. T is the positive generator of $\sum_{i=1}^{N} \mathbb{Z} \log \gamma_i$, where $\gamma_i = \sqrt{r_i \mu_i}$. Let $m_i = -T^{-1} \log \gamma_i$ for $i \in S$. Then the m_i are positive integers whose largest common divisor is 1. If we define

$$M(n) = \#\{w = w_1 w_2 \ldots w_l \in W_* : \frac{k}{r_w \mu_w} = k e^{2nT}\}$$

for $n \in \mathbb{N}$, then

$$M(n) = \#\{w = w_1 w_2 \ldots w_l \in W_* : \sum_{i=1}^{l} m_{w_i} = n\}.$$

We also define $M(0) = 1$ and $M(n) = 0$ for any negative integer n.

Lemma 4.3.7. *If* $p = e^{Tds}$, *then* $\lim_{n\to\infty} M(n)/p^n = (\sum_{i=1}^{N} m_i p^{-m_i})^{-1}$.

Note that $p^{-m_i} = e^{-m_i Tds} = \gamma_i^{ds}$. Hence $\sum_{i=1}^{N} p^{-m_i} = 1$.

Proof. Let $p_i = p^{-m_i}$. Set $\tilde{M}(n) = M(n)/p^n$ for any $n \in \mathbb{Z}$. Then this lemma follows immediately by Lemma B.4.5.   □

Proof of Theorem 4.3.4. We write $\rho(x)$ instead of $\rho_b(x, \mu)$.

(1) $\Rightarrow$ (2) : Let u be a pre-localized eigenfunction belonging to the eigenvalue k. Then, by Proposition 4.3.2, $k_n = ke^{2nT}$ is a (Dirichlet and Neumann) eigenvalue with multiplicity at least $M(n)$. Hence

$$\rho(k_n) - \rho(k_n-) \geq M(n), \qquad (4.3.1)$$

where $\rho(k-) = \lim_{x\uparrow k} \rho(x)$. As $\lim_{x\to\infty} |\rho(x)/x^{d_s/2} - G((\log x)/2)| = 0$ and $\log k_n = \log k + 2nT$, it follows that

$$\lim_{n\to\infty} (\rho(k_n) - \rho(k_n-))k_n^{-ds/2} \leq G(\alpha) - \lim_{t\uparrow\alpha} G(t), \qquad (4.3.2)$$

where $\alpha = \frac{1}{2} \log k$. Combining (4.3.1) and (4.3.2), Lemma 4.3.7 implies that $G(\alpha) - \lim_{t\uparrow\alpha} G(t) > 0$. Hence G is discontinuous.

(2) $\Rightarrow$ (3) : Let $G_n(t) = \rho(e^{2(t+nT)})/e^{(t+nT)d_s}$ for $0 \leq t \leq T$. Also let $\epsilon_n = \sup_{0\leq t\leq 1} |G_n(t) - G(t)|$. Then, by Theorem 4.1.5, $\epsilon_n \to 0$ as $n \to \infty$. Now, if G is discontinuous at α, then we can choose $a_m > 0$ and $b_m > 0$ so that $\lim_{m\to\infty} a_m = \lim_{m\to\infty} b_m = 0$ and $L = \liminf_{m\to\infty} |G(\alpha + a_m) - G(\alpha - b_m)| > 0$. It follows that

$$\liminf_{m\to\infty} |G_n(\alpha + a_m) - G_n(\alpha - b_m)| \geq L - 2\epsilon_n.$$

This implies that $e^{2(\alpha+nT)}$ is an eigenvalue whose multiplicity is no less than $2e^{(\alpha+nT)d_s}/L$ for sufficiently large n.

(3) $\Rightarrow$ (4) : This is obvious.

(4) $\Rightarrow$ (1) : Assume that $\dim E_N(k) > \#(V_0)$. Define a linear map τ : $E_N(k) \to \ell(V_0)$ by $\tau(u)(x) = u(x)$ for $x \in V_0$. As $\dim E_N(k) > \dim \ell(V_0) = \#(V_0)$, the kernel of τ is not trivial. Hence there exists a non-trivial $u \in E_N(k)$ that satisfies $u|_{V_0} = 0$, and so u is a pre-localized eigenfunction.

A similar argument works for the case of D-eigenvalues.   □

Remark. Note that the implication (4) $\Rightarrow$ (1) is true even in the non-lattice case.

Next we prove Theorem 4.3.6.

Lemma 4.3.8. *u is a pre-localized eigenfunction if and only if $u \in (\Pi_1)^{\perp}$ and u is a D-eigenfunction.*

Proof. Let $u \in E_D(\lambda) \cap E_N(\lambda)$ for some $\lambda > 0$. Also, let v be a harmonic function. Then, by the Gauss–Green's formula (Theorem 3.7.8),

$$\mathcal{E}(u, v) = \sum_{p \in V_0} u(p)(dv)_p - \int_K u \Delta_\mu v d\mu = 0$$

$$= \sum_{p \in V_0} v(p)(du)_p - \int_K v \Delta_\mu u d\mu = \lambda(u, v)_\mu.$$

Hence $(u, v)_\mu = 0$. This implies that $u \in (\Pi_1)^\perp$.

Conversely, assume that $u \in E_D(\lambda)$ and $(u, v) = 0$ for any harmonic function u. Then using the Gauss–Green's formula in a similar way to the above, we see that $(du)_p = 0$ for any $p \in V_0$. Hence u is a pre-localized eigenfunction. $\qquad \square$

Proof of Theorem 4.3.6. Step 1: $G_D(\overline{\Pi}_\infty) \subseteq \overline{\Pi}_\infty$, where $G_D = H_D{}^{-1}$ is the Green's operator defined in Proposition 3.4.8.

Proof of Step 1: Let $u \in \overline{\Pi}_\infty$. Then there exists $\{u_n\}_{n \geq 1} \subset \Pi_\infty$ such that $u_n \to u$ as $n \to \infty$ in $L^2(K, \mu)$. By Lemma 3.7.10, it follows that $G_D u_n \in \Pi_\infty$. Since G_D is a bounded operator from $L^2(K, \mu)$ to itself, $G_D u_n \to G_D u$ as $n \to \infty$. Hence $G_D u \in \overline{\Pi}_\infty$.

Step 2: $G_D(\overline{\Pi}_\infty^\perp) \subseteq \overline{\Pi}_\infty^\perp$

Proof of Step 2: Let $u \in \overline{\Pi}_\infty^\perp$. Note that G_D is symmetric. By Step 1, for any $v \in \overline{\Pi}_\infty$, $(G_D u, v)_\mu = (u, G_D v)_\mu = 0$. Hence $G_D u \in \overline{\Pi}_\infty^\perp$.

Now, by Step 1 and Step 2, it follows that $H_D(\overline{\Pi}_\infty \cap \mathrm{Dom}(H_D)) = \overline{\Pi}_\infty$ and $H_D(\overline{\Pi}_\infty^\perp \cap \mathrm{Dom}(H_D)) = \overline{\Pi}_\infty^\perp$. Since H_D is a self-adjoint operator having compact resolvent, $H_D|_{\overline{\Pi}_\infty^\perp \cap \mathrm{Dom}(H_D)}$ is a self-adjoint operator from $\overline{\Pi}_\infty^\perp$ to itself having compact resolvent as well. Hence there exists a complete orthonormal system $\{f_n\}_{n \geq 1}$ of $\overline{\Pi}_\infty^\perp$ consisting of eigenfunctions of H_D. By Proposition 4.1.2, f_n is a D-eigenfunction. As $f_n \in \overline{\Pi}_\infty^\perp \subset (\Pi_1)^\perp$, Lemma 4.3.8 implies that f_n is a pre-localized eigenfunction. Hence $\overline{\Pi}_\infty^\perp \subseteq E^{loc}$.

Next, let f be a pre-localized eigenfunction belonging to an eigenvalue $\lambda > 0$. Then, by the Gauss–Green's formula,

$$\lambda^m(f, v)_\mu = -\lambda^{m-1}(f, \Delta_\mu v)_\mu = \cdots = (-1)^m(f, (\Delta_\mu)^m v)_\mu = 0$$

for any $v \in \Pi_m$. Hence $(f, v)_\mu = 0$ for any $v \in \Pi_\infty$. This implies $E^{loc} \subseteq \overline{\Pi}_\infty^\perp$. $\qquad \square$

4.4 Existence of localized eigenfunctions

In this section, we give some sufficient conditions, in terms of the geometry of K, for the existence of localized eigenfunctions. Those conditions will require a certain kind of strong symmetry for the self-similar set, the harmonic structure and the measure. This is rather restrictive and seems a little far from a necessary condition. By using those conditions, however, we can show the existence of a pre-localized eigenfunction for the symmetric harmonic structure on (affine) nested fractals obtained in Theorem 3.8.10. See Corollary 4.4.11 for details.

Notation. If $g : K \to K$ is a bijection, and $f : K \to \mathbb{R}$, define $T_g f : K \to \mathbb{R}$ by $(T_g f)(x) = f(g^{-1}(x))$.

Definition 4.4.1. Define $\mathcal{G}$ by

$\mathcal{G} = \{g : g$ is a bijective homeomorphism from K to K that satisfies

    (1) $g : V_0 \to V_0$,

    (2) $\mu \circ g^{-1} = \mu$,

    (3) $T_g(\mathcal{F}) \subseteq \mathcal{F}$ and $\mathcal{E}(T_g u, T_g v) = \mathcal{E}(u, v)$ for any $u, v \in \mathcal{F}\}$.

Obviously, $\mathcal{G}$ is a group. $\mathcal{G}$ is called the symmetry group of $(\mathcal{L}, (D, \mathbf{r}), \mu)$.

Remark. Note that $g \in \mathcal{G}$ is not necessarily a symmetry of $\mathcal{L}$ as defined in 3.8.

Let $\{\varphi_n\}_{n \geq 1}$ be the complete orthonormal system for $L^2(K, \mu)$ consisting of the Dirichlet eigenfunctions of Δ_μ. Also, let us assume that $\varphi_n \in E_D(\lambda_n)$. (In 4.1, we used φ_n^D and λ_n^D in place of φ_n and λ_n, respectively.)

Lemma 4.4.2. *Let* $R = \sum_{i=1}^n \alpha_i T_{g_i}$, *where* $\alpha_i \in \mathbb{R}$ *and* $g_i \in \mathcal{G}$ *for* $i = 1, 2, \ldots, n$. *If* $Ru \neq 0$ *for some* $u \in L^2(K, \mu)$ *and* $R^* v \in \mathcal{F}_0$ *for all* $v \in \mathcal{F}$, *where* $R^* = \sum_{i=1}^n \alpha_i T_{g_i^{-1}}$, *then there exists a pre-localized eigenfunction.*

Proof. If $R\varphi_n = 0$ for all $n \geq 1$, then $Ru = 0$ for any $u \in L^2(K, \mu)$. Hence there exists $n \geq 1$ such that $R\varphi_n \neq 0$. Let $u = R\varphi_n$, and note that $u \in \mathcal{F}_0$. Now for all $v \in \mathcal{F}$, as $R^* v \in \mathcal{F}_0$, Proposition 4.1.2 implies

$$\mathcal{E}(u, v) = \mathcal{E}(R\varphi_n, v) = \mathcal{E}(\varphi_n, R^* v) = \lambda_n(\varphi_n, R^* v)_\mu = \lambda_n(u, v)_\mu.$$

Again, by Proposition 4.1.2, we see $u \in E_N(\lambda_n)$. Since $u \in \mathcal{F}_0$, it follows that u is a pre-localized eigenfunction. $\qquad\square$

For $g \in \mathcal{G}$, set

$$S(g) = \{x \in K : g(x) = x\}.$$

Let ι be the identity map of K.

Proposition 4.4.3. (1) *If there exists $h \in \mathcal{G}\backslash\{\iota\}$ such that $V_0 \subset \mathcal{S}(h)$, then there exist pre-localized eigenfunctions.*
(2) *If $\#(\mathcal{G})$ is infinite then there exist pre-localized eigenfunctions.*

Proof. (1) Let $R = I - T_h$. As $h \neq \iota$, there exists $x \in K$ such that $h(x) \neq x$. Since h is continuous, there exists a neighborhood A of x such that $h(A) \cap A = \emptyset$. Set $u = \chi_A$, where χ_A is the characteristic function of the set A. Then we have $Ru \neq 0$.

As $V_0 \subset \mathcal{S}(h)$, $h^{-1}(x) = x$ for all $x \in V_0$. Hence $v - T_{h^{-1}}v = R^*v \in \mathcal{F}_0$ for all $v \in \mathcal{F}$. Now the proposition follows from Lemma 4.4.2.
(2) If $\mathcal{G}$ is infinite, then, since V_0 is finite, a counting argument shows that there exist distinct elements g_1, g_2 of $\mathcal{G}$ with the same action on V_0. Hence $V_0 \subset \mathcal{S}(g_1^{-1}g_2)$, and the result is immediate from (1). □

There do exist p. c. f. self-similar sets for which $\mathcal{G}$ is infinite (the Vicsek set is one), but this is a little exceptional. We now turn to the more complicated situation when $\mathcal{G}$ is finite.

Theorem 4.4.4. *Suppose G is a finite subgroup of $\mathcal{G}$ which is vertex transitive on V_0, and that there exists $h \in \mathcal{G}$, $h \notin G$, such that*

$$\mathcal{S}_G(h) = \bigcup_{g \in G} \mathcal{S}(h^{-1}g) \neq K. \tag{4.4.1}$$

Then there exist pre-localized eigenfunctions.

Proof. Set $R_G = \sum_{g \in G} T_g = \sum_{g \in G} T_{g^{-1}}$, and $R = R_G(T_{h^{-1}} - I)$. Let $x \in K\backslash\mathcal{S}_G(h)$. Then $\{g(x) : g \in G\}$ is finite and does not contain $h(x)$. Hence there exists a neighborhood A of x such that $h(A) \cap g(A) = \emptyset$ for all $g \in G$. Set $u = \chi_{h(A)}$. If $y \in A$, then $u(g(y)) = 0$ for $g \in G$, and so

$$(Ru)(y) = \sum_{g \in G} u(h(g(y))) - \sum_{g \in G} u(g(y)) \geq u(h(y)) = 1,$$

proving that $Ru \neq 0$.

Let $v \in \mathcal{F}$. As G is vertex transitive, if $y \in V_0$ then

$$(R_G)v(y) = \sum_{g \in G} v(g(y)) = \frac{\#(G)}{\#(V_0)} \sum_{p \in V_0} v(p),$$

which is independent of y. If $x \in V_0$ then $h^{-1}(x) \in V_0$, and therefore $R^*v = (T_h - I)R_Gv \in \mathcal{F}_0$.

Now, using Lemma 4.4.2, we can complete the proof. □

As the condition (4.4.1) is a little troublesome to verify, one might hope that this symmetry argument would work under the weaker condition that

$\mathcal{G}$ contains a vertex transitive subgroup G and there exists $h \in \mathcal{G} \backslash G$. (4.4.2)

However, it is easy to see that (4.4.1) is equivalent to the statement that $R_G(T_{h^{-1}} - I) \neq 0$. See Example 4.4.9.

On the other hand, note that if (4.4.1) fails, then $\mathcal{S}(h^{-1}g)$ has an interior point for some $g \in G$. Hence (4.4.2) along with

$$\mathcal{S}(h^{-1}g) \text{ has no interior point for any } g \in G \qquad (4.4.3)$$

implies (4.4.1). In fact we will see below (Lemma 4.4.5-(3)) that for nested fractals (4.4.3) always holds, so that (4.4.2) is all that needs to be verified.

Next, we will discuss the case where K is a subset of $\mathbb{R}^d$ for some $d \in \mathbb{N}$ and the F_i are restrictions of similitudes of $\mathbb{R}^d$. In such a case, we can assume, without loss of generality, that

$$\sum_{i=1}^{M} p_i = 0, \qquad (4.4.4)$$

where $V_0 = \{p_1, p_2, \dots, p_M\}$ with $M = \#(V_0)$, and that

$$\{x - y : x, y \in K\} \quad \text{span} \quad \mathbb{R}^d. \qquad (4.4.5)$$

Under these assumptions,

Lemma 4.4.5. (1)   *If f is an affine map from $\mathbb{R}^d$ to itself with $f(V_0) = V_0$, then $f(0) = 0$.*
(2)   *Let $f_i \mathbb{R}^d \to \mathbb{R}^d$ be linear for $i = 1, 2$. If $f_1|_K = f_2|_K$ then $f_1(x) = f_2(x)$ for all $x \in \mathbb{R}^d$.*
(3)   *If $f : \mathbb{R}^d \to \mathbb{R}^d$ is a linear map with $f(K) = K$, then f is linearly conjugate to an orthogonal transformation of $\mathbb{R}^d$. Moreover, if f is not the identity map, then $\mathcal{S}(f) = \{x \in K : f(x) = x\}$ contains no interior point in the intrinsic topology K.*

Proof. (1)   Let $f(x) = Ax + b$, where A is a $d \times d$-matrix and $b \in \mathbb{R}^d$. As $\sum_{i=1}^{M} f(p_i) = \sum_{i=1}^{M} p_i = 0$, we have $A(\sum_{i=1}^{M} p_i) + Mb = 0$. Hence $b = 0$.
(2)   This is immediate from (4.4.5).
(3)   As $K \subset \text{Im } f$ and (4.4.5) holds, it follows that f is invertible. Note that K is bounded. We can easily see that $f^n(x)$ and $f^{-n}(x)$ remain bounded as $n \to \infty$ for any $x \in \mathbb{R}^d$. Hence, if we extend f to a map from $\mathbb{C}^d$ to itself, f is semisimple and the absolute values of its eigenvalues are all equal to 1. Therefore f is linearly conjugate to an orthogonal transformation.

Next suppose there exists a non-empty open subset O of K such that $f(x) = x$ for any $x \in O$. By (4.4.5), there exist $x_i, y_i \in K$ for $i = 1, 2, \ldots, d$ such that $(x_1 - y_1, \ldots, x_d - y_d)$ is a basis for $\mathbb{R}^d$. Now choose $w \in W_*$ so that $F_w(K) \subset O$, and write $z_i = F_w(x_i) - F_w(y_i)$; then $(z_1, z_2, \ldots, z_d)$ is a basis for $\mathbb{R}^d$ and $f(z_i) = z_i$ for $i = 1, 2, \ldots, d$. Thus f is the identity map. □

Hereafter, we consider a special subgroup of the symmetry group $\mathcal{G}$.

Theorem 4.4.6. *Define*

$$\mathcal{G}_0 = \{g \in \mathcal{G} : g = f|_K \text{ for some linear map } f : \mathbb{R}^d \to \mathbb{R}^d\}.$$

If there exists a proper subgroup G of $\mathcal{G}_0$ which is vertex transitive on V_0, then there exists a pre-localized eigenfunction.

Proof. As G is a proper subgroup of $\mathcal{G}_0$, there exists $h \in \mathcal{G}_0 \backslash G$. By Lemma 4.4.5-(3), we see that $\mathcal{S}(h^{-1}g)$ has no interior point for any $g \in G$. Hence we can verify (4.4.2) and (4.4.3). As is mentioned above, this implies (4.4.1). Hence, by Theorem 4.4.4, there exist localized eigenfunctions. □

The corollary below is sometimes easier to apply to examples than Theorem 4.4.6. Let P_{V_0} be the group of permutations of V_0. We define a natural restriction map $\kappa : \mathcal{G} \to P_{V_0}$ by $\kappa(g) = g|_{V_0}$.

Corollary 4.4.7. (1) *If κ is not injective then there exists a pre-localized eigenfunction.*
(2) *Set $G_0 = \kappa(\mathcal{G}_0)$. If there exists a proper subgroup of G_0 which is vertex transitive on V_0, then there exists a pre-localized eigenfunction.*

Proof. (1) If κ is not injective then there exists $g \neq \iota \in \mathcal{G}$ with $g(x) = x$ for all $x \in V_0$. The result is now immediate from Proposition 4.4.3-(1).
(2) We can find a proper subgroup of $\mathcal{G}_0$ which is vertex transitive on V_0 by using κ^{-1}. Then use Theorem 4.4.6. □

Here are some cases where we can apply Corollary 4.4.7-(2).

Example 4.4.8. (1) If $G_0 = P_{V_0}$ and $\#(V_0) \geq 3$, then the group of even permutations is a proper subgroup of P_{V_0} which is vertex transitive on V_0. This includes the case of the standard Laplacian on the Sierpinski gasket.
(2) Let V_0 be a regular n-sided polygon for $n > 2$ and suppose G_0 contains D_n, where D_n is the symmetry group of the regular n-sided polygon. We may write $V_0 = \{(\cos(2\pi j/n), \sin(2\pi j/n)) : j = 1, 2, \ldots, n\}$. Let g be the rotation by $2\pi/n$ around $(0,0)$. Then $\{g^j : j = 1, 2, \ldots, n\}$ is a proper subgroup of D_n and is vertex transitive on V_0.

Next we give an example where we can apply Corollary 4.4.7-(1).

Example 4.4.9. Set $F_1(z) = \frac{1}{2}(z+1), F_2(z) = \frac{1}{2}(z-1), F_3(z) = \frac{\sqrt{-1}}{4}(z+1)$ and $F_4(z) = \frac{\sqrt{-1}}{4}(z-1)$ for $z \in \mathbb{C}$. K is the self-similar set with respect to $\{F_i\}_{i=1,2,3,4}$. It is easy to see that $(K, S, \{F_i\}_{i \in S})$ where $S = \{1, 2, 3, 4\}$ is a p.c.f. self-similar structure. In fact, $C_{\mathcal{L}} = \{0\}$, $\mathcal{C}_{\mathcal{L}} = \pi^{-1}(0) = \{2\dot{1}, 1\dot{2}, 3\dot{2}, 4\dot{1}\}$ and $\mathcal{P}_{\mathcal{L}} = \{\dot{1}, \dot{2}\}$. As $\pi(\dot{1}) = 1$ and $\pi(\dot{2}) = -1$, $V_0 = \{-1, 1\}$.

Let $t \in (0, 1)$, and set

$$D = \begin{pmatrix} -1 & 1 \\ 1 & -1 \end{pmatrix} \quad \text{and} \quad \mathbf{r} = (\frac{1}{2}, \frac{1}{2}, t, t).$$

Then $(D, \mathbf{r})$ is a regular harmonic structure. Also let μ be a self-similar measure on K with weight $(\mu_i)_{i \in S}$, where $\mu_1 = \mu_2$ and $\mu_3 = \mu_4$.

The reflections in the real axis and the imaginary axis, denoted by g_1 and g_2 respectively, belong to $\mathcal{G}$, and $\mathcal{G}_0$ is the group generated by $\{g_1, g_2\}$. Obviously $\kappa(g_1)$ is the identity map on V_0. Hence, by Corollary 4.4.7-(1), there exists a pre-localized eigenfunction of $-\Delta_\mu$. (Note also that $\mathcal{G}$ contains infinitely many elements, and so the existence of a pre-localized eigenfunction also follows from Proposition 4.4.3-(2).)

Now let $G = \{\iota, g_2\}$, and let $h : K \to K$ be defined by $h(x) = x$ for $x \in K_1 \cup K_2$ and $h(x) = g_2(x)$ for $x \in K_3 \cup K_4$. It is not hard to check that $h \in \mathcal{G}$. Then $h \notin G$ but $R_G(T_{h^{-1}} - I) = 0$. Recall the discussion after Theorem 4.4.4.

In the rest of this section, we will consider the case of affine nested fractals. Recall Definition 3.8.2. For $x, y \in \mathbb{R}^d$, $x \neq y$, H_{xy} is the hyperplane bisecting the line segment $[x, y]$. Also $g_{xy} : \mathbb{R}^d \to \mathbb{R}^d$ is the reflection in H_{xy}.

Theorem 4.4.10. *Assume that $\#(V_0) \geq 3$. If $g_{xy} \in \mathcal{G}_0$ for all $x, y \in V_0$ with $x \neq y$, then there exists a pre-localized eigenfunction on K.*

Proof. Let $\mathcal{G}_1$ be the subgroup of $\mathcal{G}_0$ generated by $\{g_{xy} : x, y \in V_0, x \neq y\}$. In view of Lemma 4.4.5-(2) we may identify g_{xy} and $g_{xy}|_K$, and so regard $\mathcal{G}_1$ as a subgroup of $O(d)$, the d-dimensional orthogonal group. Note that, as g_{xy} is a reflection, $\det g_{xy} = -1$. Let $\mathcal{G}_2$ be the set of $g \in \mathcal{G}_1$ which are the product of an even number of the g_{xy}. Then every element g of $\mathcal{G}_2$ has $\det g = 1$, and so $\mathcal{G}_2$ is a proper subgroup of $\mathcal{G}_1$. Furthermore, if $\#(V_0) \geq 3$ then $\mathcal{G}_2$ is vertex transitive. For, if $x, y \in V_0$, let $z \in V_0 \setminus \{x, y\}$: then $g_{yz}g_{xz}(x) = y$. The result now follows from Theorem 4.4.6. $\square$

If $\#(V_0) = 2$ then examples show that both possibilities (i.e., existence or non-existence of localized eigenfunctions) can arise. Recall the case of $[0, 1]$, Example 3.7.2. If we choose a self-similar measure μ with weight $(1/2, 1/2)$ (μ is, in fact, Lebesgue measure on $[0, 1]$), then obviously g_{01} belongs to $\mathcal{G}_0$. The Laplacian is equal to d^2/dx^2 and of course has no localized eigenfunctions. On the other hand, Example 4.4.9 gives an example where $\#(V_0) = 2$ and a pre-localized eigenfunction exists.

Now let $\mathcal{L} = (K, S, \{F_i\}_{i \in S})$ be strongly symmetric. By Theorem 3.8.10, there exists a harmonic structure $(D, \mathbf{r})$, where $D \in \mathcal{L}_s(V_0)$, on $\mathcal{L}$ if $\mathbf{r}$ is $\mathcal{G}_s$-invariant. (Recall that $\mathcal{G}_s$ is the group of linear symmetries of $\mathcal{L}$ and that $\mathcal{L}_s(V_0)$ is the collection of $\mathcal{G}_s$-invariant Laplacians.) Let $(\mathcal{E}, \mathcal{F})$ be the quadratic form associated with the harmonic structure $(D, \mathbf{r})$. Then we see that $(\mathcal{E}, \mathcal{F})$ is $\mathcal{G}_s$-invariant by Theorem 3.8.10. Now let μ be a self-similar measure on K with weight $(\mu_i)_{i \in S}$. Assume that μ is $\mathcal{G}_s$-invariant, that is, $\mu_i = \mu_j$ if there exists $g \in \mathcal{G}_s$ such that $g(K_i) = K_j$. Then we see that $\mathcal{G}_s \subseteq \mathcal{G}_0$. So if $\mathcal{L}$ is an affine nested fractal, the assumption of Theorem 4.4.10 is satisfied.

Corollary 4.4.11. *Let $\mathcal{L} = (K, S, \{F_i\}_{i \in S})$ be an affine nested fractal with $\#(V_0) \geq 3$. Let $(D, \mathbf{r})$ be a $\mathcal{G}_s$-invariant harmonic structure (i.e., $(D, \mathbf{r})$ is a harmonic structure, $D \in \mathcal{L}_s(V_0)$ and $\mathbf{r}$ is $\mathcal{G}_s$-invariant.) Also, let μ be a $\mathcal{G}_s$-invariant self-similar measure on K with weight $(\mu_i)_{i \in S}$. Then the associated Laplacian Δ_μ has a pre-localized eigenfunction. Moreover, in the lattice case of Theorem 4.1.5, it follows that $\rho_b(x, \mu)/x^{d_s/2}$ does not converge as $x \to \infty$ for $b = N, D$.*

For the general strongly symmetric case, if there exists a proper subgroup of $\mathcal{G}_s$ which is vertex transitive on V_0, then the associated Laplacian Δ_μ possesses a pre-localized eigenfunction. For example,

Example 4.4.12. Recall Example 3.8.13, where V_0 is the vertices of a cube. In this case, we may write

$$V_0 = \{((-1)^i, (-1)^j, (-1)^k) : i, j, k \in \{0, 1\}\}.$$

Suppose that $\mathcal{G}^2$ is contained in $\mathcal{G}_s$. (The group $\mathcal{G}^2$ is defined in Example 3.8.13.) Let $(D, \mathbf{r})$ be a $\mathcal{G}_s$-invariant harmonic structure and let μ be a $\mathcal{G}_s$-invariant self-similar measure. Then since $\mathcal{G}^0$ is vertex transitive on V_0, we see that Δ_μ has a pre-localized eigenfunction.

Unfortunately, it is not known whether $\mathcal{G}_s$ has a proper subgroup which is vertex transitive on V_0 for the general strongly symmetric case.

4.5 Estimate of eigenfunctions

In this section, we will estimate the supremum norm of an eigenfunction of Δ_μ compared to the L^2-norm. More precisely, let $\varphi \in \mathcal{D}_\mu$ and assume that $\Delta_\mu \varphi = -\lambda\varphi$ for some $\lambda > 0$. Then, if $\alpha = \max_{i\in S}(\log \mu_i)/(\log \mu_i r_i)$, we will see that

$$||\varphi||_\infty \leq c\lambda^{\alpha/2}||\varphi||_2, \qquad (4.5.1)$$

where c is a positive constant that is independent of λ and φ, $||\varphi||_\infty$ is the supremum norm of φ and $||\varphi||_2$ is the L^2-norm of φ. Later, we will use this estimate to show continuity of the heat kernel associated with Δ_μ.

First we will study measurement of the distortion between two self-similar measures. Let μ and ν be self-similar measures on K with weights $(\mu_i)_{i\in S}$ and $(\nu_i)_{i\in S}$, respectively.

Definition 4.5.1. Define

$$\overline{h}(\lambda : \nu, \mu) = \min_{w\in\Lambda(\nu,\lambda)} \mu_w \quad \text{and} \quad \underline{h}(\lambda : \nu, \mu) = \max_{w\in\Lambda(\nu,\lambda)} \mu_w,$$

where $\Lambda(\nu, \lambda) = \{w = w_1 w_2 \ldots w_m \in W_* : \nu_{w_1\ldots w_{m-1}} > \lambda \geq \nu_w\}$.

By Definitions 1.3.9 and 1.5.6, $\Lambda(\nu, \lambda)$ is a partition for $0 < \lambda < 1$.

The quantities $\overline{h}(\lambda : \nu, \mu)$ and $\underline{h}(\lambda : \nu, \mu)$ measure the distortion between the two self-similar measures μ and ν.

Proposition 4.5.2. *Set*

$$\overline{\delta}(\nu, \mu) = \max_{i\in S} \frac{\log \mu_i}{\log \nu_i} \quad \text{and} \quad \underline{\delta}(\nu, \mu) = \min_{i\in S} \frac{\log \mu_i}{\log \nu_i}.$$

Then there exists $c > 0$ such that

$$c\lambda^{\overline{\delta}(\nu,\mu)} \leq \overline{h}(\lambda : \nu, \mu) \leq \lambda^{\overline{\delta}(\nu,\mu)} \quad \text{and} \quad c\lambda^{\underline{\delta}(\nu,\mu)} \leq \underline{h}(\lambda : \nu, \mu) \leq \lambda^{\underline{\delta}(\nu,\mu)}$$

for any $\lambda \in (0,1)$.

Obviously, by the definition, $\overline{\delta}(\nu, \mu) \geq 1 \geq \underline{\delta}(\nu, \mu)$ and any of the equalities holds if and only if $\mu = \nu$.

To show the above proposition, we need the following lemma.

Lemma 4.5.3. *Let a_i and b_i be positive numbers for $i = 1, 2, \ldots, n$. Then*

$$\min_{i=1,2,\ldots,n} \frac{b_i}{a_i} \leq \frac{b_1 + \cdots + b_n}{a_1 + \cdots + a_n} \leq \max_{i=1,2,\ldots,n} \frac{b_i}{a_i}.$$

One can easily prove this lemma by using induction on n.

We use $\overline{h}(\lambda), \underline{h}(\lambda), \overline{\delta}$ and $\underline{\delta}$ in place of $\overline{h}(\lambda : \nu, \mu), \underline{h}(\lambda : \nu, \mu), \overline{\delta}(\nu, \mu)$ and $\underline{\delta}(\nu, \mu)$ respectively.

Proof of Proposition 4.5.2. If $a = \min_{i \in S} \nu_i$, then $\nu_w \leq \lambda \leq \nu_w/a$ for any $w \in \Lambda(\nu, \lambda)$. Hence

$$(a\lambda)^{(\log \mu_w)/(\log \nu_w)} \leq \mu_w \leq \lambda^{(\log \mu_w)/(\log \nu_w)}. \tag{4.5.2}$$

Set $c = a^{\overline{\delta}}$. Then Lemma 4.5.3 implies that $c\lambda^{\overline{\delta}} \leq \mu_w \leq \lambda^{\overline{\delta}}$. Choose $k \in S$ so that $\overline{\delta} = (\log \mu_k)/(\log \nu_k)$. Then there exists $w_* \in \Lambda(\nu, \lambda)$ such that $\dot{k} \in \Sigma_{w_*}$. Hence $w_* = k \ldots k$. By (4.5.2), $c\lambda^{\overline{\delta}} \leq \mu_{w_*} \leq \lambda^{\overline{\delta}}$. Hence $c\lambda^{\overline{\delta}} \leq \overline{h}(\lambda) \leq \lambda^{\overline{\delta}}$ for any $\lambda \in (0,1)$. Exactly the same argument shows that $c\lambda^{\underline{\delta}} \leq \underline{h}(\lambda) \leq \lambda^{\underline{\delta}}$. $\qquad\square$

Now let us return to eigenfunctions of the Laplacian Δ_μ. Assume that μ is a self-similar measure on K with weight $(\mu_i)_{i \in S}$. Then, recalling Theorem 4.1.7, we have $\sum_{i \in S}(\mu_i r_i)^{d_S/2} = 1$, where d_S is the spectral exponent of $(\mathcal{E}, \mathcal{F}, \mu)$. Define $\nu_i = (\mu_i r_i)^{d_S/2}$ and let ν be the self-similar measure on K with weight $(\nu_i)_{i \in S}$. The following is the main theorem of this section.

Theorem 4.5.4. *There exist positive constants c_1 and c_2 such that*

$$||u||_\infty \leq c_1 |\lambda|^{d_S \overline{\delta}(\nu,\mu)/4} ||u||_2 \tag{4.5.3}$$

for any $u \in \mathcal{D}_\mu$ with $\Delta_\mu u = -\lambda u$ and $|\lambda| \geq c_2$. Moreover, if there exists a pre-localized eigenfunction of Δ_μ, then (4.5.3) is best possible. More precisely, there exist $\{u_n\}_{n \geq 1} \subset \mathcal{D}_\mu \backslash \{0\}$ and $c_3 > 0$ such that $\Delta_\mu u_n = -\lambda_n u_n$, $\lim_{n \to \infty} \lambda_n = \infty$ and

$$||u_n||_\infty \geq c_3 {\lambda_n}^{d_S \overline{\delta}(\nu,\mu)/4} ||u_n||_2.$$

By the definition of $\overline{\delta}(\nu, \mu)$, it is easy to see that $\alpha = d_S \overline{\delta}(\nu, \mu)/2$ and that (4.5.1) and (4.5.3) are the same.

For D-eigenfunctions (and also N-eigenfunctions), (4.5.1) is easily derived from the Nash inequality (5.3.3) along with Corollary B.3.9.

We need several lemmas to show Theorem 4.5.4.

Notation. Let Λ be a partition. For $u : K \to \mathbb{R}$, $u_\Lambda = \sum_{p \in V(\Lambda)} u(p) \psi_p^\Lambda$. In particular, if $\Lambda = \{\emptyset\}$, we write u_0 in place of u_Λ.

Recall Definition 4.2.2 for the definitions of $V(\Lambda)$ and ψ_p^Λ. u_Λ is the unique Λ-harmonic function that satisfies $u_\Lambda|_{V(\Lambda)} = u|_{V(\Lambda)}$. In particular, u_0 is the unique harmonic function that satisfies $u_0|_{V_0} = u|_{V_0}$.

Lemma 4.5.5. *There exists $c > 0$ such that*

$$c||f||_2 \geq \min_{w \in \Lambda} \sqrt{\mu_w} ||f||_\infty$$

for any partition Λ and any Λ-harmonic function f.

Proof. Since the space of harmonic functions is finite dimensional, there exists $c > 0$ such that $c\|u\|_2 \geq \|u\|_\infty$ for any harmonic function u. Now let f be a Λ-harmonic function. Then

$$c^2\|f\|_2^2 = c^2 \sum_{w\in\Lambda} \mu_w \|f \circ F_w\|_2^2 \geq \sum_{w\in\Lambda} \mu_w \|f \circ F_w\|_\infty^2$$

$$\geq \min_{w\in\Lambda} \mu_w \sum_{w\in\Lambda} \|f \circ F_w\|_\infty^2 \geq (\min_{w\in\Lambda} \mu_w) \max_{w\in\Lambda} \|f \circ F_w\|_\infty^2.$$

Since $\|f\|_\infty = \max_{w\in\Lambda} \|f \circ F_w\|_\infty$, the lemma follows from the above inequality. $\qquad\square$

By Proposition 3.4.8 and Theorem 3.6.1, it follows that the Green's operator G_μ is a bounded operator from $L^p(K,\mu)$ to $L^p(K,\mu)$ for $p = 2, \infty$. Let $\|G_\mu\|_p$ be the operator norm of G_μ as a bounded operator from $L^p(K,\mu)$ to itself for $p = 2, \infty$.

Lemma 4.5.6. *Set $a = \max\{\|G_\mu\|_2, \|G_\mu\|_\infty\}$. If $u \in \mathcal{D}_\mu$ and $\Delta_\mu u = -\lambda u$ and $|\lambda| > (2a)^{-1}$, then*

$$\|u_\Lambda\|_p/2 \leq \|u\|_p \leq 2\|u_\Lambda\|_p$$

for $p = 2, \infty$, where $\Lambda = \Lambda(\nu, (2a|\lambda|)^{-d_S/2})$.

Proof. Note that $|\lambda|\mu_w r_w a \leq 1/2$ for any $w \in \Lambda$. By Theorem 3.7.14,

$$u \circ F_w = \lambda \mu_w r_w G_\mu(u \circ F_w) + (u \circ F_w)_0.$$

This immediately implies that

$$\big| \|u \circ F_w\|_p - \|(u \circ F_w)_0\|_p \big| \leq |\lambda|\mu_w r_w \|G_\mu\|_p \|u \circ F_w\|_p$$

for $p = 2, \infty$. Hence, if $|\lambda|\mu_w r_w a \leq 1/2$,

$$\|(u \circ F_w)_0\|_p/2 \leq \|u \circ F_w\|_p \leq 2\|(u \circ F_w)_0\|_p$$

for $p = 2, \infty$. This implies the desired inequality. $\qquad\square$

Proof of Theorem 4.5.4. Let $c_2 = (2a)^{-1}$ and let $\Lambda = \Lambda(\nu, (c_2/\alpha)^{d_S/2})$. Also assume that $u \in \mathcal{D}_\mu$ and $\Delta_\mu = -\alpha u$. Then, by Lemmas 4.5.5 and 4.5.6, we have

$$\frac{\|u\|_\infty}{2} \leq \|u_\Lambda\|_\infty \leq c\overline{h}((c_2/\alpha)^{d_S/2})^{-1/2}\|u_\Lambda\|_2 \leq 2c\overline{h}((c_2/\alpha)^{d_S/2})^{-1/2}\|u\|_2,$$

where $\overline{h}(\cdot) = \overline{h}(\cdot : \nu, \mu)$. Hence, by Proposition 4.5.2, it follows that

$$\|u\|_\infty \leq c_1 \alpha^{d_S \overline{\delta}/4}\|u\|_2,$$

where $c_1 = 4c(c_2)^{-d_S \overline{\delta}/4}$ and $\overline{\delta} = \overline{\delta}(\nu, \mu)$. So we have shown (4.5.3).

Next assume that there exists a pre-localized eigenfunction u and $\Delta_\mu u = -\alpha u$. Then, as in Proposition 4.3.2, if $u_w = P_w u$, $\Delta_\mu u_w = -\frac{\alpha}{\mu_w r_w} u$. Note that $||u_w||_2 = \sqrt{\mu_w} ||u||_2$ and $||u_w||_\infty = ||u||_\infty$. Choose k so that $\overline{\delta}(\nu, \mu) = \log \mu_i / \log \nu_i$. Let $w(m) = k \ldots k \in W_m$ for all $m \geq 0$ and let $\varphi_m = u_{w(m)}$. Then $\Delta_\mu \varphi_m = -\alpha_m \varphi_m$, where $\alpha_m = \alpha(\nu_k)^{-2m/d_S}$. Also, if $c_4 = ||u||_2 / ||u||_\infty$, then $||\varphi_m||_\infty = c_4 (\mu_k)^{-m/2} ||\varphi_m||_2$. Hence we obtain

$$||\varphi_m||_\infty = c_4 (\alpha_m / \alpha)^{d_S \overline{\delta}/4} ||\varphi_m||_2.$$

$\square$

Notes and references

4.1   Theorem 4.1.5 was obtained in [93]. One of the key ideas in proving this theorem is the use of Dirichlet forms, in particular, the self-similarity of Dirichlet forms, Proposition 3.3.1. This idea was first used by Fukushima in [41], where he studied the spectral dimension of nested fractals. See also Kumagai [100].

There have been a number of related works before Theorem 4.1.5. In the late 1970s and the early 1980s, physicists studied the integrated density of states on fractal graphs, in particular, the graphs associated with the Sierpinski gasket. Although they did not give any precise definition of Laplacian, they obtained the so called "spectral dimension" of fractals, which is $\log 9 / \log 5$ for the Sierpinski gasket. The spectral dimension corresponds to the spectral exponent for the standard Laplacian on the Sierpinski gasket. They also suggested the existence of localized eigenfunctions. See Dhar [30], Alexander & Orbach [1], Rammal & Toulouse [154] and Rammal [153]. Also see the series of papers by Gefen *et al.*[47, 49, 48]. One can find a review of those works in [118] and [75].

On the mathematical side, Kusuoka [106] constructed Brownian motion on the Sierpinski gasket and obtained the value of d_S for that case. Also Barlow & Perkins [20] studied the asymptotic behavior of the heat kernel associated with Brownian motion on the Sierpinski gasket by a probabilistic approach. Later, Fukushima & Shima [44] obtained detailed information about eigenvalues and eigenfunctions for the standard Laplacian on the Sierpinski gasket by using the eigenvalue decimation method. In particular, they found localized eigenfunctions and proved that the integrated density of state is discontinuous. (This result immediately implies that the periodic function G appearing in Theorem 4.1.5 is discontinuous in this case.) Roughly speaking, the eigenvalue decimation method is based on the following property.

Eigenvalue decimation property *Let $\Phi(x) = x(5 - x)$. Define $\ell_0(V_m) = \{u \in \ell(V_m) : u|_{V_0} \equiv 0\}$. (We may identify $\ell_0(V_m)$ with $\ell(V_m \backslash V_0)$ by restricting the domain of $u \in \ell_0(V_m)$ to $V_m \backslash V_0$.) Also define $L_m^0 : \ell_0(V_m) \to \ell_0(V_m)$ by letting $L_m^0 u = (L_m u)|_{V_m \backslash V_0}$ for any $u \in \ell_0(V_m)$, where L_m is defined in Example 3.7.3. Then, if $\lambda \notin \{2, 5, 6\}$, $L_{m+1}^0 u = -\lambda u$ for $u \in \ell_0(V_{m+1})$ if and only if $L_m^0(u|_{V_m}) = -\Phi(\lambda)(u|_{V_m})$.*

See Shima [166] for the proof.

L_m^0 is the discrete version of the Dirichlet Laplacian. By virtue of the eigenvalue decimation method, we can trace all the eigenvalues of L_m^0 and the eigenfunctions starting from $m = 1, 2$. And hence we obtain the exact locations of the eigenvalues of the standard Laplacian.

In [24], Dalrymple *et al.* studied (numerically in part) eigenvalues and eigenfunctions of the standard Laplacian on the Sierpinski gasket by using the eigenvalue decimation method and obtained figures of eigenfunctions. See also [50] and [90] for detailed structure of eigenvalues and eigenfunctions of the standard Laplacian on the Sierpinski gasket.

Unfortunately, the eigenvalue decimation method only works for quite limited class of Laplacians on p. c. f. self-similar structures. See [169] for details.

In [44], Fukushima & Shima also studied the spectrum of the standard Laplacian on the infinite Sierpinski gasket and obtained the discontinuity of the integrated density of states. See also [41] and [45]. In [178], Teplyaev determined the complete structure of the spectrum of the standard Laplacian on the infinite Sierpinski gasket with Neumann boundary condition. In particular, he showed that there exists a complete orthonormal system consisting of localized eigenfunctions and that the spectrum is pure point. However, this problem is still open in the case of Dirichlet boundary condition. Recently Sabot extended Teplyaev's results to infinite nested fractals without boundary points. See [165] for details.

4.2 The results in this section was obtained in [85].

4.3 and 4.4   The results in those sections, except Theorem 4.3.6, were obtained in [18]. Theorem 4.3.6 is a new result.

4.5 The results in this section are new.

5

Heat Kernels

In this chapter, we will study the Neumann heat kernel $p_N(t, x, y)$ associated with the Neumann Laplacian H_N and the Dirichlet heat kernel $p_D(t, x, y)$ associated with the Dirichlet Laplacian H_D. Formally, the heat kernels p_D and p_N are defined as the integral kernels associated with the semigroups e^{-tH_N} and e^{-tH_D} respectively. More precisely, in the Neumann case, for example,

$$(e^{-tH_N}u)(x) = \int_K p_N(t, x, y)u(y)\mu(dy).$$

In other words, the heat kernels are the fundamental solutions of the heat equation,

$$\frac{\partial u}{\partial t} = \Delta_\mu u,$$

with corresponding boundary conditions.

We will first define p_N and p_D as formal sums with respect to the complete orthonormal systems $\{\varphi_i^N\}_{i \geq 1}$ and $\{\varphi_i^D\}_{i \geq 1}$ respectively. (See Definition 5.1.1.) Using estimates in 4.5, the formal expansions are shown to yield a continuous function on $(t, x, y) \in (0, \infty) \times K^2$. Then p_N and p_D are shown to be the integral kernels of the semigroups $e^{-H_N t}$ and $e^{-H_D t}$ respectively.

In 5.2, we will prove the parabolic maximum principle for solutions of the heat equation. This implies that $0 \leq p_D(t, x, y) \leq p_N(t, x, y)$ when $(D, \mathbf{r})$ is a regular harmonic structure. (As we will explain in 5.2, the restriction that $(D, \mathbf{r})$ is a regular harmonic structure is purely technical. In fact, by using probabilistic methods or the general theory of Dirichlet forms, it is known that this holds even if $(D, \mathbf{r})$ is not regular.)

In 5.3, we will study the asymptotic behavior of $p_N(t, x, x)$ and $p_D(t, x, x)$ as $t \to 0$. If $(D, \mathbf{r})$ is regular and $\mu = \mu_*$, where μ_* is the self-similar

157

measure defined in Theorem 4.2.1, then our estimate will be sharp. See Corollary 5.3.2.

As in the last chapter, we suppose that $(K, S, \{F_i\}_{i \in S})$ is a connected p. c. f. self-similar structure with $S = \{1, 2, \ldots, N\}$, $(D, \mathbf{r})$ is a harmonic structure on $\mathcal{L}$ and μ is a self-similar measure with weight $(\mu_i)_{i \in S}$ on K with $r_i \mu_i < 1$ for any $i \in S$. We use $(\mathcal{E}, \mathcal{F})$ to denote the resistance form associated with $(D, \mathbf{r})$. Also, d is a metric on K which is compatible with the original topology of K. (Note that (K, d) is compact.) We will use the convention from the previous chapters that the symbol b represents the boundary conditions, i.e., $b \in \{N, D\}$, where N represents the Neumann boundary condition and D represents the Dirichlet boundary condition.

5.1 Construction of heat kernels

In this section, we will construct heat kernels $p_D(t, x, y)$ and $p_N(t, x, y)$ corresponding to Dirichlet and Neumann boundary conditions respectively.

By the results in 4.1, for $b = D, N$, there exist $\{\lambda_n^b\}_{n \geq 1}$ and $\{\varphi_n^b\}_{n \geq 1} \subset C(K, d) \cap \mathcal{D}_{b,\mu}$ such that $\varphi_n^b \in E_b(\lambda_n^b)$ for all $n \geq 1$,

$$0 \leq \lambda_1^b \leq \cdots \leq \lambda_m^b \leq \lambda_{m+1}^b \leq \cdots$$

and $\{\varphi_n^b\}_{n \geq 1}$ is a complete orthonormal system for $L^2(K, \mu)$.

Definition 5.1.1 (Heat kernel). For $b = D, N$, we define the b-heat kernel $p_b(t, x, y)$ for $(t, x, y) \in (0, \infty) \times K \times K$ by

$$p_b(t, x, y) = \sum_{n \geq 1} e^{-\lambda_n^b t} \varphi_n^b(x) \varphi_n^b(y). \tag{5.1.1}$$

The right-hand side of the above definition is only a formal sum so far. We will show, however, that the sum converges uniformly on $[T, \infty) \times K \times K$ for any $T > 0$. Hence $p_b(t, x, y)$ is continuous on $(0, \infty) \times K \times K$.

The following gives some important properties of the heat kernels. In the rest of this section, the symbol b always represents D or N.

Proposition 5.1.2. (1) $p_b(t, x, y)$ *is non-negative and continuous on* $(0, \infty) \times K \times K$.
(2) *For any* $(t, x) \in (0, \infty) \times K$, *define* $p_b^{t,x}$ *by* $p_b^{t,x}(y) = p_b(t, x, y)$ *for any* $y \in K$. *Then* $p_b^{t,x} \in \mathrm{Dom}(H_b) \cap \mathcal{D}_\mu = \mathcal{D}_{b,\mu}$ *for any* $(t, x) \in (0, \infty) \times K$.
(3) *For any* $(x, y) \in K \times K$, $p_b(\cdot, x, y) \in C^1((0, \infty))$.
(4) *For any* $(t, x, y) \in (0, \infty) \times K \times K$,

$$\frac{\partial p_b(t, x, y)}{\partial t} = (\Delta_\mu p_b^{t,x})(y).$$

(5) *For any $s, t > 0$ and any $x, y \in K$,*

$$p_b(t + s, x, y) = \int_K p_b(t, x, x') p_b(s, x', y) \mu(dx').$$

We need several lemmas to prove this proposition.

Lemma 5.1.3. *There exist positive constants c_1 and c_2 such that*

$$c_1 n^{2/d_S} \le \lambda_n^b \le c_2 n^{2/d_S}$$

for $n \ge 2$.

Remark. Recall that H_D is invertible and that $\lambda_1^D > 0$. Therefore, if $b = D$, then the above inequality holds for $n \ge 1$ as well. However, this is not the case if $b = N$, since $\lambda_1^N = 0$.

Proof. By Theorem 4.1.5, there exist positive constants c_3 and c_4 such that

$$c_3 x^{d_S/2} \le \rho_b(x, \mu) \le c_4 x^{d_S/2}$$

for sufficiently large $x > 0$. As $n \le \rho_b(\lambda_n^b, \mu) \le c_4 \lambda_n^{b\,d_S/2}$, we have that $c_1 n^{2/d_S} \le \lambda_n^b$, where $c_1 = c_4^{-2/d_S}$. For any $\epsilon > 0$, $c_3 (\lambda_n^b - \epsilon)^{d_S/2} \le \rho_b(\lambda_n^b - \epsilon, \mu) \le n$. Letting $\epsilon \downarrow 0$, we see that $\lambda_n^b \le c_2 n^{2/d_S}$, where $c_2 = c_3^{-2/d_S}$. $\square$

Lemma 5.1.4. *For any $\alpha, \beta > 0$ and any $T > 0$, $\sum_{n \ge 1} n^\alpha e^{-n^\beta t}$ is uniformly convergent on $[T, \infty)$.*

Proof. Choose $m \in \mathbb{N}$ so that $\beta m - \alpha > 1$. Let $\gamma = \beta m - \alpha$. Since $n^{-\alpha} e^{n^\beta t} = \sum_{k \ge 0} n^{\beta k - \alpha} t^k / k!$, we see that $n^\alpha e^{-n^\beta t} \le A n^{-\gamma}$ for any $t \ge T$ and $n \ge 1$, where $A = m! T^{-m}$. As $\sum_{n \ge 1} n^{-\gamma} < \infty$, the lemma follows immediately. $\square$

Now we prove the proposition except the non-negativity of $p_b(t, x, y)$.

Proof of Proposition 5.1.2. (1)   By Theorem 4.5.4 and Lemma 5.1.3,

$$\|\varphi_n^b\|_\infty \le c(\lambda_n^b)^{d_S \bar{\delta}/4} \le c' n^{\bar{\delta}/2}, \tag{5.1.2}$$

where $\bar{\delta} = \bar{\delta}(\nu, \mu)$. Again, using Lemma 5.1.3, we obtain

$$|e^{-\lambda_n^b t} \varphi_n^b(x) \varphi_n^b(y)| \le (c')^2 n^{\bar{\delta}} e^{-c'' n^\beta t},$$

where $\beta = 2/d_S$ and c'' is a positive constant which is independent of n. Now, by Lemma 5.1.4, $\sum_{n \ge 1} (c')^2 n^{\bar{\delta}} e^{-c'' n^\beta t}$ converges uniformly on $[T, \infty)$ for any $T > 0$. Therefore, the right-hand side of (5.1.1) converges uniformly on $[T, \infty)$. So $p_b(t, x, y)$ is continuous on $(0, \infty) \times K \times K$.

(2) Note that $f = \sum_{n \geq 1} a_n \varphi_n^b \in \mathrm{Dom}(H_b)$ if and only if $\sum_{n \geq 1} |\lambda_n^b a_n|^2 < \infty$. Also, if $f \in \mathrm{Dom}(H_b)$, then $H_b f = \sum_{n \geq 1} \lambda_n^b a_n \varphi_n^b$. Now, by Lemmas 5.1.3 and 5.1.4, $\sum_{n \geq 1} (\lambda_n^b e^{-\lambda_n^b t} \varphi_n^b(x))^2 < \infty$ for any $(t, x) \in (0, \infty) \times K$. Therefore, $p_b^{t,x} \in \mathrm{Dom}(H_b)$ and

$$(H_b p_b^{t,x})(y) = \sum_{n \geq 1} \lambda_n^b e^{-\lambda_n^b t} \varphi_n^b(x) \varphi_n^b(y). \tag{5.1.3}$$

The same argument as in the proof of (1) implies that the right-hand side of (5.1.3) converges uniformly on $[T, \infty) \times K \times K$ for any $T > 0$. Therefore $H_b p_b^{t,x} \in C(K, d)$. Corollary 3.7.13 implies that $p_b^{t,x} \in \mathrm{Dom}(H_b) \cap \mathcal{D}_\mu = \mathcal{D}_{b,\mu}$.

(3) As the right-hand side of (5.1.3) converges uniformly on $[T, \infty) \times K \times K$ for any $T > 0$, we see that

$$\int_{t_1}^{t_2} -(H_b p_b^{t,x})(y) dt = p_b(t_2, x, y) - p_b(t_1, x, y) \tag{5.1.4}$$

for any $t_1 < t_2$ and any $x, y \in K$. Therefore $p_b(t, x, y) \in C^1((0, \infty))$.

(4) Obvious by (5.1.3) and (5.1.4).

(5) Since the infinite sum in the right-hand side of (5.1.1) converges uniformly, a direct calculation shows the desired equality. □

Let $H_R = \{z \in \mathbb{C} : \mathrm{Re}\, z > 0\}$. Then, we can extend the heat kernel $p_b(t, x, y)$ to a holomorphic function on H_R as follows.

Proposition 5.1.5. *Define $p_b(z, x, y)$ by substituting $z \in H_R$ for t in (5.1.1). Then $p_b(z, x, y)$ is a holomorphic function on H_R for any $x, y \in K$.*

Later, this proposition will play a key role in proving Proposition 5.1.10, where $p_b(t, x, y)$ is shown to be strictly positive for any $t > 0$.

Proof. Note that

$$\sum_{n \geq 1} |e^{-\lambda_n^b z} \varphi_n^b(x) \varphi_n^b(y)| \leq \sum_{n \geq 1} e^{-\lambda_n^b t} |\varphi_n^b(x)| |\varphi_n^b(y)|,$$

where $t = \mathrm{Re}\, z$ for $z \in H_R$. The same discussion as in the proof of Proposition 5.1.2-(1) implies that $\sum_{n \geq 1} e^{-\lambda_n^b z} \varphi_n^b(x) \varphi_n^b(y)$ converges uniformly on $\{z \in \mathbb{C} : \mathrm{Re}\, z \geq T\} \times K \times K$ for any $T > 0$. Since $e^{-\lambda_n^b z}$ is holomorphic, it follows that $p(z, x, y)$ is holomorphic on H_R. □

Next we construct the heat semigroup associated with $p_b(t, x, y)$. See B.2 for a definition and basic properties of semigroups.

Definition 5.1.6. For $u \in L^1(K, \mu)$, define

$$(T_t^b u)(x) = \int_K p_b(t, x, y) u(y) \mu(dy).$$

for $t > 0$ and $x \in K$.

Note that $L^1(K, \mu) \supseteq L^p(K, \mu)$ for $p \in [1, \infty]$.

Theorem 5.1.7. (1) T_t^b *is a bounded operator from* $L^p(K, \mu) \to C(K, d)$ *for any* $t > 0$ *and any* $p \in [1, \infty]$.
(2) $T_t^b \circ T_s^b = T_{t+s}^b$ *for any* $t, s > 0$.
(3) $T_t^b(L^1(K, \mu)) \subseteq \mathcal{D}_{b,\mu}$ *for any* $t > 0$.
(4) *Let* $u \in L^1(K, \mu)$. *Set* $u(t, x) = (T_t^b u)(x)$ *for* $(t, x) \in (0, \infty) \times K$. *Then* $u(\cdot, x) \in C^\infty((0, \infty))$ *for any* $x \in K$. *Moreover,*

$$\frac{\partial u(t, x)}{\partial t} = (\Delta_\mu u_t)(x) \tag{5.1.5}$$

for any $(t, x) \in (0, \infty) \times K$, *where* $u_t = T_t^b u$.
(5) $\{T_t^b\}_{t>0}$ *is a strongly continuous semigroup on* $L^2(K, \mu)$. *The generator of* $\{T_t^b\}_{t>0}$ *is* $-H_b$.

Proof. (1) This is obvious since $p_b(t, x, y)$ is continuous on $(0, \infty) \times K \times K$.
(2) This follows immediately by Proposition 5.1.2-(5).
(3) and (4) Let $u \in L^1(K, \mu)$. Choose t_1 so that $0 < t_1 < t$. Set $s = t - t_1$. Since $u_1 = T_{t_1}^b u \in C(K, d) \subset L^2(K, \mu)$, we see that $u_1 = \sum_{n \geq 1} a_n \varphi_n^b$ for some $\{a_n\}_{n \geq 1}$ with $\sum_{n \geq 1} |a_n|^2 < \infty$. Therefore $T_t^b u = T_s^b u_1 = \sum_{n \geq 1} a_n e^{-\lambda_n^b s} \varphi_n^b$. Making use of Lemmas 5.1.3 and 5.1.4, we obtain that $\sum_{n \geq 1} (a_n e^{-\lambda_n^b s} \lambda_n^b)^2 < \infty$. This implies that $T_t^b u \in \text{Dom}(H_b)$ and

$$H_b(T_t^b u) = \sum_{n \geq 1} \lambda_n^b a_n e^{-\lambda_n^b s} \varphi_n^b.$$

By Lemmas 5.1.3 and 5.1.4 along with (5.1.2), $H_b(T_t^b u) \in C(K, d)$. Now, Corollary 3.7.13 implies that $T_t^b u \in \mathcal{D}_{b,\mu}$. Also, it follows that

$$-\int_{t_1}^t (H_b(T_t^b u))(x) = (T_t^b u)(x) - (T_{t_1}^b u)(x).$$

Hence $u(t, x)$ is a C^1-class function and (5.1.5) holds for $t > t_1$.
(5) If $u \in L^2(K, \mu)$, then $u = \sum_{n \geq 1} a_n \varphi_n^b$, where $\sum_{n \geq 1} |a_n|^2 < \infty$. So $T_t^b u = \sum_{n \geq 1} a_n e^{-\lambda_n^b t}$. Hence $\|T_t^b u\|_2 \leq \|u\|_2$ and $\|T_t^b u - u\|_2 \to 0$ as $t \to 0$. Therefore $\{T_t^b\}_{t>0}$ is a strongly continuous semigroup on $L^2(K, \mu)$. Now let A be the generator of $\{T_t^b\}_{t>0}$. Then, by the definition of the generator in

Theorem B.2.2, it follows that $\varphi_n^b \in \mathrm{Dom}(A)$ and $A\varphi_n^b = -\lambda_n^b \varphi_n^b = -H_b \varphi_n^b$. Since $\{\varphi_n^b\}_{n\geq 1}$ is a complete orthonormal system for $L^2(K,\mu)$, we deduce that $A = -H_b$.   □

As the generator of $\{T_t^b\}_{t>0}$ is $-H_b$, the corresponding closed form on $L^2(K,\mu)$ is the Dirichlet form $(\mathcal{E},\mathcal{F})$ if $b = N$ or $(\mathcal{E},\mathcal{F}_0)$ if $b = D$. Theorem B.3.4 implies that $\{T_t^b\}_{t>0}$ has the Markov property. One important consequence of this fact is the non-negativity of the heat kernel. See Lemma B.3.5 and Proposition B.3.6 for further properties of strong continuous semigroups with the Markov property.

Proof of Proposition 5.1.2-(1). The remaining property is that $p_b(t,x,y)$ is non-negative. By Theorem B.3.4, $\{T_t^b\}_{t>0}$ has the Markov property. Therefore, if $u \in C(K,d)$ and $u(x) \geq 0$ for any $x \in K$, then $(T_t u)(x) \geq 0$ for any $x \in K$. This implies that $p_b(t,x,y) \geq 0$ for any $(t,x,y) \in (0,\infty) \times K \times K$.   □

Here we have shown non-negativity of the heat kernels by using the Markov property of $\{T_t^b\}_{t>0}$, more precisely, by Theorem B.3.4. However in this book we will not give a proof of Theorem B.3.4. There is another way of showing non-negativity of the heat kernels by using the (parabolic) maximum principle, as in the classical theory of the heat equation on $\mathbb{R}^n$. See the next section for details. Unfortunately, however, so far this alternative proof only works in the case of a regular harmonic structure.

From Theorem 5.1.7, the heat kernel $p_b(t,x,y)$ is the fundamental solution of the heat equation (5.1.5).

Definition 5.1.8. Let $u : (0,\infty) \times K \to \mathbb{R}$. For $t \in (0,\infty)$, set $u_t(x) = u(t,x)$ for any $x \in K$. u is called a solution of the heat equation (5.1.5) on $(0,\infty)$ if and only if u is continuous on $(0,\infty) \times K$, $u(\cdot,x) \in C^1((0,\infty))$ for any $x \in K$, $u_t \in \mathcal{D}_\mu$ for any $t \in (0,\infty)$ and u satisfies (5.1.5) for any $(t,x) \in (0,\infty) \times K$.

Corollary 5.1.9. (1) *Set $u(t,x) = (T_t^D \varphi)(x)$ for $\varphi \in L^2(K,\mu)$. Then $u(t,x)$ is the unique solution of the heat equation (5.1.5) on $(0,\infty)$ with*

$$\lim_{t\downarrow 0} ||u_t - \varphi||_2 = 0 \quad and \quad u(t,p) = 0$$

for any $t > 0$ and any $p \in V_0$.
(2) *Set $u(t,x) = (T_t^N \varphi)(x)$ for $\varphi \in L^2(K,\mu)$. Then $u(t,x)$ is the unique solution of the heat equation (5.1.5) on $(0,\infty)$ with*

$$\lim_{t\downarrow 0} ||u_t - \varphi||_2 = 0 \quad and \quad (du_t)_p = 0$$

for any $t > 0$ and any $p \in V_0$.

Proof. We only need to show uniqueness of the solution, which follows from Theorem B.2.6. □

If φ is a continuous function on K, we naturally expect that $\|T_t^b \varphi - \varphi\|_\infty \to 0$ as $t \downarrow 0$. (If $b = D$, we need to assume that $\varphi|_{V_0} = 0$ as well.) However, such continuity of a solution up to $t = 0$ is not straightforward in general. In the next section, we can show such continuity of a solution only when $(D, \mathbf{r})$ is a regular harmonic structure. In fact, this problem is closely related to the proof of the non-negativity of the heat kernels through the parabolic maximum principle mentioned above. See the next section for details.

By Proposition 5.1.2, we know that $p_b(t, x, y) \geq 0$ for any $(t, x, y) \in (0, \infty) \times K \times K$. In fact, we can show a stronger result as in the next proposition.

Proposition 5.1.10. (1) $p_N(t, x, y) > 0$ *for any* $(t, x, y) \in (0, \infty) \times K \times K$.
(2) $p_D(t, x, y) > 0$ *for any* $t > 0$ *if* x *and* y *belong to the same connected component of* $K \backslash V_0$.

In this section, we only give a proof of (1) of the proposition above, while (2) will be proven in 5.3.

Proof. Step 1: For any $x, y \in K$, there exists $t > 0$ such that $p_N(t, x, y) > 0$.

Proof of Step 1. Since $\lambda_1^N = 0$ and $\varphi_1^N = 1$, it follows that $p_N(s, a, a) \geq 1$ for any $(s, a) \in (0, \infty) \times K$. Let us fix $s > 0$. For $a \in K$, if $U(a) = \{b \in K : p_N(s, a, b) > 0\}$, then $U(a)$ is a non-empty open subset of K. As K is compact and connected, we can choose $x_0, \ldots, x_m$ so that $x_0 = x$, $x_m = y$ and $x_{i+1} \in U(x_i)$ for all $i = 0, 1, \ldots, m$. Applying Proposition 5.1.2-(5), we obtain

$$p_N(ms, x, y) =$$
$$\int p_N(s, x, y_1) p_N(s, y_1, y_2) \cdots p_N(s, y_{m-1}, y) \mu(dy_1) \mu(dy_2) \cdots \mu(dy_{m-1}).$$

Since $x_{i+1} \in U(x_i)$ for any i, $p_N(s, x, x_1) p_N(s, x_1, x_2) \cdots p_N(s, x_{m-1}, y) > 0$. Hence $p_N(ms, x, y) > 0$. □

Step 2: If $p_N(t, x, y) > 0$ for $t > 0$, then $p_N(s, x, y) > 0$ for any $s \geq t$.

Proof of Step 2. If $s > t$, Proposition 5.1.2-(5) implies that

$$p_N(s, x, y) = \int_K p_N(s - t, x, x') p_N(t, x', y) \mu(dx').$$

Since $p_N(s - t, x, x') p_N(t, x', y) > 0$ for $x' = x$, it follows that $p_N(s, x, y) > 0$. $\qquad\square$

Now, by Step 1 and Step 2, for any $x, y \in K$, there exists $T \geq 0$ such that $p_N(t, x, y) = 0$ for $t \in (0, T)$ and $p_N(t, x, y) > 0$ for $t \in (T, \infty)$. Suppose $T > 0$. By Proposition 5.1.5, $p_N(z, x, y)$ is a holomorphic function on H_R. Hence, if $p_N(t, x, y) = 0$ for $t \in (0, T)$, then $p_N(z, x, y) = 0$ for any $z \in H_R$. Obviously this contradicts the fact that $p_N(t, x, y) > 0$ for $t > T$. Therefore $T = 0$ and $p_N(t, x, y) > 0$ for any $t > 0$. $\qquad\square$

5.2 Parabolic maximum principle

In 5.1, we defined the notion of solutions of the heat equation

$$\frac{\partial u(t, x)}{\partial t} = (\Delta_\mu u_t)(x) \qquad (5.2.1)$$

on $(0, \infty)$ in Definition 5.1.8. In this section, we study the continuity of the solution to (5.2.1) at $t = 0$. More precisely,

Definition 5.2.1. Let $u : [0, \infty) \times K \to \mathbb{R}$. u is called an L^∞-solution of the heat equation (5.2.1) on $[0, \infty)$ if and only if u is continuous on $[0, \infty) \times K$ and is a solution of the heat equation (5.2.1) on $(0, \infty)$.

A remarkable property of an L^∞-solution of the heat equation on $[0, \infty)$ is the parabolic maximum principle. In the classical theory of heat equations on $\mathbb{R}^n$, the parabolic maximum principle plays an important role. For example, it is often used to show uniqueness of a solution of a certain heat equation. See, for example, [77] and [152] for details.

Theorem 5.2.2 (Parabolic maximum principle).
Let $u : [0, \infty) \times K \to \mathbb{R}$ be an L^∞-solution of the heat equation (5.2.1) on $[0, \infty)$. Then for any $T > 0$,

$$\max_{(t,x) \in U_T} u(t, x) = \max_{(t,x) \in \partial U_T} u(t, x) \quad and \quad \min_{(t,x) \in U_T} u(t, x) = \min_{(t,x) \in \partial U_T} u(t, x),$$

where $U_T = [0, T] \times K$ and $\partial U_T = \{0\} \times K \cup [0, T] \times V_0$. Moreover, if $u_t \in \mathcal{D}_{N,\mu}$ for any $t > 0$, where $u_t(x) = u(t, x)$ for any $t \in [0, \infty)$ and any $x \in K$, then,

$$\max_{(t,x) \in [0,\infty) \times K} u(t, x) = \max_{x \in K} u(0, x) \quad and \quad \min_{(t,x) \in (0,\infty) \times K} u(t, x) = \min_{x \in K} u(0, x).$$

To prove the above theorem, we need the following lemmas.

Lemma 5.2.3. *Let $u \in \mathcal{D}_\mu$. If $(\Delta_\mu u)(x) > 0$ for any $x \in K \backslash V_0$, then $u(x) < \max_{p \in V_0} u(p)$ for any $x \in K$.*

Proof. By Theorem 3.7.16, we obtain

$$(H_m u)(p) = \int_K \psi_p^m \Delta_\mu u \, d\mu \qquad (5.2.2)$$

for any $p \in V_m \backslash V_0$. By the assumption of the lemma, the right-hand side of (5.2.2) is positive. Since $H_m \in \mathcal{LA}(V_m)$, there exists $q \in V_m$ such that $u(q) > u(p)$. Hence, if $u(p_*) = \max_{q \in V_m} u(q)$ for $p_* \in V_m$, then $p_* \in V_0$. Therefore $u(p) < \max_{q \in V_0} u(q)$ for any $p \in V_m \backslash V_0$.

Now, let $x \notin V_*$. Choose $w \in W_*$ so that $x \in K_w$ and $F_w(V_0) \cap V_0 = \emptyset$. Using the same argument as above, we have $u(p) < \max_{q \in F_w(V_0)} u(q)$ for any $p \in K_w \cap V_*$. Therefore it follows that $u(x) \leq \max_{q \in F_w(V_0)} u(q)$. Again the above argument implies that $\max_{q \in F_w(V_0)} u(q) < \max_{p \in V_0} u(p)$. Hence we obtain $u(x) < \max_{p \in V_0} u(p)$ for any $x \in K \backslash V_0$. $\qquad \square$

Lemma 5.2.4. *Let $u \in \mathcal{D}_\mu$. If $u(x) = \max_{y \in K} u(y)$ for $x \in K \backslash V_0$, then $(\Delta_\mu u)(x) \leq 0$. Moreover, if $u \in \mathcal{D}_{N,\mu}$, then the statement above is true for $x \in V_0$ as well.*

Proof. Suppose that $(\Delta_\mu u)(x) > 0$. First assume that $x \in V_*$. Then, using (5.2.2) in the same way as in the proof of Lemma 5.2.3, we deduce that there exists $y \in V_*$ such that $u(y) > u(x)$. This is a contradiction. Hence $x \notin V_*$. Then choose $w \in W_*$ so that $x \in K_w$ and $(\Delta_\mu u)(y) > 0$ for any $y \in K_w$. Set $v = u \circ F_w$. Then obviously v satisfies the assumptions of Lemma 5.2.3. Therefore, $u(x) < \max_{p \in F_w(V_0)} u(p)$. This contradicts the assumption of the lemma.

Next assume that $u \in \mathcal{D}_{N,\mu}$, $p \in V_0$ and $(\Delta_\mu u)(p) > 0$. Recalling the proof of Lemma 3.6.3, we see that $\mathcal{E}(\psi_m^p, u) = -(H_m u)(p)$. The Gauss–Green's formula (Theorem 3.7.8) implies that (5.2.2) holds for any $p \in V_0$. Therefore, $(H_m u)(p) > 0$ for sufficiently large m. Then there exists $q \in V_m$ such that $u(q) > u(p)$. Hence $u(p) < \max_{y \in K} u(y)$. $\qquad \square$

Proof of Theorem 5.2.2. Let $u : [0, \infty) \times K \to \mathbb{R}$ be an L^∞-solution of the heat equation (5.2.1) on $[0, \infty)$. For $n \geq 1$, define u_n by $u_n(t, x) = u(t, x) - t/n$ for any $(t, x) \in [0, \infty) \times K$. First, we show the following claim.

$$\max_{(t,x) \in U_T} u_n(t, x) = \max_{(t,x) \in \partial U_T} u_n(t, x).$$

Proof of the claim. Suppose that $(t_*, p) \in U_T \backslash \partial U_T$ and that

$$u_n(t_*, p) = \max_{(t,x) \in U_T} u_n(t, x).$$

This implies that $\frac{\partial u_n(t_*, p)}{\partial t} \geq 0$. Note that $\Delta_\mu u_n = \Delta_\mu u$ and that $\frac{\partial u_n}{\partial t} = \frac{\partial u}{\partial t} - 1/n$. Since $\Delta_\mu u = \frac{\partial u}{\partial t}$, it follows that $(\Delta_\mu u_n)(t_*, p) \geq 1/n > 0$. Hence, Lemma 5.2.4 implies that $u_n(t_*, p) < \max_{x \in K} u_n(t_*, x)$. This contradicts the fact that $u_n(t_*, p) = \max_{(t,x) \in U_T} u_n(t, x)$. □

Now, letting $n \to \infty$, we have

$$\max_{(t,x) \in U_T} u(t, x) = \max_{(t,x) \in \partial U_T} u(t, x).$$

The minimum also follows if we apply the maximum case to $-u$.

If $u_t \in \mathcal{D}_{N,\mu}$ for any $t > 0$, we may repeat the above argument by replacing U_T and ∂U_T by $[0, \infty) \times K$ and $\{0\} \times K$ respectively. □

Later in this section, we will apply the parabolic maximum principle to show non-negativity of the heat kernels $p_D(t, x, y)$ and $p_N(t, x, y)$. For that purpose, we need to know whether or not the solutions of the heat equations constructed by the heat kernels in 5.1 can be extended to L^∞-solutions on $[0, \infty)$. More precisely, the problem is whether or not $T_t^b \varphi$ is a L^∞-solution of the heat equation on $[0, \infty)$. This is still an open problem in general: we have only a partial answer if the harmonic structure is regular.

Theorem 5.2.5. *Suppose that $(D, \mathbf{r})$ is a regular harmonic structure.*
(1) *Let $\varphi \in \{f \in C(K, d) : f|_{V_0} \equiv 0\}$. Define $u(t, x) = (T_t^D \varphi)(x)$ for any $(t, x) = (0, \infty) \times K$ and $u(0, x) = \varphi(x)$ for any $x \in K$. Set $u_t(x) = u(t, x)$ for any $(t, x) \in [0, \infty) \times K$. Then u is an L^∞-solution of the heat equation (5.2.1) on $[0, \infty)$ with $u_t|_{V_0} \equiv 0$ for any $t \geq 0$. In particular, $||u_t - u_0||_\infty \to 0$ as $t \to 0$.*
(2) *Let $\varphi \in C(K, d)$. Define $u(t, x) = (T_t^N \varphi)(x)$ for any $(t, x) = (0, \infty) \times K$ and $u(0, x) = \varphi(x)$ for any $x \in K$. Set $u_t(x) = u(t, x)$ for any $(t, x) \in [0, \infty) \times K$. Then u is an L^∞-solution of the heat equation (5.2.1) on $[0, \infty)$ with $(du_t)_p = 0$ for any $(t, p) \in (0, \infty) \times V_0$. In particular, $||u_t - u_0||_\infty \to 0$ as $t \to 0$.*

By Corollary 5.1.9, the following proposition immediately implies this theorem.

Proposition 5.2.6. *Suppose that $(D, \mathbf{r})$ is a regular harmonic structure.*
(1) *Let $\varphi \in \{f \in C(K, d) : f|_{V_0} \equiv 0\}$. Then $||T_t^D \varphi - \varphi||_\infty \to 0$ as $t \to 0$.*
(2) *Let $\varphi \in C(K, d)$. Then $||T_t^N \varphi - \varphi||_\infty \to 0$ as $t \to 0$.*

We need several results to prove this proposition. For the moment, we show the following special case.

Lemma 5.2.7. *Suppose that $(D, \mathbf{r})$ is a regular harmonic structure.*
(1) *Let $\varphi \in \mathcal{F}_0$. Then $\|T_t^D \varphi - \varphi\|_\infty \to 0$ as $t \to 0$.*
(2) *Let $\varphi \in \mathcal{F}$. Then $\|T_t^N \varphi - \varphi\|_\infty \to 0$ as $t \to 0$.*

To prove Lemma 5.2.7, we need the following lemma.

Lemma 5.2.8. *Suppose that $(D, \mathbf{r})$ is a regular harmonic structure. Then there exists $c > 0$ such that*

$$\|u\|_\infty \leq c\sqrt{\mathcal{E}(u, u)} + c\|u\|_2$$

for any $u \in \mathcal{F}$.

Proof. Since $(D, \mathbf{r})$ is a regular harmonic structure, Theorem 3.3.4 implies that $\Omega = K$ and that (K, R) is a compact metric space. Hence, by (2.2.6),

$$|f(x) - f(y)|^2 \leq \mathcal{E}(f, f)R(x, y) \tag{5.2.3}$$

for any $x, y \in K$ and any $f \in \mathcal{F}$.

Let $v \in \mathcal{F}_0$. Then, by (5.2.3), it follows that

$$|v(x)|^2 \leq \mathcal{E}(v, v)R(x, p)$$

for any $x \in K$, where $p \in V_0$. Therefore, there exists $c > 0$ such that

$$\|v\|_\infty \leq c\sqrt{\mathcal{E}(v, v)} \tag{5.2.4}$$

for any $v \in \mathcal{F}_0$.

Let $u \in \mathcal{F}$. Define $u_0 = \sum_{p \in V_0} u(p)\psi_p$. Then, by (5.2.4),

$$\|u - u_0\|_\infty \leq c\sqrt{\mathcal{E}(u - u_0, u - u_0)} \leq c\sqrt{\mathcal{E}(u, u)}.$$

Also, there exists $c' > 0$ such that $\|u_0\|_\infty \leq c'\|u_0\|_2$ for any $u \in \mathcal{F}$. Also note that $\|u - u_0\|_2 \leq \|u - u_0\|_\infty$. Combining all the inequalities, we obtain

$$\|u\|_\infty \leq \|u - u_0\|_\infty + \|u_0\|_\infty \leq c\sqrt{\mathcal{E}(u, u)} + c'\|u_0\|_2$$
$$\leq c\sqrt{\mathcal{E}(u, u)} + c'\|u - u_0\|_2 + c'\|u\|_2 \leq c(1 + c')\sqrt{\mathcal{E}(u, u)} + c'\|u\|_2.$$

$\square$

Proof of Lemma 5.2.7. Let $b = D$ or N. By Proposition B.2.4,

$$\mathcal{E}_*(T_t^b \varphi - \varphi, T_t^b \varphi - \varphi) = \mathcal{E}(T_t^b \varphi - \varphi, T_t^b \varphi - \varphi) + \|T_t^b \varphi - \varphi\|_2^2 \to 0$$

as $t \downarrow 0$. Hence, by Lemma 5.2.8, it follows that $\|T_t^b \varphi - \varphi\|_\infty \to 0$ as $t \to 0$. $\square$

By Lemma 5.2.7, $T_t^b \varphi$ is a L^∞-solution of the heat equation on $[0, \infty)$ under the corresponding assumption on φ in Lemma 5.2.7. Applying the parabolic maximum principle to those solutions, we can obtain non-negativity of the heat kernels without using Theorem B.3.4.

Theorem 5.2.9. *For any* $(t, x, y) \in (0, \infty) \times K \times K$,

$$0 \le p_D(t, x, y) \le p_N(t, x, y).$$

In the following, we will prove this theorem assuming that $(D, \mathbf{r})$ is regular. For the general case, the fact that $p_D(t, x, y) \le p_N(t, x, y)$ can be shown by considering the Markov processes associated with $p_D(t, x, y)$ and $p_N(t, x, y)$. See A.3 for details.

Proof. First we show that $p_D(t, x, y) \ge 0$. Suppose that $p_D(s, a, b) < 0$ for some (s, a, b). As p_D is continuous, we can choose ψ_p^m so that $p_D(s, a, b)\psi_p^m(b) < 0$ and $p_D(s, a, y)\psi_p^m(y) \le 0$ for any $y \in K$. Define $u(t, x) = (T_t^D \psi_p^m)(x)$ for any $(t, x) \in (0, \infty) \times K$ and $u(0, x) = \psi_p^m(x)$ for any $x \in K$. Then, as $\psi_p^m \in \mathcal{F}_0$, Lemma 5.2.7 implies that u is an L^∞-solution of the heat equation on $[0, \infty)$. Note that $u|_{\partial U_T} \ge 0$ for any $T > 0$. Hence, by the parabolic maximum principle, we see that $u(t, x) \ge 0$ for any $(t, x) \in [0, \infty) \times K$. This contradicts the fact that $u(s, a) = \int_K p_D(s, a, y)\psi_p^m(y)\mu(dy) < 0$. Therefore, $p_D(t, x, y) \ge 0$ for any $(t, x, y) \in (0, \infty) \times K \times K$.

Next let $\varphi \in \mathcal{F}$. Define $u(t, x) = (T_t^N \varphi)(x)$ for $(t, x) \in (0, \infty) \times K$ and $u(0, x) = \varphi(x)$ for any $x \in K$. Then u is an L^∞-solution of the heat equation on $[0, \infty)$. Moreover, by Corollary 5.1.9, $(du_t)_p = 0$ for any $(t, p) \in (0, \infty) \times V_0$. Hence $u_t \in \mathcal{D}_{N,\mu}$ for any $t > 0$. Therefore, we can apply Theorem 5.2.2 to $u(t, x)$ and obtain that $u(t, x) \ge 0$ for any $(t, x) \in [0, \infty) \times K$ if $\varphi(x) \ge 0$ for any $x \in K$. Using entirely the same argument as in proving non-negativity of $p_D(t, x, y)$, we can verify that $p_N(t, x, y) \ge 0$ for any $(t, x, y) \in (0, \infty) \times K \times K$.

Finally, we show that $p_D(t, x, y) \le p_N(t, x, y)$ for any $(t, x, y) \in (0, \infty) \times K \times K$. Let $\varphi \in \mathcal{F}_0$. Define $v(t, x) = (T_t^N \varphi)(x) - (T_t^D \varphi)(x)$ for any $(t, x) \in [0, \infty) \times K$ and $v(0, x) = 0$ for any $x \in K$. Since

$$v(x, t) = \int_K (p_N(t, x, y) - p_D(t, x, y))\varphi(y)\mu(dy),$$

v is an L^∞-solution of the heat equation on $[0, \infty)$. Suppose that $\varphi(x) \ge 0$ for any $x \in K$. Then the above argument implies that $T_t^N \varphi \ge 0$. Since $T_t^D \varphi|_{(0,T) \times V_0} \equiv 0$, it follows that $v|_{\partial U_T} \ge 0$ for any $T > 0$. Hence, by the parabolic maximum principle, $v(t, x) \ge 0$ for any $(t, x) \in [0, \infty) \times$

K. Taking this fact into account, the same argument as in proving non-negativity of $p_D(t, x, y)$ implies that $p_N(t, x, y) - p_D(t, x, y) \geq 0$ for any $(t, x, y) \in (0, \infty) \times K \times K$. □

Now we are ready to give a proof of Proposition 5.2.6.

Lemma 5.2.10. *For any* $(t, x) \in (0, \infty) \times K$,

$$||p_D^{t,x}||_1 \leq ||p_N^{t,x}||_1 = 1.$$

Proof. Theorem 5.2.9 implies that $||p_D^{t,x}||_1 \leq ||p_N^{t,x}||_1 = \int_K p_N(t, x, y)\mu(dy)$. Since $\{\varphi_n^N\}_{n \geq 1}$ is an orthonormal system of $L^2(K, \mu)$ and $\varphi_1^N \equiv 1$, we see that $\int_K \varphi_n^N d\mu = 0$ for $n \geq 2$. Therefore, by the definition of $p_N(t, x, y)$,

$$\int_K p_N(t, x, y)\mu(dy) = \sum_{n \geq 1} e^{-\lambda_n^N t} \varphi_n^N(x) \int_K \varphi_n^N(y)\mu(dy) = 1.$$

□

Remark. As in Theorem 5.2.9, this lemma holds even if $(D, \mathbf{r})$ is not a regular harmonic structure.

Proof of Proposition 5.2.6. Let $\varphi \in C(K, d)$ and assume that $\varphi|_{V_0} \equiv 0$. As $\mathcal{F}_0$ is dense in $\{u \in C(K, d) : u|_{V_0} \equiv 0\}$ with respect to the supremum norm, there exists $\{u_n\}_{n \geq 1} \subset \mathcal{F}_0$ such that $||u_n - \varphi||_\infty \to 0$ as $n \to \infty$. Note that

$$||T_t^D \varphi - \varphi||_\infty \leq ||T_t^D(\varphi - u_n)||_\infty + ||T_t^D u_n - u_n||_\infty + ||u_n - \varphi||_\infty. \tag{5.2.5}$$

On the other hand, by Lemma 5.2.10,

$$|(T_t^D(u_n - \varphi))(x)| \leq ||p_D^{t,x}||_1 ||u_n - \varphi||_\infty \leq ||u_n - \varphi||_\infty.$$

Hence $||T_t^D(u_n - \varphi)||_\infty \leq ||u_n - \varphi||_\infty$. Combining this with (5.2.5), we obtain

$$||T_t^D \varphi - \varphi||_\infty \leq 2||u_n - \varphi||_\infty + ||T_t^D u_n - u_n||_\infty.$$

By Lemma 5.2.7, it follows that $||T_t^D u_n - u_n||_\infty \to 0$ as $t \to 0$. Therefore, using a routine argument, it follows that $||T_t^D \varphi - \varphi||_\infty \to 0$ as $t \to 0$.

We can repeat exactly the same argument in the Neumann case. □

Remark. The key element of the above proof is that $||T_t^b \varphi||_\infty \leq ||\varphi||_\infty$ for any $\varphi \in C(K, d)$. Even if $(D, \mathbf{r})$ is not a regular harmonic structure, this follows from Proposition B.3.6, which is a consequence of Theorem B.3.4.

Using Theorem 5.2.9, we obtain the following corollary of Theorem 5.2.5.

Corollary 5.2.11. *Let $(D, \mathbf{r})$ be regular. For any $u \in C(K, d)$, $T_t^D u$ converges to u uniformly on compacts in $K \backslash V_0$ as $t \downarrow 0$.*

To prove this corollary, we need the following lemma.

Lemma 5.2.12. *Let U be a closed subset of (K, d) and let $u \in C(K, d)$. If $\operatorname{supp}(u) \cap U = \emptyset$, then $T_t^D u$ converges to 0 uniformly on U as $t \downarrow 0$.*

Proof. First assume that u is non-negative on K. Then, by Theorem 5.2.9, $(T_t^N u)(x) \geq (T_t^D u)(x) \geq 0$ for any $x \in K$. Theorem 5.2.5 implies that $T_t^N u \to 0$ uniformly on U as $t \downarrow 0$. Therefore, $T_t^D u \to 0$ uniformly on U as $t \downarrow 0$. For general u, let $u_+(x) = \max\{u(x), 0\}$ and $u_-(x) = -\min\{u(x), 0\}$ for any $x \in K$. Then $u = u_+ - u_-$ and $\operatorname{supp}(u_+) \cap U = \operatorname{supp}(u_-) \cap U = \emptyset$. So the general case follows on applying the result on non-negative u to u_+ and u_-. $\qquad\square$

Proof of Corollary 5.2.11. Let U be any compact subset of K contained in $K \backslash V_0$. Define $u_m = \sum_{x \in V_0} u(x) \psi_x^m$. Then $\operatorname{supp}(u_m) \cap U = \emptyset$ for sufficiently large m. By Lemma 5.2.12, $T_t^D u_m \to 0$ uniformly on U as $t \downarrow 0$. Now $T_t^D u = T_t^D(u - u_m) + T_t^D u_m$. By Theorem 5.2.5, $T_t^D(u - u_m) \to u - u_m$ uniformly on K as $t \downarrow 0$. Since $u_m(x) = 0$ on U, we see that $T_t^D u \to u$ uniformly on U as $t \downarrow 0$. $\qquad\square$

Using Corollary 5.2.11, we obtain an explicit expression for the difference $p_N(t, x, y) - p_D(t, x, y)$, which was shown to be non-negative in Theorem 5.2.9. Recall that $p_D^{t,x}(y) = p_D(t, x, y)$. Note that $p_D(t, x, y) \geq 0$ and $p_D(t, x, y) = 0$ if $x \in V_0$ or $y \in V_0$. Hence it follows that $(dp_D^{t,x})_q \leq 0$ for any $q \in V_0$.

Theorem 5.2.13. *Let $(D, \mathbf{r})$ be regular.*
(1) *For any $x, y \in K \backslash V_0$ and $t > 0$,*

$$p_N(t, x, y) - p_D(t, x, y) = -\sum_{q \in V_0} \int_0^t p_N(t - s, x, q)(dp_D^{s,y})_q \, ds,$$

where $\int_0^t p_N(t - s, x, q)(dp_D^{s,y})_q \, ds$ means, in fact, the following limit:

$$\lim_{\epsilon \to 0} \int_\epsilon^{t-\epsilon} p_N(t - s, x, q)(dp_D^{s,y})_q \, ds.$$

(2) *For any $q \in V_0$, any $x \in K \backslash V_0$ and any $t > 0$,*

$$\int_K p_D(t, x, y) \psi_q(y) \mu(dy) = \psi_q(x) - \left(-\int_0^t (dp_D^{s,x})_q \, ds \right),$$

where $\int_0^t (dp_D^{s,x})_q \, ds$ means $\lim_{\epsilon \to 0} \int_\epsilon^t (dp_D^{s,x})_q \, ds$.

Proof. First we prove (1). Using the Gauss–Green's formula, we obtain

$$(T_\epsilon^N(p_D^{t-\epsilon,y}))(x) - (T_\epsilon^D(p_N^{t-\epsilon,x}))(y)$$

$$= \Big[\int_K p_N(t-s,x,z)p_D(s,z,y)\mu(dz) \Big]_\epsilon^{t-\epsilon}$$

$$= \int_\epsilon^{t-\epsilon} \Big(\int_K \big(p_N(t-s,x,z)\frac{\partial p_D}{\partial t}(s,z,y)$$

$$- \frac{\partial p_N}{\partial t}(t-s,x,z)p_D(s,z,y)\big)\mu(dz)\big)ds$$

$$= \sum_{q\in V_0} \int_\epsilon^{t-\epsilon} p_N(t-s,x,q)(dp_D^{s,y})_q ds.$$

Applying Theorem 5.2.5 and Corollary 5.2.11,

$$(T_\epsilon^N(p_D^{t-\epsilon,y}))(x) - (T_\epsilon^D(p_N^{t-\epsilon,x}))(y) \to p_D(t,x,y) - p_N(t,x,y)$$

as $\epsilon \downarrow 0$. Thus we have shown (1).

For (2), by the Gauss–Green's formula,

$$\int_K \psi_q(z)(\Delta_\mu p_D^{s,x})(z)\mu(dz) = (dp_D^{s,x})_q.$$

By integrating this,

$$-\int_\epsilon^t (dp_D^{s,x})_q = (T_\epsilon^D\psi_q)(x) - (T_t^D\psi_q)(x).$$

Applying Corollary 5.2.11, we can immediately obtain (2). □

Summing the equalities in Theorem 5.2.13-(2) for all $q \in V_0$, we obtain the following corollary.

Corollary 5.2.14. *Let $(D,\mathbf{r})$ be regular. Then for any $x \in K\backslash V_0$ and any $t > 0$,*

$$\|p_D^{t,x}\|_1 = \int_K p_D(t,x,y)\mu(dy) = 1 - \big(-\sum_{q\in V_0}\int_0^t (dp_D^{s,x})_q ds\big).$$

5.3 Asymptotic behavior of the heat kernels

In this section, we will study asymptotic behavior of the heat kernel $p_b(t,x,y)$ as $t \downarrow 0$. Recall that ν is the self-similar measure with weight $((\mu_i r_i)^{d_S/2})_{i\in S}$, where d_S is the spectral exponent defined in (4.1.2). Let $\overline{\delta} = \overline{\delta}(\nu,\mu)$ and $\underline{\delta} = \underline{\delta}(\nu,\mu)$, where $\overline{\delta}(\nu,\mu)$ and $\underline{\delta}(\nu,\mu)$ are defined in Proposition 4.5.2.

Theorem 5.3.1. (1)   *For $b = D, N$, there exists $c_1 > 0$ such that*

$$\sup_{x,y \in K} p_b(t, x, y) \leq c_1 t^{-d_S \overline{\delta}/2} \tag{5.3.1}$$

for any $t \in (0, 1]$.
(2)   *There exists $c_2 > 0$ such that*

$$c_2 t^{-d_S \underline{\delta}/2} \leq p_N(t, x, x) \tag{5.3.2}$$

for any $t \in (0, 1]$ and any $x \in K$.

The following corollary is immediate.

Corollary 5.3.2. *If $\mu_i = \nu_i$ for any $i \in S$, then there exist positive constants c_1 and c_2 such that*

$$c_2 t^{-d_S/2} \leq p_N(t, x, x) \leq c_1 t^{-d_S/2}$$

for any $t \in (0, 1]$ and $x \in K$.

Since $\nu_i = (\mu_i r_i)^{d_S/2}$, we see that $\mu_i = \nu_i$ for any $i \in S$ if and only if the harmonic structure $(D, \mathbf{r})$ is regular and $\mu_i = (r_i)^{d_H}$ for any $i \in S$, where d_H is defined by (4.2.1). By Theorem 4.2.1, $d_S = 2d_H/(d_H + 1)$. Also, in such a case, the spectral dimension of (K, R), $d_S(K, R)$ equals $d_S(\mu)$. See (4.2.3).

First, we prove the upper bound (Theorem 5.3.1-(1)) through the following Nash inequality (5.3.3).

Theorem 5.3.3. *Let $\theta = d_S \overline{\delta}$. Then there exists $c > 0$ such that*

$$\|u\|_2^{2+4/\theta} \leq c(\mathcal{E}(u, u) + \|u\|_2^2)\|u\|_1^{4/\theta} \tag{5.3.3}$$

for any $f \in \mathcal{F}$.

Lemma 5.3.4. *There exists $c > 0$ such that*

$$\|u\|_2^2 \leq c(\mathcal{E}(u, u) + \|u\|_1^2)$$

for any $u \in \mathcal{F}$.

Proof. Letting $m = 0$ in (3.4.2), we see that there exists $c_1 > 0$ such that

$$\|u - u_0\|_2 \leq c_1 \sqrt{\mathcal{E}(u - u_0, u - u_0)}$$

for any $u \in \mathcal{F}$, where $u_0 = \sum_{p \in V_0} u(p)\psi_p$. Also, since the dimension of the space of harmonic functions is finite, there exists $c_2 > 0$ such that $\|u_0\|_2 \leq c_2 \|u_0\|_1$ for any $u \in \mathcal{F}$. Also note that $\|u\|_1 \leq \|u\|_2$ for any

$u \in L^2(K, \mu)$ and that $\mathcal{E}(u, u) = \mathcal{E}(u - u_0, u - u_0) + \mathcal{E}(u_0, u_0)$. Combining those inequalities, we obtain

$$\begin{aligned}
||u||_2 &\leq c_2||u_0||_1 + c_1\sqrt{\mathcal{E}(u - u_0, u - u_0)} \\
&\leq c_2(||u||_1 + ||u - u_0||_1) + c_1\sqrt{\mathcal{E}(u - u_0, u - u_0)} \\
&\leq c_2||u||_1 + (c_1 c_2 + c_1)\sqrt{\mathcal{E}(u - u_0, u - u_0)} \\
&\leq c_3||u||_1 + c_3\sqrt{\mathcal{E}(u, u)},
\end{aligned}$$

where $c_3 = \max\{c_2, c_1 c_2 + c_1\}$. This implies the desired inequality. $\square$

Proof of Theorem 5.3.3. Let Λ be a partition of Σ. Then, Lemma 5.3.4 along with the self-similarity of μ and $\mathcal{E}$ implies that

$$\begin{aligned}
||u||_2^2 &= \sum_{w \in \Lambda} \mu_w \int_K |u \circ F_w|^2 d\mu \\
&\leq c \sum_{w \in \Lambda} \mu_w (\mathcal{E}(u \circ F_w, u \circ F_w) + ||u \circ F_w||_1^2) \\
&\leq c \sum_{w \in \Lambda} (\mu_w r_w) r_w^{-1} \mathcal{E}(u \circ F_w, u \circ F_w) + c \sum_{w \in \Lambda} \mu_w^{-1}\Big(\mu_w \int_K |u \circ F_w| d\mu\Big)^2 \\
&\leq c(\max_{w \in \Lambda} \mu_w r_w)\mathcal{E}(u, u) + c(\min_{w \in \Lambda} \mu_w)^{-1}||u||_1^2.
\end{aligned}$$

Let $\Lambda = \Lambda(\nu, \lambda)$, which is defined in Definition 4.5.1. Then, by Proposition 4.5.2,

$$||u||_2^2 \leq c'\lambda^{2/d_s}\mathcal{E}(u, u) + c'\lambda^{-\overline{\delta}}||u||_1^2 \qquad (5.3.4)$$

for any $u \in \mathcal{F}$ and any $\lambda \in (0, 1)$.

Now, suppose that $\mathcal{E}(u, u) > ||u||_1^2$ and choose λ so that $\lambda^{2/d_s + \overline{\delta}} = ||u||_1^2/\mathcal{E}(u, u)$. Then, by (5.3.4),

$$||u||_2^{2+4/\theta} \leq c''\mathcal{E}(u, u)||u||_1^{4/\theta}$$

for any $u \in \mathcal{F}$. This implies (5.3.3).

If $\mathcal{E}(u, u) \leq ||u||_1^2$, then, by Lemma 5.3.4, it follows that

$$||u||_2^2 \leq c(\mathcal{E}(u, u) + ||u||_1^2) \leq 2c||u||_1^2.$$

Obviously this suffices for proving (5.3.3) in this case. $\square$

Proof of Theorem 5.3.1-(1). By Theorems 5.3.3 and B.3.7, we know that there exists $c > 0$ such that $||T_t^b||_{1 \to \infty} \leq ct^{-\theta/2}$ for $t \in (0, 1]$. Since $||T_t^b||_{1 \to \infty} = \sup_{x, y \in K} p_b(t, x, y)$, we can immediately verify (5.3.1). $\square$

Now we show Proposition 5.1.10-(2). First, recall that $g^{(m)}(x, y)$ is a continuous function for $m > d_S \overline{\delta}/2$ by Corollary 3.6.6.

Lemma 5.3.5. *Let* $m > d_S \bar{\delta}/2$. *Then, for any* $x, y \in K$,

$$g^{(m)}(x, y) = \frac{1}{\Gamma(m)} \int_0^\infty t^{m-1} p_D(t, x, y) dt,$$

where Γ *is the gamma function defined by* $\Gamma(s) = \int_0^\infty x^{s-1} e^{-x} dx$.

Proof. Note that $\lambda_1^D > 0$. Applying Lemmas 5.1.3 and 5.1.4 to the infinite sum $\sum_{n \geq 1} e^{-(\lambda_n^D - \lambda_1^D)t} \varphi_n^D(x) \varphi_n^D(y)$, we see that

$$\sup_{(t,x,y) \in [1,\infty) \times K \times K} e^{\lambda_1^D t} p_D(t, x, y) < \infty.$$

Moreover, by Theorem 5.3.1-(1), $\sup_{x,y \in K} t^{m-1} p_D(t, x, y) \leq ct^{-d_S \bar{\delta}/2 - 1 + m}$ for $t \in (0, 1]$. Therefore, if

$$F(t) = c \begin{cases} t^{m-1} e^{-\lambda_1^D t} & \text{for } t > 1, \\ t^{-d_S \bar{\delta}/2 - 1 + m} & \text{for } 0 < t \leq 1, \end{cases}$$

for sufficiently large $c > 0$, then $\int_0^\infty F(t) dt < \infty$ and $t^{m-1} p_D(t, x, y) \leq F(t)$ for any $(t, x, y) \in (0, \infty) \times K \times K$. If $h(x, y) = \Gamma(m)^{-1} \int_0^\infty t^{m-1} p_D(t, x, y) dt$, then Lebesgue's convergence theorem implies that $h(x, y)$ is continuous on $K \times K$. Now, by Fubini's theorem and the definition of $g^{(m)}$,

$$\int_K h(x, y) \varphi_n^D(y) \mu(dy) = \varphi_n^D / (\lambda_n^D)^m = \int_K g^{(m)}(x, y) \varphi_n^D(y) \mu(dy)$$

for any $n \geq 1$. As both $g^{(m)}$ and h are continuous on $K \times K$, it follows that $g^{(m)}(x, y) = h(x, y)$ for any $x, y \in K$.   □

Proof of Proposition 5.1.10-(2). Assume that x and y belong to the same connected component of $K \backslash V_0$. Let $m > d_S \bar{\delta}/2$. Then, by Corollary 3.6.7, $g^{(m)}(x, y) > 0$. Hence, Lemma 5.3.5 implies that $p_D(t, x, y) > 0$ for some $t > 0$, and in particular that $p_D(t, x, x) > 0$. This means that $\varphi_n^D(x)^2 > 0$ for some $n \geq 1$. Therefore we have that $p_D(t, x, x) > 0$ for any $t > 0$. We have therefore proved the following.

(i) $p_D(t, x, y) > 0$ for some $t > 0$.

(ii) $p_D(t, x, x) > 0$ for any $(t, x) \in (0, \infty) \times K \backslash V_0$.

Using these, we can repeat the same argument as in the proof of the Neumann case after Step 2 to deduce the desired conclusion.   □

Next we prove the lower estimate, Theorem 5.3.1-(2).

Definition 5.3.6. *Let* Λ *be a partition of* Σ *and let* A *be a subset of* K. *Write*

(1) $\Lambda_A = \{w : w \in \Lambda, K_w \cap A \neq \emptyset\}$.

(2) $N_\Lambda(A) = \cup_{w \in \Lambda_A} K_w$.

(3) $\tilde{\Lambda}_A = \Lambda_{N_\Lambda(A)} \backslash \Lambda_A$.

(4) $U_\Lambda(A) = N_\Lambda(N_\Lambda(A))$.

(5) $\partial U_\Lambda(A) = U_\Lambda(A) \cap \overline{K \backslash U_\Lambda(A)}$.

For $x \in K$, we use Λ_x, $N_\Lambda(x)$ and so on instead of $\Lambda_{\{x\}}$, $N_\Lambda(\{x\})$, etc.

Now let us choose m so that $\{K_{m,p}\}_{p \in V_0}$ are mutually disjoint. (Note that $K_{m,p} = N_{W_m}(p)$.) Set $K_0 = \overline{K \backslash \cup_{p \in V_0} K_{m,p}}$. Then K_0 is a non-empty compact subset of K. Also define ∂K_0 by $\partial K_0 = K_0 \cap (\cup_{p \in V_0} K_{m,p})$. Note that if $p, q \in V_0$ and $p \neq q$, then p and q belong to different connected components of $K \backslash \partial K_0$.

Lemma 5.3.7. *For $t > 0$, define*

$$q(t) = \inf_{x \in K_0} \int_K p_D(t, x, y) \mu(dy).$$

Then $q(t) > 0$ for any $t > 0$ and $q(\cdot)$ is monotonically non-increasing.

Proof. Let $u(t, x) = \int_K p_D(t, x, y) \mu(dy)$. Then $u(t, x)$ is continuous on $(0, \infty) \times K$. By Proposition 5.1.10, we also see that $u(t, x) > 0$ for any $x \in K_0$. Since K_0 is compact, it follows that $q(t) = \inf_{x \in K_0} u(t, x) > 0$.

Since $T_{t-s}^D p_D^{s,x} = p_D^{t,x}$ for $t > s$, Proposition B.3.6 implies that $u(s, x) = \|p_D^{s,x}\|_1 \geq \|p_D^{t,x}\|_1 = u(t, x)$. (This can be deduced from Lemma 5.2.10 as well.) Hence, $u(t, x)$ is a monotonically non-increasing function of t. Therefore $q(\cdot)$ is monotonically non-increasing. $\square$

To prove Theorem 5.3.1-(2), we also need results from A.2, where we study general boundary conditions. Let B be a finite subset of V_*. By Theorem A.2.1, if

$$\mathcal{F}_B = \{u \in \mathcal{F} : u|_B = 0\},$$

then $(\mathcal{E}, \mathcal{F}_B)$ is a local regular Dirichlet form on $L^2(K \backslash B, \mu)$. As $\mu(B) = 0$, $L^2(K \backslash B, \mu)$ can be identified with $L^2(K, \mu)$. If $B = \emptyset$, then $\mathcal{F}_B = \mathcal{F}$ and this is just Theorem 3.4.6. Also, if $B = V_0$, then this is Corollary 3.4.7. Exactly the same way as for $p_b(t, x, y)$, we can define the heat kernel associated with $(\mathcal{E}, \mathcal{F}_B)$ and μ, which is denoted by $p_B(t, x, y)$. See Definition A.2.12 for details. For example, the Neumann heat kernel $p_N(t, x, y)$ coincides with $p_\emptyset(t, x, y)$ and the Dirichlet heat kernel coincides with $p_{V_0}(t, x, y)$.

Let Λ be a partition of Σ. We define $p_\Lambda(t, x, y) = p_{V(\Lambda)}(t, x, y)$.

Lemma 5.3.8.

$$p_\Lambda(t, x, y) = \begin{cases} \frac{1}{\mu_w} p_D(t/(\mu_w r_w), F_w{}^{-1}(x), F_w{}^{-1}(y)) & \text{if } \{w\} = \Lambda_x = \Lambda_y, \\ 0 & \text{otherwise.} \end{cases}$$

Proof. Let H_Λ be the non-negative self-adjoint operator on $L^2(K, \mu)$ associated with the closed form $(\mathcal{E}, \mathcal{F}_B)$, where $B = V(\Lambda)$. Then, the same argument as in the proof of Lemma 4.1.9 implies that $H_\Lambda u = \lambda u$ if and only if $H_D(u \circ F_w) = \mu_w r_w \lambda(u \circ F_w)$ for any $w \in \Lambda$. Comparing Definition A.2.12 with (5.1.1), we immediately obtain this lemma. $\qquad\square$

For $x \in K$, we define $p_{\Lambda,x}(t, z, y) = p_{\partial U_\Lambda(x)}(t, z, y)$.

Lemma 5.3.9.

$$\int_{U_\Lambda(x)} p_{\Lambda,x}(t, x, y)\mu(dy) \geq \min_{w \in \tilde{\Lambda}_x} \inf_{z \in F_w(K_0)} \int_K p_\Lambda(t, z, y)\mu(dy).$$

We will give an analytical proof of this lemma. However, so far, the proof works only when the harmonic structure is regular. Using probabilistic methods, we may prove this lemma even if the harmonic structure is not regular. See A.3 for details.

Proof. Assume that $(D, \mathbf{r})$ is regular. Let $A = \cup_{w \in \tilde{\Lambda}_x} F_w(\partial K_0)$ and let $B = A \cup \partial U_\Lambda(x)$. Then for any $q \in \partial U_\Lambda(x)$, x and q belong to different connected components of $K \backslash A$. Hence, by Theorem A.2.19, there exists a neighborhood, U, of $q \in \partial U_\Lambda(x)$ such that $p_B(t, x, y) = 0$ for any $y \in U$. This implies that $(dp_B^{s,x})_q = 0$ for any $q \in \partial U_\Lambda(x)$ and any $s > 0$.

Since $\Lambda \supseteq \partial U_\Lambda(x)$, by Theorem A.2.16,

$$p_\Lambda(t, z, y) \leq p_{\Lambda,x}(t, z, y)$$

for any $t > 0$ and any $z, y \in K$. Also note that, for any $z \in K$, $\|p_\Lambda^{t,z}\|_1$ is monotonically decreasing and is no greater that 1 for any $t > 0$. Combining those facts with Theorem A.2.18 and (A.2.10), we have the following assertion.

$$\int_K p_{\Lambda,x}(t,x,y)\mu(dy)$$

$$= \int_K p_B(t,x,y)\mu(dy) + \int_K (p_{\Lambda,x}(t,x,y) - p_B(t,x,y))\mu(dy)$$

$$= 1 - \sum_{q \in B} \int_0^t (-dp_B^{s,x})_q ds + \sum_{q \in A} \int_0^t \int_K p_{\Lambda,x}(t-s,q,y)\mu(dy)(-dp_B^{s,x})_q ds$$

$$\geq 1 - \sum_{q \in A} \int_0^t (-dp_B^{s,x})_q ds + \sum_{q \in A} \int_0^t \int_K p_\Lambda(t-s,q,y)\mu(dy)(-dp_B^{s,x})_q ds$$

$$\geq 1 + \sum_{q \in A} \Big(\int_K p_\Lambda(t,q,y)\mu(dy) - 1 \Big) \int_0^t (-dp_B^{s,x})_q ds$$

$$\geq 1 + (\min_{q \in A} \int_K p_\Lambda(t,q,y)\mu(dy) - 1) \sum_{q \in A} \int_0^t (-dp_B^{s,x})_q ds.$$

By (A.2.10), $0 \leq \sum_{q \in A} \int_0^t (-dp_B^{s,x})_q ds \leq 1$. Hence

$$\int_K p_{\Lambda,x}(t,x,y)\mu(dy) \geq \min_{q \in A} \int_K p_\Lambda(t,q,y)\mu(dy).$$

Since $p_{\Lambda,x}(t,x,y) = 0$ if $y \notin U_\Lambda(x)$, we obtain the desired inequality. □

Proposition 5.3.10. *There exists $c > 0$ such that*

$$q\Big(\frac{t}{\min_{w \in \bar{\Lambda}_x} \mu_w r_w}\Big)^2 \leq c\Big(\max_{w \in \Lambda(N_\Lambda(x))} \mu_w\Big)p_N(2t,x,x) \qquad (5.3.5)$$

for any $x \in K$ and any partition Λ.

Proof. By Lemma 5.3.8,

$$\int_K p_\Lambda(t,z,y)\mu(dy) = \int_{K_w} (\mu_w)^{-1} p_D(t/(\mu_w r_w), F_w^{-1}(z), F_w^{-1}(y))\mu(dy)$$

$$= \int_K p_D(t/(\mu_w r_w), F_w^{-1}(z), y)\mu(dy) \geq q(t/(\mu_w r_w))$$

for any $z \in F_w(K_0)$. Also, by Theorem A.2.16, it follows that

$$\int_{U_\Lambda(x)} p_N(t,x,y)\mu(dy) \geq \int_{U_\Lambda(x)} p_{\Lambda,x}(t,x,y)\mu(dy).$$

Therefore, Lemma 5.3.9 along with Lemma 5.3.7 implies that

$$\int_{U_\Lambda(x)} p_N(t,x,y)\mu(dy) \geq \min_{w \in \bar{\Lambda}_x} q(t/(\mu_w r_w)) \geq q\Big(\frac{t}{\min_{w \in \bar{\Lambda}_x} \mu_w r_w}\Big). \qquad (5.3.6)$$

On the other hand, by Schwarz's inequality, it follows that

$$\Big(\int_{U_\Lambda(x)} p_N(t,x,y)\mu(dy)\Big)^2 \le \mu(U_\Lambda(x))\int_K p(t,x,y)^2\mu(dy).$$

Now by a similar argument to the one in the proof of Lemma 4.2.3, we can see that there exists $c > 0$ such that $\#(\Lambda_{N_\Lambda(x)}) \le c$ for any $x \in K$ and any partition Λ. Also, by Proposition 5.1.2-(5), we obtain that $\int_K p_N(t,x,y)^2\mu(dy) = p_N(2t,x,x)$. Combining those facts with (5.3.6), we obtain (5.3.5). $\qquad\square$

Proof of Theorem 5.3.1-(2). Since $q(\cdot)$ is monotonically non-increasing, Proposition 5.3.10 implies that

$$q\Big(\frac{t}{\min_{w\in\Lambda}\mu_w r_w}\Big)^2 \le c(\max_{w\in\Lambda}\mu_w)p(2t,x,x)$$

for any $x \in K$ and any partition Λ. Let $\Lambda = \Lambda(\nu,\lambda)$ and let $t = \min_{w\in\Lambda}\mu_w r_w/2$. Then we have

$$q(1/2)^2 \le c\underline{h}(\lambda:\nu,\mu)p_N\Big(\min_{w\in\Lambda(\nu,\lambda)}\mu_w r_w,x,x\Big)$$

for any $x \in K$ and any $\lambda > 0$. Using Proposition 4.5.2, we see that there exists $c' > 0$ such that $c' \le s^{d_S\underline{\ell}/2}p_N(s,x,x)$ for any $x \in K$ and any $s > 0$. $\qquad\square$

Notes and references

In [20], Barlow & Perkins studied the heat kernel associated with Brownian motion on the Sierpinski gasket, which is the Neumann heat kernel associated with the standard Laplacian in our context. They obtained a uniform off-diagonal (Aronson-type) estimate,

$$c_1 t^{-d_S/2}\exp\big(-c_2(|x-y|^{d_w}/t)^{1/(d_w-1)}\big) \le p_N(t,x,y)$$
$$\le c_3 t^{-d_S/2}\exp\big(-c_4(|x-y|^{d_w}/t)^{1/(d_w-1)}\big),$$

where we think that the Sierpinski gasket is embedded in $\mathbb{R}^2$, $d_S = \log 9/\log 5$ is the spectral dimension and $d_w = \log 5/\log 2$ is the walk dimension. This result was extended to nested fractals by Kumagai in [99] and then to affine nested fractals by Fitzsimmons, Hambly & Kumagai in [37] as follows. Let K be an affine nested fractal. Let $(D,\mathbf{r})$ be the harmonic structure on K obtained in Theorem 3.8.10. Suppose that $(D,\mathbf{r})$ is regular. Also let $\mu = \mu_*$ be the self-similar measure defined in Theorem 4.2.1. Recall that $(r_i\mu_i)^{d_S/2} = \mu_i$ and that $d_S = 2d_H/(d_H+1)$, where

$d_H = \dim_H(K, R)$. Let $p_N(t, x, y)$ be the Neumann heat kernel associated with $(D, \mathbf{r})$ and μ. Then there exist constants c_i such that

$$c_1 t^{-d_S/2} \exp\left(-c_2\left(\frac{R(x, y)^{d_H+1}}{t}\right)^{1/(d_w-1)}\right) \le p_N(t, x, y)$$
$$\le c_3 t^{-d_S/2} \exp\left(-c_4\left(\frac{R(x, y)^{d_H+1}}{t}\right)^{1/(d_w-1)}\right)$$

for any $(t, x, y) \in (0, 1] \times K \times K$, where d_w is the walk dimension with respect to the shortest path metric. See [99] and [37] for the definition of the shortest path metric. The Aronson-type estimate above is obtained by a probabilistic approach. In [6], one can find a good presentation of the probabilistic methods including the proof of the estimate above.

In [60], Hambly and Kumagai have shown that a uniform off-diagonal estimate as above fails without the strong symmetry that affine nested fractals possess. Moreover, if $\mu \ne \mu_*$, then $\lim_{t\downarrow 0} \log p_N(t, x, x)/\log t$ depends sensitively on x and exhibits multi-fractal nature. See [19] and [58]. For example, by those results, it follows that the heat kernel estimates in Theorem 5.3.1 are best possible. Specifically, define

$$\underline{d}(x) = -2 \liminf_{t\downarrow 0} \frac{\log p_N(t, x, x)}{\log t}$$

for any $x \in K$. Replacing the limit infimum by the limit supremum, we also define $\overline{d}(x)$. Then

$$d_S \underline{\delta} = \inf_{x \in K} \overline{d}(x) \le \sup_{x \in K} \underline{d}(x) = d_S \overline{\delta}.$$

Note that Proposition 5.3.10 essentially suffices to show that $\sup_{x \in K} \underline{d}(x) = d_S \overline{\delta}$.

5.1   The expression of the heat kernel by using the eigenfunctions of the Laplacian has been known in many places. See, for example, [6, Theorem 3.44]. The proof of the positivity of the heat kernel, Proposition 5.1.10, is essentially based on that in [14].

5.3   The essential idea in proving the upper estimate of the heat kernel, Theorem 5.3.1-(1), is the use of the Nash inequality. This idea is originally used in [37] in proving the counterpart of Theorem 5.3.1-(1) for the symmetric Laplacian (corresponding to the harmonic structures obtained in Theorem 3.8.10) on an affine nested fractal. See also [6, Theorem 8.3].

Appendix A
Additional Facts

A.1 Second eigenvalue of A_i

Let $(D, \mathbf{r})$ be a harmonic structure on a connected p.c.f. self-similar structure $(K, S, \{F_i\}_{i \in S})$, where $S = \{1, 2, \ldots, N\}$ and $\mathbf{r} = (r_1, r_2, \ldots, r_N)$. In this section, we study the eigenvalues of A_i, which were defined in (3.2.2). Recall (3.2.3). We have

$$A_i u|_{V_0} = u|_{F_i(V_0)}$$

for any harmonic function u and any $i \in S$. So the matrices $\{A_i\}_{i \in S}$ completely determine the structure of harmonic functions. The following proposition follows easily from (3.2.3).

Proposition A.1.1. *If k is an eigenvalue of A_i, then $|k| \leq 1$. Moreover, if $|k| = 1$ then $k = 1$ and $A_i u = u$ if and only if u is a constant on V_0.*

We will focus on the second eigenvalue of A_i in the rest of this section. It has been suggested in Example 3.2.6 that the second eigenvalue of A_i equals the resistance scaling ratio r_i. We will show that this is true under some assumptions. Indeed, Strichartz [175] has proved that if $K \backslash V_0$ is connected, then this is true. In this section, we will give a complete answer in the general case.

By the proposition above, if we set

$$\lambda_i = \max\{|k| : |k| < 1, k \text{ is an eigenvalue of } A_i\},$$

then $\lambda_i < 1$.

For $U \subset K$, define

$$m(U) = \text{the number of connected components of } K \backslash U.$$

U is said to be a non-focal set if $m(U) = 1$. If $U = \{p\}$ for $p \in K$, we use $m(p)$ instead of $m(\{p\})$. We say that p is a focal point if $m(p) \geq 2$.

Now let p be the fixed point of F_i for some i. Then, by Proposition 1.6.7, $m(p) \le \#(V_0)$. Let $\{J_j\}_{j=1,\dots,m(p)}$ be the collection of all connected components of $K \backslash \{p\}$. Then, again by Proposition 1.6.7, there exists a permutation of $\{1, \dots, m(p)\}$, τ, such that $F_i(J_j) = K_i \cap J_{\tau(j)}$. Define $k(p)$ by $k(p) = \min\{k : \tau^k = e\}$, where e is the identity.

The following is the main result on the second eigenvalue of A_i, which includes information on corresponding eigenvectors and multiplicity of the eigenvalue as well.

Theorem A.1.2. *Let p be the fixed point of F_i.*
(1) *If $p \in V_0$, then $\lambda_i = r_i$ and r_i is an eigenvalue of A_i. There exists a non-negative (non-trivial) eigenvector of A_i belonging to the eigenvalue r_i. Also, $r_i^{k(p)}$ is the unique eigenvalue of $A_i^{k(p)}$ whose absolute value equals $\lambda_i^{k(p)}$ and*

$$\dim E(r_i^{k(p)}, A_i^{k(p)}) = m(p),$$

where $E(k, A) = \{u \in \ell(V_0) : Au = ku\}$ for any linear map $A : \ell(V_0) \to \ell(V_0)$.
(2) *Assume $p \notin V_0$. If p is non-focal point, then $\lambda_i < r_i$. If p is a focal point, then $\lambda_i = r_i$ and $\dim E(r_i^{k(p)}, A_i^{k(p)}) = m(p) - 1$.*

Remark. In the case (1) of this theorem, one may obtain that $\dim E(r_i, A_i)$ equals the number of orbits of the action of τ. Specifically, let τ be the permutation appearing in the definition of $k(p)$. We write $i \sim j$ for $i, j \in \{1, \dots, m(p)\}$ if and only if there exists $l \in \mathbb{N}$ such that $\tau^l(i) = j$. Then $\sim$ is an equivalence relation on $\{1, \dots, m(p)\}$. Each equivalence class of $\sim$ is called an orbit of the action of τ. So, what we mean is

$$\dim E(r_i, A_i) = \#(\{1, \dots, m(p)\}/\sim).$$

Corollary A.1.3. *Let p be the fixed point of F_i. If $p \in V_0$ and p is a non-focal point, then $\lambda_i = r_i$ and r_i is the unique eigenvalue of A_i whose absolute value is λ_i. Moreover, $\dim E(r_i, A_i) = 1$ and there exists $u \in E(r_i, A_i)$ that satisfies $u(q) > 0$ if $q \ne p$ and $u(p) = 0$.*

To prove the above theorem, we need several lemmas. In the following lemmas, we assume that $p = F_i(p) \in V_0$.

Lemma A.1.4. *For any $m \ge 0$ and any $u \in \ell(V_0)$,*

$$(D(A_i - kI)^m u)(p) = (r_i - k)^m (Du)(p),$$

where I is the identity matrix.

Proof. Let u be a harmonic function on K. By the definitions of H_m and A_i, $(H_m u)(p) = (r_i)^{-m}(D(A_i)^m u)(p)$. Also $(H_m u)(p) = (Du)(p) = -(du)_p$. Hence $r_i{}^m (Du)(p) = (D(A_i)^m u)(p)$. This immediately implies the lemma. □

By this lemma, we can immediately obtain

Lemma A.1.5. *If $A_i u = ku$ for $u \in \ell(V_0)$ and $(Du)(p) \neq 0$, then $k = r_i$.*

Lemma A.1.6. *Suppose that $A_i u = ku$ for $u \in \ell(V_0)$ and that $u(p) \neq 0$. Then u is a constant on V_0 and $k = 1$.*

Proof. As p is a fixed point of F_i, we have $(A_i u)(p) = u(p)$. Hence, by Proposition A.1.1, if $u(p) \neq 0$ then $k = 1$ and u is a constant on V_0. □

Recall the notion of an H_m-path defined in Definition 3.2.9.

Lemma A.1.7. *Define*

$$U_{m,x} = \{y \in V_0 : \text{there exists an } H_m\text{-path between } x \text{ and } y\}$$

for $x \in V_m$. Suppose $F_w(q) \in V_m \backslash V_0$ for $q \in V_0$ and $w \in W_m$. Then $(A_w)_{qq_*} > 0$ for $q_* \in V_0$ if and only if $q_* \in U_{m,F_w(q)}$.

Using Lemma 3.2.13, we see that if $x \in V_m \backslash V_0$, then $U_{m,x} = V_0 \cap \overline{C}$, where C is the connected component of $K \backslash V_0$ containing q.

Proof. Let C be the connected component of $K \backslash V_0$ containing $F_w(q)$. Then $U_{m,F_w(q)} = \overline{C} \cap V_0$. Hence, by Theorem 3.2.14, if u is a harmonic function, then $u(F_w(q))$ is determined by $u|_{U_{m,F_w(q)}}$. Since $(A_w u)(q) = u(F_w(q))$, we can verify the claim of this lemma. □

We also assume that $F_i(p) = p \in V_0$ in the following three lemmas.

Lemma A.1.8. *Set $U_p = \{q : q \in V_0, D_{pq} > 0\}$. Then, for sufficiently large m, $U_{m,q} \subseteq U_p \cup \{p\}$ for any $q \in F_i{}^m(V_0)$.*

Proof. Choose m so that $F_i{}^m(V_0) \cap V_0 = \{p\}$. Then $p \in U_{m,q}$ if $q \in F_i{}^m(V_0)$ and $q \neq p$. If $p_* \in U_{m,q}$ and $p_* \neq p$, then combining two H_m-paths between p_* and q and between q and p, we see that $p_* \in U_{m,p}$. By Theorem 3.2.11 and Lemma 3.2.13, it follows that $U_{m,p} \subseteq U_p$. □

By Theorem 3.2.11, we obtain the following.

Lemma A.1.9. *p is a non-focal point if and only if, for any $q_1, q_2 \in V_0 \backslash \{p\}$, there exists $\{p_i\}_{i=1,2,\ldots,m} \subset V_0 \backslash \{p\}$ such that $p_1 = q_1$, $p_m = q_2$ and $D_{p_i p_{i+1}} \neq 0$ for $i = 1, 2, \ldots, m-1$.*

Lemma A.1.10. *If p is a non-focal point, then for sufficiently large m, $U_{m,q} = U_p \cup \{p\}$ for any $q \in F_i{}^m(V_0) \backslash \{p\}$.*

Proof. Choose m so that $F_i{}^m(V_0) \cap V_0 = \{p\}$. If $q \in F_i{}^m(V_0)$ and $q \neq p$, then $q \in V_m \backslash V_0$. Hence, by Lemma 3.2.13, $U_{m,q} = V_0 \cap \overline{C}$, where C is the connected component of $K \backslash V_0$ containing q. Since there exists an H_m-path between q and p, Lemma 3.2.13 implies that $p \in \overline{C}$. Now, by Proposition 1.6.8-(3), $\#(C(p, V_0)) = m(p) = 1$. Therefore $C(p, V_0) = \{C\}$. Hence, by Theorem 3.2.11, we see that $U_p \cup \{p\} = V_0 \cap \overline{C}$. $\square$

Proof of Theorem A.1.2. First assume that $p \in V_0$. Let $U = \{u \in \ell(V_0) : u(p) = 0\}$. We identify U as $\ell(V_0 \backslash \{p\})$. Note that $A_i U \subseteq U$. Set $B_i = A_i|_U$. Then, by Lemma A.1.6, the maximum of the absolute value of the eigenvalues of B_i equals λ_i. Now, B_i is a non-negative matrix (i.e., $(B_i)_{pq} \geq 0$). Hence, by the Perron–Frobenius Theorem, λ_i is an eigenvalue of B_i and there exists a non-negative (non-trivial) eigenvector v belonging to the eigenvalue λ_i. By Lemma A.1.7 and Lemma A.1.8, we see that $(A_i{}^m)_{qq_*} = 0$ for $q_* \in Z_p$, where $Z_p = V_0 \backslash (U_p \cup \{p\})$. Hence, $A_i{}^m$ is expressed by

$$A_i{}^m = \begin{pmatrix} 1 & 0 & 0 \\ * & C_m & 0 \\ * & D_m & 0 \end{pmatrix}, \tag{A.1.1}$$

where $B_m : \ell(U_p) \to \ell(U_p)$ and $C_m : \ell(U_p) \to \ell(Z_p)$ are non-negative matrices. Hence,

$$B_i{}^m = \begin{pmatrix} C_m & 0 \\ D_m & 0 \end{pmatrix}. \tag{A.1.2}$$

Note that $C_m u = k u$ for $u \in \ell(U_p)$ and $k \neq 0$ if and only if $B_i^m v = kv$, where $v \in \ell(V_p \cup Z_p)$ is defined by $v|_{U_p} = u$ and $v|_{Z_p} = \frac{1}{k} D_m u$. Now, if v is the non-negative (non-trivial) eigenvector of B_i belonging to the eigenvalue λ_i, then $v|_{U_p} \neq 0$. Hence $(Dv)(p) = \sum_{q \in U_p} D_{pq} v(q) > 0$. By Lemma A.1.5, we see that $\lambda_i = r_i$.

Next assume that p is a non-focal point. Then, by Lemma A.1.10, $(A_i{}^m)_{qq_*} > 0$ for $q_* \in U_p \cup \{p\}$ and $q \in V_0 \backslash \{p\}$. Therefore C_m and D_m in (A.1.1) are strongly positive matrices. Hence the Perron-Frobenius theorem implies that if k is an eigenvalue of C_m and $|k| = r_i^m$, then $k = r_i^m$ and $\dim E(r_i{}^m, A_i{}^m) = 1$. Since this holds for sufficiently large m, we deduce that if k is an eigenvalue of A_i with $|k| = r_i$, then $k = r_i$ and $\dim E(r_i, A_i) = 1$.

Next suppose $m(p) > 1$. Let $\{J_j\}_{j=1,2,\dots,m(p)}$ be the connected components of $K \backslash \{p\}$ and define $I_j = J_j \cap V_0$, $U_p^j = U_p \cap I_j$ and $Z_p^j = Z_p \cap I_j$

for any $j = 1, 2, \ldots, m(p)$. By Proposition 1.6.7, $I_j \neq \emptyset$ for any $j = 1, 2, \ldots, m(p)$. In fact, $U_p^j \neq \emptyset$. By analogy with Lemma A.1.10, it follows that $U_{k(p)m,q} = U_p^j \cup \{p\}$ for $q \in F_i^{k(p)m}(I_j)$ for sufficiently large m. Hence,

$$B_i^{k(p)m} = \begin{pmatrix} \tilde{B}_{m,1} & 0 & 0 \\ 0 & \ddots & 0 \\ 0 & 0 & \tilde{B}_{m,m(p)} \end{pmatrix} \quad \text{and} \quad \tilde{B}_{m,j} = \begin{pmatrix} C_{m,j} & 0 \\ D_{m,j} & 0 \end{pmatrix},$$

where $\tilde{B}_{m,j} : \ell(I_j) \to \ell(I_j)$, $C_{m,j} : \ell(U_p^j) \to \ell(U_p^j)$ and $D_{m,j} : \ell(U_p^j) \to \ell(Z_p^j)$ are strongly positive for any $j = 1, 2, \ldots, m(p)$. Applying the same discussion as in the non-focal case to $\tilde{B}_{m,j}$, we see that if k is an eigenvalue of $\tilde{B}_{m,j}$ with $|k| = r_i^{k(p)m}$, then $k = r_i^{k(p)m}$ and $\dim E(r_i^{k(p)m}, \tilde{B}_{m,j}) = 1$ for all j. Hence, if k is an eigenvalue of $B_i^{k(p)m}$ with $|k| = r_i^{k(p)m}$, then $k = r_i^{k(p)m}$ and $\dim E(r_i^{k(p)m}, B_i^{k(p)m}) = m(p)$. Since this holds for sufficiently large m, it follows that if k is an eigenvalue of $A_i^{k(p)}$ with $|k| = r_i^{k(p)}$, then $k = r_i^{k(p)}$ and $\dim E(r_i^{k(p)}, A_i^{k(p)}) = m(p)$.

Next, suppose $p \notin V_0$. Define $\tilde{V}_0 = V_0 \cup \{p\}$ and $\tilde{V}_m = \cup_{w \in W_m} F_w(\tilde{V}_0)$. Then $\tilde{V}_m \subseteq \tilde{V}_{m+1}$ for any $m \geq 0$. There exists a Laplacian $\tilde{H}_m$ on $\tilde{V}_m$ such that

$$\mathcal{E}_{\tilde{H}_m}(u,u) = \min\{\mathcal{E}(v,v) : v \in \mathcal{F}, v|_{\tilde{V}_m} = u\}$$

for any $u \in \ell(\tilde{V}_m)$. We write $\tilde{D} = \tilde{H}_0$. Then the $\tilde{V}_0$-harmonic function with boundary value $\rho \in \ell(\tilde{V}_0)$ is the unique $u \in \mathcal{F}$ with $u|_{\tilde{V}_0} = \rho$ that attains the following minimum:

$$\min\{\mathcal{E}(u,u) : u \in \mathcal{F}, u|_{\tilde{V}_0} = \rho\}.$$

In this case, also, there exists $\tilde{A}_i : \ell(\tilde{V}_0) \to \ell(\tilde{V}_0)$ such that

$$u|_{F_i(\tilde{V}_0)} = \tilde{A}_i u|_{\tilde{V}_0},$$

where we identify $\ell(F_i(\tilde{V}_0))$ with $\ell(\tilde{V}_0)$ through the one-to-one mapping F_i. Note that a $\tilde{V}_0$-harmonic function u is a harmonic function if and only if $(\tilde{D}u)(p) = 0$. Hence, for any k and n,

$$E(k, A_i^n) = \{u|_{V_0} : u \in E(k, \tilde{A}_i^n), (\tilde{D}u)(p) = 0\}. \tag{A.1.3}$$

Therefore, if $\tilde{\lambda}_i$ is the maximum of the second largest absolute value of the eigenvalues of $\tilde{A}_i$, then $\tilde{\lambda}_i \geq \lambda_i$.

We can apply exactly the same arguments for $\tilde{A}_i$ as for A_i, and get analogous results. So if p is non-focal, then r_i is the unique eigenvalue of $\tilde{A}_i$ whose absolute value is $\tilde{\lambda}_i$ and $\dim E(r_i, \tilde{A}_i) = 1$. Also, we can choose a non-negative generator of $E(r_i, \tilde{A}_i)$ that satisfies $(\tilde{D}u)(p) > 0$. Hence, there is no eigenvalue of A_i whose absolute value equals r_i. Hence $\lambda_i < r_i$.

Next, if p is a focal point, it follows that $\dim E(r_i{}^{k(p)}, \tilde{A}_i{}^{k(p)}) = m(p)$. By (A.1.3), $\dim E(r_i{}^{k(p)}, A_i{}^{k(p)}) = m(p)-1$. This also implies that $\lambda_i = r_i$. $\square$

A.2 General boundary conditions

In Chapters 3, 4 and 5, we always think of V_0 as the boundary of the self-similar set K. This choice of a boundary is quite natural from the topological point of view; see, for example, Proposition 1.3.5. From the analytical point of view, however, we may choose any "small" subset of K as a boundary and wish to extend all the results in Chapters 3, 4 and 5 to this situation. In this section, we choose an arbitrary (finite) subset of V_* as a boundary of K and present such extensions. One advantage of such extensions is that we can unify the cases of the Dirichlet boundary condition, which was represented by the symbol "D", and the Neumann boundary condition, represented by the symbol "N". More precisely, in the scheme of this section, if we choose V_0 as the boundary, then we get the statements for the Dirichlet boundary condition, and if we choose the empty set $\emptyset$ as the boundary, then we get the ones for the Neumann boundary condition.

Let $\mathcal{L} = (K, S, \{F_i\}_{i \in S})$ be a connected, p. c. f. self-similar structure and let $(D, \mathbf{r})$ be a harmonic structure on $\mathcal{L}$. Also let $(\mathcal{E}, \mathcal{F})$ be the resistance form associated with $(D, \mathbf{r})$. Throughout this section, we assume that μ is a self-similar measure on K with weight $(\mu_i)_{i \in S}$ and that $\mu_i r_i < 1$ for all $i \in S$. Let d be a distance on K which is compatible with the original topology of K. Note that (K, d) is compact.

Let B be a finite subset of V_*. We will regard B as the boundary of K hereafter and define the Dirichlet form, the Laplacian and the heat kernel associated with the boundary B. Most of the results in this section are straightforward extensions of the corresponding ones in Chapters 3, 4 and 5. The proofs can be obtained by slight modifications of the original proofs. So we only present a sketch of the proofs in this section and leave the details to the reader.

First we introduce the Dirichlet form associated with the boundary B. The following theorem is an extension of Theorem 3.4.6.

Theorem A.2.1. *Define $\mathcal{F}_B = \{u \in \mathcal{F} : u|_B \equiv 0\}$ and $\mathcal{E}_B = \mathcal{E}|_{\mathcal{F}_B \times \mathcal{F}_B}$. Then $(\mathcal{E}_B, \mathcal{F}_B)$ is a local regular Dirichlet form on $L^2(K \backslash B, \mu)$. The cor-*

responding non-negative self-adjoint operator H_B on $L^2(K\backslash B, \mu)$ has compact resolvent.

Remark. Since $\mu(B) = 0$, we may naturally identify $L^2(K\backslash B, \mu)$ with $L^2(K, \mu)$. In this manner, we think of H_B as a self-adjoint operator on $L^2(K, \mu)$.

The proof of this theorem is similar to that of Theorem 3.4.6. If $B = \emptyset$, then this theorem coincides with the Neumann boundary case, Theorem 3.4.6, while if $B = V_0$, then this theorem coincides with the Dirichlet boundary case, Corollary 3.4.7.

Note that if $B \neq \emptyset$, then H_B is invertible and the inverse $(H_B)^{-1}$ is a compact operator on $L^2(K, \mu)$. We write $G_B = (H_B)^{-1}$.

Next we introduce the Laplacian associated with the boundary B.

Definition A.2.2. Define

$$\mathcal{D}_\mu^B = \{u \in C(K, d) : \text{there exsits } \varphi \in C(K, d) \text{ such that}$$
$$\lim_{m\to\infty} \max_{p\in V_m\backslash B} |(\mu_{m,p})^{-1}(H_m u)(p) - \varphi(p)| = 0\}.$$

Furthermore, we define a linear operator $\Delta_{B,\mu} : \mathcal{D}_\mu^B \to C(K, d)$ by $\Delta_{B,\mu}u = \varphi$, where $\varphi \in C(K, d)$ is the function appearing in the definition of $\mathcal{D}_\mu^B$. $\Delta_{B,\mu}$ is called the Laplacian associated with B.

Lemma A.2.3. *Let $p \in F_w(V_0)$ for some $w \in W_m$. Define $(H_{n,w}f)(p) = (r_w)^{-1}(H_{n-m}(f\circ F_w))((F_w)^{-1}(p))$ for any $n \geq m$. Assume that $B\cap K_w \subseteq F_w(V_0)$. Then, for any $f \in \mathcal{D}_\mu^B$,*

$$\lim_{n\to\infty} -(H_{n,w}f)(p) = -(H_{m,w}f)(p) + \int_{K_w} \psi_p^m(\Delta_{B,\mu}f)d\mu. \quad (A.2.1)$$

Note that, for any $p \in V_m$,

$$(H_m f)(p) = \sum_{w:w\in W_m, p\in F_w(V_0)} (H_{m,w}f)(p). \quad (A.2.2)$$

Hence, the quantity $\lim_{n\to\infty} -(H_{n,w}f)(p)$, which is denoted by $(df)_{w,p}$, is a kind of directional (Neumann) derivative of f at p.

Proof. Write $h = f\circ F_w$ and $q = (F_w)^{-1}(p)$. Then since $f \in \mathcal{D}_\mu^B$, we see that $h \in \mathcal{D}_\mu$. Hence, by Lemma 3.7.5,

$$\lim_{k\to\infty} -(H_k h)(q) = -(Dh)(q) + \int_K \psi_q \Delta_\mu h d\mu.$$

Since $(H_k h)(q) = r_w(H_{k+m,w}f)(p)$ for any $k \geq 0$, the above equation immediately implies (A.2.1). $\square$

Choose m so that $B \subseteq V_m$. Then $B \cap K_w \subseteq F_w(V_0)$ for any $w \in W_m$. Recall (A.2.2). Then, for any $p \in V_*$ and any $f \in \mathcal{D}_\mu^B$,

$$\lim_{n \to \infty} -(H_n f)(p) = \sum_{w:w \in W_m, p \in F_w(V_0)} (df)_{w,p}.$$

Definition A.2.4. For any $f \in \mathcal{D}_\mu^B$ and any $p \in V_*$, define

$$(df)_p = \lim_{n \to \infty} -(H_n f)(p).$$

$(df)_p$ is called the Neumann derivative of f at p.

By the above argument,

$$(df)_p = \sum_{w:w \in W_m, p \in F_w(V_0)} (df)_{w,p}.$$

Also, we immediately obtain the following proposition.

Proposition A.2.5. *Assume that $B \subseteq V_m$ and that $p \in V_m$. Then, for any $f \in \mathcal{D}_\mu^B$,*

$$(df)_p = -(H_m f)(p) + \int_K \psi_p^m \Delta_{B,\mu} f \, d\mu.$$

It is easy to see that $(df)_p = 0$ if $p \notin B$. Therefore, for any $p \in V_* \backslash B$, if $B \subseteq V_m$,

$$(H_m f)(p) = \int_K \psi_p^m \Delta_{B,\mu} f \, d\mu. \tag{A.2.3}$$

Conversely, if $(df)_p = 0$, (A.2.3) implies that $\lim_{m \to \infty} (\mu_{m,p})^{-1} (H_m f)(p) = (\Delta_{B,\mu} f)(p)$. Hence we have the following fact.

Proposition A.2.6. *If $B_1 \subseteq B_2 \subseteq V_m$, then $\mathcal{D}_\mu^{B_1} \subseteq \mathcal{D}_\mu^{B_2}$. More precisely,*

$$\mathcal{D}_\mu^{B_1} = \{u \in \mathcal{D}_\mu^{B_2} : (du)_p = 0 \text{ for all } p \in B_2 \backslash B_1\}.$$

If $u \in \mathcal{D}_\mu^{B_1}$, then $\Delta_{B_1,\mu} u = \Delta_{B_2,\mu} u$.

Next we present an extension of Theorem 3.7.8, the Gauss–Green's formula.

Proposition A.2.7. *$\mathcal{D}_\mu^B \subset \mathcal{F}$. Moreover, for any $u \in \mathcal{F}$ and any $v \in \mathcal{D}_\mu^B$,*

$$\mathcal{E}(u,v) = \sum_{p \in B} u(p)(dv)_p - \int_K u \Delta_{B,\mu} v \, d\mu. \tag{A.2.4}$$

The proof of this proposition is analogous to that of Theorem 3.7.8. We also obtain the counterpart of Theorem 3.7.9.

Theorem A.2.8. *Define* $\mathcal{D}_{B,\mu}$ *by* $\mathcal{D}_{B,\mu} = \{u \in \mathcal{D}_\mu^B : u|_B \equiv 0\}$. *Suppose that* $B \neq \emptyset$. *Then* $\mathcal{D}_{B,\mu} = G_B(C(K,d)) \subset \mathrm{Dom}(H_B)$ *and* $\Delta_{B,\mu}|_{\mathcal{D}_{B,\mu}} = -H_B|_{\mathcal{D}_{B,\mu}}$. *Furthermore,* H_B *is the Friedrichs extension of* $-\Delta_{B,\mu}|_{\mathcal{D}_{B,\mu}}$.

Remark. If $B = \emptyset$, then $\mathcal{D}_{B,\mu} = \mathcal{D}_{\mu,N}$ and $H_B = H_N$. So, by Theorem 3.7.9, we still see that $\Delta_{B,\mu}|_{\mathcal{D}_{B,\mu}} = -H_B|_{\mathcal{D}_{B,\mu}}$ and that H_B is the Friedrichs extension of $-\Delta_{B,\mu}|_{\mathcal{D}_{B,\mu}}$.

The key fact in proving the above theorem is the following lemma corresponding to Lemma 3.7.10.

Lemma A.2.9. *Assume that* $B \neq \emptyset$. *For any* $\varphi \in C(K,d)$, $G_B\varphi \in \mathcal{D}_{B,\mu}$ *and* $\Delta_{B,\mu}(G_B\varphi) = -\varphi$.

Proof. Let $u = G_B\varphi$ and assume $B \subseteq V_k$. Then, for any $f \in \mathcal{F}_B$, $\mathcal{E}(f,u) = (f,\varphi)$. If $m \geq k$ and $p \in V_m\backslash B$, then $\psi_p^m \in \mathcal{F}_B$. Therefore we obtain $\mathcal{E}(\psi_p^m, u) = \int_K \psi_p^m \varphi d\mu$. On the other hand, $\mathcal{E}(\psi_p^m, u) = \mathcal{E}_m(\psi_p^m, u) = -(H_m u)(p)$. Hence it follows that

$$(H_m u)(p) = -\int_K \psi_p^m \varphi d\mu. \qquad (A.2.5)$$

(This equation is analogous to that of Lemma 3.6.4.) Using a similar argument to that in the proof of Lemma 3.7.10, we see that

$$\lim_{m\to\infty} \max_{p \in V_m\backslash B} |(\mu_{m,p})^{-1}(H_m u)(p) + \varphi(p)| = 0.$$

Let $w \in W_k$. Define $u_w = u \circ F_w$ and $\varphi_w = \varphi \circ F_w$. Then, by (A.2.5), it follows that

$$(H_m u_w)(p) = -r_w \mu_w \int_K \psi_p^m \varphi_w d\mu$$

for any $m \geq 1$ and any $p \in V_m\backslash V_0$. Therefore, using Theorem 3.7.16, we obtain

$$u_w = \sum_{p \in V_0} u_w(p)\psi_p + r_w \mu_w G_\mu \varphi_w \qquad (A.2.6)$$

and that $u_w \in \mathcal{D}_\mu \subset C(K,d)$. Hence $u \in C(K,d)$ and therefore $u \in \mathcal{D}_{B,\mu}$ and $\Delta_{B,\mu}(G_B\varphi) = -\varphi$. $\qquad \square$

Given this lemma, a similar argument as in the proof of Theorem 3.7.9 suffices to show Theorem A.2.8.

Using (A.2.6), we may define the Green's function $g_B(x,y)$ as follows.

Proposition A.2.10. *Assume that $B \neq \emptyset$ and that $B \subseteq V_k$. Divide H_k into four parts, and write*

$$H_k = \begin{pmatrix} T_B & {}^t J_B \\ J_B & X_B \end{pmatrix},$$

where $T_B : \ell(B) \to \ell(B)$, $J_B : \ell(B) \to \ell(V_k \backslash B)$ and $X_B : \ell(V_k \backslash B) \to \ell(V_k \backslash B)$. Set $G_B^{(k)} = -(X_B)^{-1}$ and define

$$g_B(x,y) = \sum_{p,q \in V_k \backslash B} (G_B^{(k)})_{pq} \psi_p^k(x) \psi_q^k(y) + \sum_{w \in W_k} r_w g_w(x,y),$$

where

$$g_w(x,y) = \begin{cases} g((F_w)^{-1}(x), (F_w)^{-1}(y)) & \text{if both } x \text{ and } y \in K_w, \\ 0 & \text{otherwise,} \end{cases}$$

where g is the Green's function defined in 3.5. Then, for $\varphi \in C(K,d)$, the following four conditions are equivalent:

(1) $u \in \mathcal{D}_{B,\mu}$ *and* $\Delta_{B,\mu} u = \varphi$,

(2) $u = -G_B \varphi$,

(3) $u(x) = -\int_K g_B(x,y) \varphi(y) d\mu$ *for any* $x \in K$,

(4) $u \in \mathcal{F}_B$ *and, for any $m \geq k$ and any $p \in V_m \backslash B$,*

$$(H_m u)(p) = \int_K \psi_p^m \varphi d\mu.$$

g_B is called the Green's function associated with the boundary B.

Remark. The original Green's function g equals g_{V_0}.

Using Lemma A.2.9 and Proposition A.2.10, we can prove Theorem A.2.8 by arguments similar to those in the proof of Theorem 3.7.9.

Next, we discuss the eigenvalues and the eigenfunctions of $\Delta_{B,\mu}$. By Proposition A.2.10, we can verify the counterpart of Corollary 3.6.6, which was the key fact in proving Proposition 4.1.2. So, in the same manner, we may obtain the following statement corresponding to Proposition 4.1.2.

Proposition A.2.11. *For $\lambda \in \mathbb{R}$, define $E_B(\lambda) = \{\varphi \in \mathcal{D}_{B,\mu} : \Delta_{B,\mu} u = -\lambda \varphi\}$. Then the following three conditions are equivalent.*

(1) $\varphi \in \text{Dom}(H_B)$ *and* $H_B \varphi = \lambda \varphi$,

(2) $\varphi \in \mathcal{F}_B$ *and* $\mathcal{E}(\varphi, u) = -\lambda(\varphi, u)_\mu$ *for any* $u \in \mathcal{F}_B$.

(3) $\varphi \in E_B(\lambda)$.

Combining this proposition with the fact that H_B has compact resolvent, we see that there exist a complete orthonormal system $\{\varphi_i^B\}_{i \geq 1}$ of $L^2(K, \mu)$

and a sequence of non-negative numbers, $\{\lambda_i^B\}_{i\geq 1}$, such that $\lambda_i^B \leq \lambda_{i+1}^B$ for every $i \geq 1$, $\lim_{n\to\infty} \lambda_n^B = \infty$ and $\varphi_i^B \in E_B(\lambda_i^B)$ for every $i \geq 1$. So, if we define the eigenvalue counting function of $-\Delta_{B,\mu}$, $\rho_B(x,\mu)$, by $\rho_B(x,\mu) = \sum_{\lambda\leq x} \dim E_B(\lambda)$, then $\rho_B(x,\mu) = \#\{n : \lambda_n^B \leq x\}$. Note that $\rho_\emptyset(x,\mu) = \rho_N(x,\mu)$ and $\rho_{V_0}(x,\mu) = \rho_D(x,\mu)$.

Since $\mathcal{F}_B \subseteq \mathcal{F}$, Corollary 4.1.8 implies that

$$\rho_N(x,\mu) - \#(B) \leq \rho_B(x,\mu) \leq \rho_N(x,\mu).$$

Therefore, the asymptotic behavior of $\rho_B(x,\mu)$ as $x \to \infty$ turns out to be exactly the same as that of $\rho_N(x,\mu)$. More precisely, replacing $\rho_b(x,\mu)$ by $\rho_B(x,\mu)$, we can immediately verify all the statements of Theorem 4.1.5.

Finally, we discuss the heat kernel associated with the Laplacian $\Delta_{B,\mu}$.

Definition A.2.12. We define the heat kernel associated with the Laplacian $\Delta_{B,\mu}$ by

$$p_B(t,x,y) = \sum_{n\geq 1} e^{-\lambda_n^B t} \varphi_n^B(x)\varphi_n^B(y).$$

Note that the Neumann heat kernel $p_N(t,x,y)$ equals $p_\emptyset(t,x,y)$ and the Dirichlet heat kernel $p_D(t,x,y)$ equals $p_{V_0}(t,x,y)$. Since we have the counterpart of Lemma 5.1.3 for $\Delta_{B,\mu}$, which is shown by Theorem 4.1.5, it follows that $p_B(t,x,y)$ is a non-negative, continuous function on $(0,\infty) \times K \times K$, like the Neumann and Dirichlet heat kernels. Moreover, replacing $p_b(t,x,y)$, Δ_μ and $\mathcal{D}_{b,\mu}$ by $p_B(t,x,y)$, $\Delta_{B,\mu}$ and $\mathcal{D}_{B,\mu}$ respectively, we can verify Propositions 5.1.2 and 5.1.5. Also define the heat semigroup $\{T_t^B\}_{t>0}$ by

$$(T_t^B u)(x) = \int_K p_B(t,x,y)u(y)\mu(dy)$$

for $u \in L^1(K,\mu)$. Then we immediately have the counterpart of Theorem 5.1.7 by using exactly the same arguments as in the proof of the original theorem.

Definition A.2.13. (1)   Let $u : (0,\infty) \times K \to \mathbb{R}$. For any $t > 0$, define $u_t : K \to \mathbb{R}$ by $u_t(x) = u(t,x)$ for any $x \in K$. If $u_t \in \mathcal{D}_\mu^B$ for all $t > 0$, $u(\cdot,x) \in C^1((0,\infty))$ for all $x \in K$ and

$$\frac{\partial u(t,x)}{\partial t} = (\Delta_{B,\mu} u_t)(x) \tag{A.2.7}$$

holds for any $(t,x) \in (0,\infty) \times K$, then u is called a solution of the heat equation (A.2.7) (associated with the Laplacian $\Delta_{B,\mu}$) on $(0,\infty)$.

(2) Let $u : [0, \infty) \times K$. For $t \geq 0$, define u_t exactly the same as above. u is called an L^∞-solution of the heat equation (A.2.7) (associated with the Laplacian $\Delta_{B,\mu}$) on $[0, \infty)$ if u is a solution of the heat equation (A.2.7) on $(0, \infty)$ and u is continuous on $[0, \infty) \times K$.

Remark. Suppose that u is a solution of the heat equation (A.2.7) on $(0, \infty)$. Then u is an L^∞-solution of the heat equation (A.2.7) if and only if $\|u_t - u_0\|_\infty \to 0$ as $t \downarrow 0$.

The next theorem is the counterpart of Corollary 5.1.9 and Theorem 5.2.5 describing the relation between the heat semigroup and solutions of the heat equation.

Theorem A.2.14. (1) *Let $\varphi \in L^2(K, \mu)$. Define $u(t, x) = (T_t^B u)(x)$ for all $(t, x) \in (0, \infty) \times K$. Then u is the unique solution of the heat equation (A.2.7) on $(0, \infty)$ with*

$$\lim_{t \downarrow 0} \|u_t - \varphi\|_2 = 0 \quad \text{and} \quad u_t|_B \equiv 0 \text{ for any } t > 0.$$

(2) *Suppose $(D, \mathbf{r})$ is a regular harmonic structure. Let $\varphi \in \{f \in C(K, d) : f|_B \equiv 0\}$. Define $u(t, x) = (T_t^B u)(x)$ for any $(t, x) \in (0, \infty) \times K$ and $u(0, x) = \varphi(x)$ for any $x \in K$. Then u is the unique L^∞-solution of the heat equation (A.2.7) on $[0, \infty)$ with $u_t|_B \equiv 0$ for all $t \geq 0$ and $u_0 = \varphi$.*

From this theorem, the heat kernel $p_B(t, x, y)$ may be thought of as the fundamental solution of the heat equation (A.2.7).

We can prove Theorem A.2.14-(1) by using the counterpart of Theorem 5.1.7. We need the following version of parabolic maximum principle to show Theorem A.2.14-(2).

Theorem A.2.15. *Let $u : [0, \infty) \times K \to \mathbb{R}$ be an L^∞-solution of the heat equation (A.2.7) on $[0, \infty)$. Then, for any $T > 0$,*

$$\max_{(t,x) \in U_T} u(t, x) = \max_{(t,x) \in \partial U_T^B} u(t, x) \quad \text{and} \quad \min_{(t,x) \in U_T} u(t, x) = \min_{(t,x) \in \partial U_T^B} u(t, x),$$

where $U_T = [0, T] \times K$ and $\partial U_T^B = \{0\} \times K \cup [0, T] \times B$.

The proof of this theorem is analogous to that of Theorem 5.2.2.

The following comparison theorem for heat kernels is the counterpart of Theorem 5.2.9 and Theorem 5.2.13-(1).

Theorem A.2.16. *Let B_1 and B_2 be finite subsets of V_*. If $B_1 \subseteq B_2$, then*

$$0 \leq p_{B_2}(t, x, y) \leq p_{B_1}(t, x, y) \tag{A.2.8}$$

for any $(t, x, y) \in (0, \infty) \times K \times K$. Moreover, if $(D, \mathbf{r})$ is regular, then

$$p_{B_1}(t, x, y) - p_{B_2}(t, x, y) = -\sum_{q \in B_2 \backslash B_1} \int_0^t p_{B_1}(t - s, x, q)(dp_{B_2}^{s,y})_q ds$$

$$(A.2.9)$$

for any $t > 0$ and any $x, y \in K \backslash B_2$.

Remark. As in Theorem 5.2.13, the integral $\int_0^t p_{B_1}(t - s, x, q)(dp_{B_2}^{s,y})_q ds$ actually means

$$\lim_{\epsilon \to 0} \int_\epsilon^{t-\epsilon} p_{B_1}(t - s, x, q)(dp_{B_2}^{s,y})_q ds.$$

If $(D, \mathbf{r})$ is regular, then this theorem can be proven in an analogous way to 5.2 by using the parabolic maximum principle, Theorem A.2.15. Even if $(D, \mathbf{r})$ is not regular, (A.2.8) is still true. However, to prove it, we must employ probabilistic arguments on the diffusion processes whose transition densities are p_{B_1} and p_{B_2} respectively. See Corollary A.3.4.

Also, we can verify the following theorem, which is an extension of Corollary 5.2.14.

Theorem A.2.17. *Let B be a finite subset of V_*. Suppose that $(D, \mathbf{r})$ is regular. Then, for any $x \in K \backslash B$,*

$$\|p_B^{t,x}\|_1 = \int_K p_B(t, x, y)\mu(dy) = 1 - (-\sum_{q \in B} \int_0^t (dp_B^{s,x})_q ds). \qquad (A.2.10)$$

Moreover, if $B \neq \emptyset$, then

$$\sum_{q \in B} -\int_0^\infty (dp_B^{s,x})_q ds = 1. \qquad (A.2.11)$$

Proof. The arguments are essentially the same as in the proof of Theorem 5.2.13. To obtain (A.2.11), let $B_1 = \emptyset$ and $B_2 = B$ in (A.2.9). Then integrate it on K with respect to the measure μ. Using Lemma 5.2.10, we have (A.2.11). Next, if $B \neq \emptyset$, then $\ker H_B = \{0\}$. Therefore all the eigenvalues of H_B are positive. Recalling Definition A.2.12, we see that $\lim_{t \to \infty} \int_K p_B(t, x, y)\mu(dy) = 0$. Then (A.2.11) is immediate from (A.2.10). $\qquad \square$

Next, we present an upper estimate for the asymptotic behavior of the heat kernel $p_B(t, x, y)$ as $t \downarrow 0$.

Theorem A.2.18. *There exists $c > 0$ such that*

$$\sup_{x,y \in K} p_B(t,x,y) \le ct^{-d_s \bar{\delta}/2}$$

for any $t > 0$.

To prove this theorem, we first establish the counterpart of the Nash inequality, Theorem 5.3.3, and apply Theorem B.3.7 as in the proof of Theorem 5.3.1-(1).

Theorem A.2.19. *Let $x, y \in K \backslash B$ and let $t > 0$. Then $p_B(t, x, y) > 0$ if and only if x and y belong to the same connected component of $K \backslash B$.*

This theorem is the counterpart of Proposition 5.1.10. To prove this theorem, first establish the counterpart of Lemma 5.3.5 by using Theorem A.2.18. Then note the following lemma and proceed with the same argument as in 5.3.

Lemma A.2.20. *Let $x, y \in K \backslash B$. Then $g_B(x, y) > 0$ if and only if x and y belong to the same connected component of $K \backslash B$.*

Proof. We present only a brief sketch of the proof. It is enough to prove that $\sum_{p,q \in V_k \backslash B}(G_B^{(k)})_{pq}\psi_p^k(x)\psi_q^k(y) > 0$ if x and y belong to the same connected component of $K \backslash B$. This can be shown by an argument similar to that in the proof of Proposition 3.5.5. $\square$

A.3 Probabilistic approach

In this section, we will briefly present the probabilistic approach to the heat kernels, $p_B(t, x, y)$. In particular, we will give probabilistic proofs of Theorem 5.2.9 and Lemma 5.3.9. Recall that our analytical proofs only work when the harmonic structure is regular.

Let $\mathcal{L} = (K, S, \{F_i\}_{i \in S})$ be a p. c. f. self-similar set with $S = \{1, 2, \dots, N\}$ and let $(D, \mathbf{r})$ be a harmonic structure on K. Also let μ be a self-similar measure on K with weight $(\mu_i)_{i \in S}$. We assume that $\mu_i r_i < 1$ for all $i \in S$. Then, by Theorem 3.4.6, the resistance form $(\mathcal{E}, \mathcal{F})$ derived from $(D, \mathbf{r})$ is a local regular Dirichlet form on $L^2(K, \mu)$. Set

$$\Omega = \{u : u \text{ is a continuous function from } [0, \infty) \text{ to } K\}.$$

Also, for $t \ge 0$, define $X_t : \Omega \to K$ by $X_t(u) = u(t)$. By the general theory of Dirichlet forms (see [43, Theorem 7.2.1] for example), there exists a diffusion process associated with the local regular Dirichlet form $(\mathcal{E}, \mathcal{F})$. One may also find a concise exposition on Dirichlet forms and diffusion processes in [6, Section 4], in particular, [6, Theorem 4.8]. See also [6, Section 7]. Using those discussions, we can prove the following theorem.

Theorem A.3.1. *There exists a diffusion process* $(\Omega, \{X_t\}_{t\geq 0}, \{P_x\}_{x\in K})$ *such that, for any* $x \in K$ *and any* $f \in C(K)$,

$$E_x(f(X_t)) = \int_K p_N(t,x,y)f(y)\mu(dy),$$

where p_N *is the Neumann heat kernel defined in Definition 5.1.1.*

$E_x(\cdot)$ is the expectation with respect to the measure P_x, where x is the starting point of the process.

Remark. By [43, Theorem 7.2.1] and [43, Theorem 7.2.2], there exists a diffusion process associated with $(\mathcal{E}, \mathcal{F})$. This diffusion process is, in general, defined only on q. e. starting points. However, since the heat kernel $p_N(t,x,y)$ is continuous on $(0,\infty) \times K \times K$, we can obtain the above theorem.

Definition A.3.2. For any $A \subseteq K$, define the hitting time of A, σ_A, by

$$\sigma_A(u) = \inf\{t \geq 0 : X_t(u) \in A\}$$

for any $u \in \Omega$.

Theorem A.3.3. *Let* B *be a finite subset of* V_*. *Then, for any Borel set* A *of* K,

$$\int_A p_B(t,x,y)\mu(dy) = P_x(\sigma_B > t, X_t \in A),$$

where p_B *is the heat kernel associated with the boundary* B *defined in Definition A.2.12.*

By this theorem, we immediately obtain the following fact, which is (A.2.8).

Corollary A.3.4. *Let* B_1 *and* B_2 *be finite subsets of* V_* *with* $B_1 \subseteq B_2$. *Then, for any* $t > 0$ *and any* $x, y \in K$,

$$0 \leq p_{B_2}(t,x,y) \leq p_{B_1}(t,x,y).$$

If $B_1 = \emptyset$ and $B_2 = V_0$, then the theorem above is Theorem 5.2.9.

Proof. Since $\sigma_{B_1} \geq \sigma_{B_2}$, it follows that

$$P_x(\sigma_{B_1} > t, X_t \in A) \geq P_x(\sigma_{B_2} > t, X_t \in A)$$

for any $t > 0$, any A and any $x \in K$. Hence Theorem A.3.3 implies the desired inequality. $\qquad\qquad\qquad\square$

Next we give a probabilistic proof of Lemma 5.3.9.

Proof of Lemma 5.3.9. Let $A = \cup_{w \in \tilde{\Lambda}_x} F_w(\partial K_0)$. By Definition A.2.12,

$$\lim_{t \to \infty} \int_K p_A(t, x, y)\mu(dy) = 0.$$

This implies that

$$\sum_{q \in A} P_x(X_{\sigma_A} = q) = P_x(\sigma_A < \infty) = 1. \tag{A.3.1}$$

Recall that for any $p \in \partial U_\Lambda(x)$, x and p belong to different connected component of $K \backslash A$. Therefore, any continuous path between p and x must intersect with A. Hence

$$\sigma_A \leq \sigma_{\partial U_\Lambda(x)}. \tag{A.3.2}$$

Using the results above, we have

$$1 - \int_K P_{\Lambda,x}(t, x, y)\mu(dy) = P_x(\sigma_{\partial U_\Lambda(x)} \leq t)$$

$$= \sum_{q \in A} P_x(X_{\sigma_A} = q, \sigma_{\partial U_\Lambda(x)} \leq t)$$

$$\leq \sum_{q \in A} P_x(X_{\sigma_A} = q, \sigma_{\partial U_\Lambda(x)} \leq t + \sigma_A)$$

$$= \sum_{q \in A} P_x(X_{\sigma_A} = q)P_q(\sigma_{\partial U_\Lambda(x)} \leq t)$$

$$\leq \sum_{q \in A} P_x(X_{\sigma_A} = q)P_q(\sigma_{V(\Lambda)} \leq t)$$

The equality in the fourth line is deduced by the Markov property of the process and (A.3.2). Again by (A.3.1),

$$1 - \int_K P_{\Lambda,x}(t, x, y)\mu(dy) \leq \max_{q \in A} P_q(\sigma_{V(\Lambda)} \leq t)$$

$$= \max_{q \in A}(1 - \int_K p_\Lambda(t, q, y)\mu(dy)).$$

Since $\int_K p_{\Lambda,x}(t, x, y)\mu(dy) = \int_{U_\Lambda(x)} p_{\Lambda,x}(t, x, y)\mu(dy)$, this immediately implies Lemma 5.3.9. $\qquad\square$

Comparing the proof above with the one in 5.3, we have the following.

Theorem A.3.5. *Suppose $(D, \mathbf{r})$ is regular. Let B be a finite subset of V_*. Then, for any $x \in K$, any $q \in B$ and any $t > 0$,*

$$P_x(X_{\sigma_B} = q, \sigma_B \leq t) = \int_0^t -(dp_B^{s,x})_q ds.$$

Appendix B

Mathematical Background

B.1 Self-adjoint operators and quadratic forms

In this section, we will introduce basic concepts and results on self-adjoint operators and non-negative quadratic forms on a Hilbert space. One can find detailed accounts and proofs of those subjects in Davies [26]. Also one can refer to classical textbooks, for example, Kato [81], Reed & Simon [156].

Let $\mathcal{H}$ be a (real or complex) separable Hilbert space with an inner product $(\cdot, \cdot)$. A linear map H is called a linear operator on $\mathcal{H}$ if and only if the domain of H, denoted by $\mathrm{Dom}(H)$, is a dense subspace of $\mathcal{H}$ and $H : \mathrm{Dom}(H) \to \mathcal{H}$. In this appendix, we assume that $\mathcal{H}$ is a real Hilbert space for simplicity.

Definition B.1.1. Let H be a linear operator on $\mathcal{H}$.
(1) H is called symmetric if and only if $(Hf, g) = (f, Hg)$ for any $f, g \in \mathrm{Dom}(H)$.
(2) H is said to be a self-adjoint operator if and only if H is symmetric and

$$\mathrm{Dom}(H) = \{g \in \mathcal{H} : \text{there exists } h \in \mathcal{H} \text{ such that}$$
$$(Hf, g) = (f, h) \text{ for all } f \in \mathrm{Dom}(H)\}.$$

(3) A symmetric operator H is called non-negative if $(Hf, f) \geq 0$ for all $f \in \mathrm{Dom}(H)$.

Proposition B.1.2. *Let H be a non-negative self-adjoint operator. Then there exists a unique non-negative self-adjoint operator G on $\mathcal{H}$ such that $\mathrm{Dom}(H) \subseteq \mathrm{Dom}(G)$, $\mathrm{Dom}(H) = \{f : f \in \mathrm{Dom}(G) \text{ and } Gf \in \mathrm{Dom}(G)\}$ and $G^2 = H$. We write $G = H^{1/2}$.*

Definition B.1.3. Let H be a non-negative self-adjoint operator on $\mathcal{H}$.

The associated quadratic form Q_H with domain $\mathrm{Dom}(H^{1/2}) \times \mathrm{Dom}(H^{1/2})$ is defined by $Q_H(f,g) = (H^{1/2}f, H^{1/2}g)$ for $f,g \in \mathrm{Dom}(H^{1/2})$.

The following lemma is an immediate consequence of the definition of a self-adjoint operator and the definition of $H^{1/2}$.

Lemma B.1.4. *Let H be a non-negative self-adjoint operator on $\mathcal{H}$. Then*

$$\mathrm{Dom}(H) = \{g \in \mathrm{Dom}(H^{1/2}) : \text{ there exists } h \in \mathcal{H} \text{ such that}$$
$$Q_H(f,g) = (f,h) \text{ for all } f \in \mathrm{Dom}(H^{1/2})\}.$$

In particular, in the above situation, $h = Hg$.

Definition B.1.5. $Q(\cdot, \cdot)$ is called a non-negative quadratic form on $\mathcal{H}$ if
(QF1) $Q : \mathrm{Dom}(Q) \times \mathrm{Dom}(Q) \to \mathbb{R}$, where the domain of Q, $\mathrm{Dom}(Q)$ is a dense subspace of $\mathcal{H}$.
(QF2) Q is bilinear and symmetric: $Q(af + bg, h) = aQ(f,h) + bQ(g,h) = aQ(h,f) + bQ(h,g)$ for any $f,g,h \in \mathcal{D}$ and $a,b \in \mathbb{R}$.
(QF3) Q is non-negative definite: $Q(f,f) \geq 0$ for any $f \in \mathcal{D}$.

Theorem B.1.6. *Let Q be a non-negative quadratic form on $\mathcal{H}$ with dense domain $\mathrm{Dom}(Q)$. Then the following conditions (1) and (2) are equivalent.*
(1) $\mathrm{Dom}(Q) = \mathrm{Dom}(H^{1/2})$ and $Q = Q_H$ for some non-negative self-adjoint operator H on $\mathcal{H}$.
(2) Define $Q_*(f,g) = Q(f,g) + (f,g)$ for any $f,g \in \mathrm{Dom}(Q)$. Then $(\mathrm{Dom}(Q), Q_*)$ is a Hilbert space.

A non-negative quadratic form Q on $\mathcal{H}$ is said to be a closed form if Q satisfies the conditions in Theorem B.1.6.

Definition B.1.7. (1) Let Q and Q' be non-negative quadratic forms on $\mathcal{H}$. Q is called an extension of Q' if and only if $\mathrm{Dom}(Q) \supseteq \mathrm{Dom}(Q')$ and $Q|_{\mathrm{Dom}(Q') \times \mathrm{Dom}(Q')} = Q'$.
(2) A non-negative quadratic form on $\mathcal{H}$ is said to be closable if there exists a closed extension of Q.

If a non-negative quadratic form Q is closable, then there exists a minimal closed extension of Q. Precisely, let $(U, \overline{Q_*})$ be the completion of $(\mathrm{Dom}(Q), Q_*)$. Then define $\overline{Q}(u,v) = \overline{Q_*}(u,v) - (u,v)$ for any $u,v \in U$. Then $\overline{Q}$, whose domain is U, is the minimal closed extension of Q. This $\overline{Q}$ is called the closure of Q.

Theorem B.1.8 (Friedrichs). *Let H be a non-negative symmetric operator on $\mathcal{H}$. Define a symmetric bilinear form $Q : \mathrm{Dom}(H) \times \mathrm{Dom}(H) \to \mathbb{R}$ by $Q(f,g) = (Hf, g)$ for $f,g \in \mathrm{Dom}(H)$. Then Q is closable.*

Let $\overline{Q}$ be the closure of the symmetric form Q in Theorem B.1.8. Then, by Theorem B.1.6, there exists a non-negative self-adjoint operator $\overline{H}$ on $\mathcal{H}$ such that $\overline{Q} = Q_{\overline{H}}$. It is known that $\overline{H}$ is an extension of H: $\mathrm{Dom}(H) \subseteq \mathrm{Dom}(\overline{H})$ and $\overline{H}|_{\mathrm{Dom}(H)} = H$. This $\overline{H}$ is called the Friedrichs extension of H. In fact, $\mathrm{Dom}(\overline{Q})$ is the completion of $\mathrm{Dom}(H)$ with respect to the inner product Q_*. By this fact, it follows that $\overline{H}$ is the minimal self-adjoint extension of H: if $\widetilde{H}$ is a self-adjoint extension of H, then $\mathrm{Dom}(\overline{H}) \subseteq \mathrm{Dom}(\widetilde{H})$ and $\widetilde{H}|_{\mathrm{Dom}(\overline{H})} = \overline{H}$.

Next we will introduce a non-negative self-adjoint operator with compact resolvent. For this purpose, we need the notion of compact operators.

Definition B.1.9. Let $(B_1, \|\cdot\|_1)$ and $(B_2, \|\cdot\|_2)$ be Banach spaces.
(1) A linear operator $A : B_1 \to B_2$ is called a compact operator if and only if $A(U)$ is relatively compact in B_2 for any bounded subset $U \subset B_1$.
(2) A linear operator $A : B_1 \to B_2$ is called a finite rank operator if $A(B_1)$ is a finite dimensional vector space.

Remark. A compact operator is bounded and hence it is continuous.

The following are important facts about compact operators.

Proposition B.1.10. *A finite rank operator is a compact operator.*

Theorem B.1.11. *Let $A : B_1 \to B_2$ be a bounded operator. If there exists a sequence of compact operators $\{A_n\}_{m \geq 1}$ such that $\|A_n - A\| \to 0$ as $n \to \infty$, where $\|\cdot\|$ is the operator norm, then A is a compact operator.*

Definition B.1.12. A non-negative self-adjoint operator H on a Hilbert space $\mathcal{H}$ is said to have compact resolvent if the resolvent $(H + I)^{-1}$ is a compact operator, where I is the identity map.

Theorem B.1.13. *Let H be a non-negative self-adjoint operator on $\mathcal{H}$. Then the following three conditions are equivalent.*
(1) H has compact resolvent.
(2) There exists a complete orthonormal basis of $\mathcal{H}$, $\{\varphi_n\}_{n \geq 1}$ such that $H\varphi_n = \lambda_n \varphi_n$ for all $n \geq 0$, $0 \leq \lambda_1 \leq \cdots \leq \lambda_n \leq \lambda_{n+1} \leq \cdots$ and $\lim_{n \to \infty} \lambda_n = \infty$.
(3) Define id : $\mathrm{Dom}(H^{1/2}) \to \mathcal{H}$ by $id(x) = x$ for all $x \in \mathrm{Dom}(H^{1/2})$. Then id is a compact operator from $(\mathrm{Dom}(H^{1/2}), (Q_H)_*) \to (\mathcal{H}, (\cdot, \cdot))$, where $(Q_H)_*$ is the inner product defined in Theorem B.1.6.

Remark. By Theorem B.1.6, $(\mathrm{Dom}(H^{1/2}), (Q_H)_*)$ is a Hilbert space.

Theorem B.1.14 (Variational formula). *Let H be a self-adjoint operator on $\mathcal{H}$. Suppose that H is non-negative and has compact resolvent. Let L be any finite-dimensional subspace of $\mathrm{Dom}(H^{1/2})$. Define*

$$\lambda(L) = \sup\{Q_H(f,f) : f \in L, \|f\| = 1\},$$

where $\|f\|$ is the $\mathcal{H}$-norm of $f \in \mathcal{H}$. Also let $\{\lambda_n\}_{n \geq 1}$ be the set of eigenvalues of H appearing in Theorem B.1.13. Then

$$\lambda_n = \inf\{\lambda(L) : L \subseteq \mathrm{Dom}(H^{1/2}), \dim L = n\}$$
$$= \inf\{\lambda(L) : L \subseteq \mathrm{Dom}(H), \dim L = n\}.$$

B.2 Semigroups

In this section, we will briefly introduce basic notions and results on one parameter semigroup of symmetric bounded operators on a Hilbert space. See Fukushima, Oshima and Takeda [43] for details and proofs. In the following, $\mathcal{H}$ is a Hilbert space with the inner product $(\cdot, \cdot)$.

Definition B.2.1 (Semigroup). A family of bounded symmetric operators $\{T_t\}_{t>0}$ from $\mathcal{H}$ to itself is called a semigroup (of symmetric operators) on $\mathcal{H}$ if it satisfies the following two properties.
Semigroup property: For any $t, s > 0$, $T_{t+s} = T_t T_s$.
Contraction property: $(T_t u, T_t u) \leq (u, u)$ for any $u \in \mathcal{H}$ and any $t > 0$.
   Moreover, if a semigroup $\{T_t\}_{t>0}$ satisfies

$$\lim_{t \downarrow 0}(T_t u - u, T_t u - u) = 0$$

for any $u \in \mathcal{H}$, then $\{T_t\}_{t>0}$ is called a strongly continuous semigroup.

Remark. In [186], the family $\{T_t\}_{t>0}$, where the T_t are not necessarily symmetric, but with the semigroup property and strong continuity, is called a semigroup of class (C_0). In addition, if $\{T_t\}_{t>0}$ has the contraction property, it is called a contraction semigroup of class (C_0).

Theorem B.2.2. *Let $\{T_t\}_{t>0}$ be a strongly continuous semigroup on $\mathcal{H}$. Set*

$$\mathrm{Dom}(A) = \{u \in \mathcal{H} : \text{there exists } f \in \mathcal{H} \text{ such that } \lim_{t \downarrow 0} \|f - \frac{T_t u - u}{t}\| = 0\}$$

and define $Au = \lim_{t \downarrow 0}(T_t u - u)/t$ for any $u \in \mathrm{Dom}(A)$. Then A is a densely defined non-positive self-adjoint operator on $\mathcal{H}$ and $\mathrm{Im}\, T_t \subseteq \mathrm{Dom}(A)$ for any $t > 0$. A is called the generator of the strongly continuous semigroup $\{T_t\}_{t>0}$. Moreover, for a non-positive self-adjoint operator A

on $\mathcal{H}$, there exists a unique strongly continuous semigroup $\{T_t\}_{t>0}$ whose generator is A.

By the above theorem, we see that there is a one-to-one correspondence between strongly continuous semigroups and non-positive self-adjoint operators. Also, by Theorems B.1.6 and B.1.8, there is a one-to-one correspondence between non-negative self-adjoint operators and closed forms. Combining those two correspondences, we obtain a one-to-one correspondence between closed forms and strongly continuous semigroups as follows.

$$\{(\mathcal{E},\mathcal{F}) : (\mathcal{E},\mathcal{F}) \text{ is a closed form on } \mathcal{H}\}$$
$$\updownarrow \mathcal{E} = Q_H$$
$$\{H : H \text{ is a non-negative self-adjoint operator on } \mathcal{H}\}$$
$$\updownarrow -H \text{ is a generator of } \{T_t\}_{t>0}.$$
$$\{\{T_t\}_{t>0} : \{T_t\}_{t>0} \text{ is a strongly continuous semigroup on } \mathcal{H}\}.$$

Now let $(\mathcal{E},\mathcal{F})$ be a closed form on a Hilbert space $\mathcal{H}$. Let H be the self-adjoint operator associated with $(\mathcal{E},\mathcal{F})$. Also let $\{T_t\}_{t>0}$ be the strongly continuous semigroup associated with $(\mathcal{E},\mathcal{F})$. We will assume these throughout the rest of this section.

Lemma B.2.3. *Let f and g belong to* $\mathrm{Dom}(H)$.
(1) $-HT_tf = -T_tHf = \lim_{h\to 0}(T_{t+h}f - T_tf)/h$ *for any $t > 0$, where the limit is the strong limit in $\mathcal{H}$.*
(2) *For any $t > 0$, $\mathcal{E}(T_tf,g) = \mathcal{E}(f,T_tg)$.*
(3) *$\mathcal{E}(T_tf,T_tf)$ is monotonically decreasing on $(0,\infty)$, and $\mathcal{E}(T_tf,T_tf) \uparrow \mathcal{E}(f,f)$ as $t \downarrow 0$.*

Proof. (1)   Set $f_t = T_tf$. Then

$$(T_{t+h}f - T_tf)/h = T_t(T_hf - f)/h = (T_hf_t - f_t)/h.$$

Note that $f_t = T_tf \in \mathrm{Dom}(H)$ by Theorem B.2.2 and $-H$ is the generator of $\{T_t\}_{t>0}$. Letting $t \to 0$, the equality follows immediately.
(2) Let $(\cdot,\cdot)$ be the inner product of $\mathcal{H}$. Then $\mathcal{E}(T_tf,g) = (T_tf,Hg) = (HT_tf,g) = (T_tHf,g) = (Hf,T_tg) = \mathcal{E}(f,T_tg)$.
(3) Let $u(t) = \mathcal{E}(T_{t/2}f,T_{t/2}f)$. Then $u(t) = \mathcal{E}(f,T_tf) = (Hf,T_tf)$. Hence $u'(t) = -(Hf,HT_tf) = -(Hf,T_tHf) = -(T_{t/2}Hf,T_{t/2}Hf) \le 0$. Therefore $u(t)$ is monotonically decreasing. Since T_t is strongly continuous, $u(t) = (Hf,T_tf) \to (Hf,f) = \mathcal{E}(f,f)$ as $t \downarrow 0$. □

Let $\mathcal{E}_*(u,v) = \mathcal{E}(u,v) + (u,v)$. Then, recalling Theorem B.1.6, we see that $(\mathcal{F},\mathcal{E}_*)$ is a Hilbert space.

Proposition B.2.4. $\{T_t\}_{t>0}$ *is a strongly continuous semigroup on* $(\mathcal{F}, \mathcal{E}_*)$.

Proof. The semigroup property is obvious. Note that $\mathrm{Dom}(H)$ is dense in $\mathcal{F}$ with respect to $\mathcal{E}_*$. Let $\{f_n\}_{n\geq 1} \subset \mathrm{Dom}(H)$. If $f_n \to f$ as $n \to \infty$ in $\mathcal{E}_*$ for some $f \in \mathcal{F}$, then, by Lemma B.2.3-(3), $\{T_t f_n\}_{n\geq 1}$ is an $\mathcal{E}_*$-Cauchy sequence. Since $\{T_t f_n\}_{n\geq 1}$ is an $\mathcal{H}$-Cauchy sequence as well and $T_t f_n \to T_t f$ as $n \to \infty$ in $\mathcal{H}$, it follows that $T_t f_n \to T_t f$ as $n \to \infty$ in $\mathcal{E}_*$. Hence $\mathcal{E}(T_t f, T_t f) \leq \mathcal{E}(f, f)$ for any $f \in \mathcal{F}$.

If $f \in \mathrm{Dom}(H)$, $\mathcal{E}(T_t f - f, T_t f - f) = \mathcal{E}(T_t f, T_t f) - 2\mathcal{E}(T_{t/2} f, T_{t/2} f) + \mathcal{E}(f, f)$. Again by Lemma B.2.3-(3), we have $\mathcal{E}(T_t f - f, T_t f - f) \to 0$ as $t \to 0$. If $f \in \mathcal{F}$, an ordinary approximation argument by a sequence in $\mathrm{Dom}(H)$ gives strong continuity. $\qquad\square$

Next we give a characterization of a strongly continuous semigroup $\{T_t\}_{t>0}$ as a solution of a formal differential equation on a Hilbert space $\mathcal{H}$.

Definition B.2.5. A map $u : [0, \infty) \to \mathcal{H}$ is called a solution of

$$\frac{du}{dt} = -Hu \qquad\qquad\qquad (\mathrm{B.2.1})$$

if and only if $u : [0, \infty) \to \mathcal{H}$ is strongly continuous (i.e., $\|u(t+h) - u(t)\| \to 0$ as $h \to 0$ for any $t \geq 0$), $u(t) \in \mathrm{Dom}(H)$ for any $t > 0$ and $(u(t+h) - u(t))/h$ converges to $-Hu(t)$ as $h \to 0$ with respect to the $\mathcal{H}$-norm for any $t > 0$.

Theorem B.2.6. *For any* $\varphi \in \mathcal{H}$, *there exists a unique solution* $u : [0, \infty) \to \mathcal{H}$ *of* (B.2.1) *that satisfies* $u(0) = \varphi$. *Moreover, the unique solution* $u(t)$ *is given by* $u(t) = T_t \varphi$.

Proof. If $u(t) = T_t \varphi$, it immediately follows that $u(t)$ is a solution of (B.2.1) with $u(0) = \varphi$. Now suppose $u(t)$ and $v(t)$ are solutions of (B.2.1) with $u(0) = v(0) = \varphi$. Let $f(t) = (u(t) - v(t), u(t) - v(t))$. Then we see that $f'(t) = -2\mathcal{E}(u(t) - v(t), u(t) - v(t)) \leq 0$ for $t > 0$. Hence $f(t)$ is a non-negative monotone non-increasing function for $t \geq 0$. Since $f(0) = 0$, $f(t) = 0$ for any $t > 0$. Therefore $u(t) = v(t)$ for any $t \geq 0$. $\qquad\square$

Corollary B.2.7. *If* $\varphi \in \mathrm{Dom}(H)$ *is an eigenfunction of* H *belonging to an eigenvalue* λ *(i.e.,* $H\varphi = \lambda\varphi$*), then* $T_t \varphi = e^{-\lambda t} \varphi$.

Proof. As $e^{-\lambda t} \varphi$ is a solution of (B.2.1), the above theorem implies that $T_t \varphi = e^{-\lambda t} \varphi$. $\qquad\square$

B.3 Dirichlet forms and the Nash inequality

In this section, we introduce the notion of Dirichlet forms. One can find a more detailed and well organized account in [43]. We will also introduce the Nash inequality to study the asymptotic behavior of a strongly continuous semigroup associated with a Dirichlet form as $t \downarrow 0$.

Let X be a locally compact metric space. Also let μ be a σ-finite Borel measure on X that satisfies $\mu(A) < \infty$ for any compact set A and $\mu(O) > 0$ for any non-empty open set O. Define $C_0(X)$ by

$$C_0(X) = \{f : X \to \mathbb{R}, f \text{ is continuous and } \operatorname{supp}(f) \text{ is compact}\}.$$

Definition B.3.1. Let $\mathcal{E}$ be a closed form on $L^2(X, \mu)$. Set $\mathcal{F} = \operatorname{Dom}(\mathcal{E})$.
(1) Markov property: We say that the unit contraction operates on $\mathcal{E}$ if and only if $\bar{u} \in \mathcal{F}$ and $\mathcal{E}(\bar{u}, \bar{u}) \leq \mathcal{E}(u, u)$ for any $u \in \mathcal{F}$, where $\bar{u}$ is defined in the same way as (DF3) in Definition 2.1.1. $(\mathcal{E}, \mathcal{F})$ is said to have the Markov property if and only if the unit contraction operates on $\mathcal{E}$.
(2) Core: A subspace $\mathcal{C}$ of $\mathcal{F} \cap C_0(X)$ is called a core of $(\mathcal{E}, \mathcal{F})$ if and only if $\mathcal{C}$ is dense in $\mathcal{F}$ with respect to the $\mathcal{E}_*$-norm and in $C_0(X)$ with respect to the supremum norm.
(3) Local property: $(\mathcal{E}, \mathcal{F})$ is said to have the local property if $\mathcal{E}(u, v) = 0$ whenever $u, v \in \mathcal{F}$, $\operatorname{supp}(u)$ and $\operatorname{supp}(v)$ are compact and $\operatorname{supp}(u) \cap \operatorname{supp}(v) = \emptyset$.

Definition B.3.2 (Dirichlet form). Let $\mathcal{E}$ be a closed form on $L^2(X, \mu)$. Set $\mathcal{F} = \operatorname{Dom}(\mathcal{E})$. $(\mathcal{E}, \mathcal{F})$ is called a Dirichlet form on $L^2(X, \mu)$ if and only if it has the Markov property. Moreover, a Dirichlet form is called regular if and only if it possesses a core. Also a Dirichlet form which has the local property is called a local Dirichlet form.

If $(\mathcal{E}, \mathcal{F})$ possesses a core, $\mathcal{F} \cap C_0(X)$ is also a core. Hence, a Dirichlet form $(\mathcal{E}, \mathcal{F})$ is regular if and only if $\mathcal{F} \cap C_0(X)$ is a core.

In the rest of this section, we assume that $(\mathcal{E}, \mathcal{F})$ is a Dirichlet form on $L^2(X, \mu)$ and H is the associated non-negative self-adjoint operator. Then Theorem B.2.2 implies that there exists a corresponding strongly continuous semigroup $\{T_t\}_{t>0}$ whose generator is equal to $-H$. Since $(\mathcal{E}, \mathcal{F})$ satisfies the Markov property in addition to being a closed form, $\{T_t\}_{t>0}$ also has the following additional property.

Definition B.3.3. A strongly continuous semigroup $\{T_t\}$ on $L^2(X, \mu)$ is said to have the Markov property if and only if it satisfies the following condition: if $u \in L^2(X, \mu)$ and $0 \leq u(x) \leq 1$ for μ-a. e. $x \in X$, then $0 \leq (T_t u)(x) \leq 1$ for μ-a. e. $x \in X$ and any $t > 0$.

Theorem B.3.4. *There is a one-to-one correspondence between Dirichlet forms on $L^2(X, \mu)$ and strongly continuous semigroups on $L^2(X, \mu)$ with the Markov property. The correspondence is given by the natural one between closed forms and strongly continuous semigroups described in B.2.*

One can also find details on semigroups associated with Dirichlet forms (including a proof of the above theorem) in Davies [25].

Next we show that T_t can be extended to a bounded operator on $L^p(X, \mu)$ for $p = 1, \infty$. For μ-measurable functions f and g on X, we write $f \geq g$ if and only if $f(x) \geq g(x)$ for μ-almost every $x \in X$. Also, if no confusion can occur, we use L^p in place of $L^p(X, \mu)$.

Lemma B.3.5. *Let $f \in L^2$.*
(1) *If $f \geq 0$, then $T_t f \geq 0$.*
(2) *$T_t|f| \geq |T_t f|$.*

Proof. (1)   Let $X_n = f^{-1}([0, n])$ and let $f_n = f \cdot \chi_{X_n}$, where χ_{X_n} is the characteristic function of X_n. Then $f_n \to f$ as $n \to \infty$ in L^2. Since $0 \leq f_n \leq n$, the Markov property of $\{T_t\}_{t>0}$ implies that $0 \leq T_t f_n \leq n$. Letting $n \to \infty$, we obtain $0 \leq T_t f$.
(2) Since $-|f| \leq f \leq |f|$, this follows immediately from (1).   □

Proposition B.3.6. *For $p = 1, \infty$, $T_t(L^2 \cap L^p) \subseteq L^p$ for any $t > 0$ and $||T_t f||_p \leq ||f||_p$ for any $f \in L^2 \cap L^p$ and for any $t > 0$, where $|| \cdot ||_p$ is the L^p-norm.*

Proof. First, we will show the case $p = 1$. As μ is σ-finite, there exists $\{E_n\}_{n \geq 1} \subset \mathcal{B}(X)$ such that $\cup_{n \geq 1} E_n = X$, $\mu(E_n) < \infty$ and $E_n \subseteq E_{n+1}$ for any $n \geq 1$. Let $f \in L^2 \cap L^1$. By Lemma B.3.5, $(|T_t f|, \chi_{E_n}) \leq (T_t |f|, \chi_{E_n}) = (|f|, T_t \chi_{E_n}) \leq ||f||_1$, where $(\cdot, \cdot)$ is the L^2-inner product. (The last inequality comes from the fact that $0 \leq T_t \chi_{E_n} \leq 1$.) Hence, we see that $T_t f \in L^1$ and $||T_t f||_1 \leq ||f||_1$.

Next let $p = \infty$. By the Markov property, $0 \leq T_t |f| \leq ||f||_\infty$ for $f \in L^2 \cap L^\infty$. Hence Lemma B.3.5 implies that $0 \leq |T_t f| \leq T_t |f| \leq ||f||_\infty$. Therefore $T_t f \in L^\infty$ and $||T_t f||_\infty \leq ||f||_\infty$.   □

By the above proposition, for $p = 1, \infty$, $T_t : L^2 \cap L^p \to L^p$ naturally extends to a bounded operator $T_t^{(p)} : L^p \to L^p$ with $||T_t^{(p)}||_p \leq 1$, where $||A||_p$ is the operator norm of a bounded operator $A : L^p \to L^p$. (If $\mu(X) = \infty$, then $T_t^{(\infty)}$ is a bounded operator from $\widetilde{L}^\infty$ to L^∞, where $\widetilde{L}^\infty$ is the closure of $L^\infty \cap L^2$ in L^∞.) Moreover, by the Riesz–Thorin interpolation theorem ([155, Theorem IX.17], [25]), we can extend this to any $p \in [1, \infty]$. If no confusion can occur, we write T_t in place of $T_t^{(p)}$. If $A : L^p \to L^q$

is a bounded operator, we use $||A||_{p \to q}$ to denote the operator norm of $A : L^p \to L^q$.

Now we consider the asymptotic behavior of T_t as $t \downarrow 0$ by using an idea originally due to Nash.

Theorem B.3.7. *The following four conditions are equivalent.*
(1) *There exist positive constants c_1 and θ such that, for any $f \in \mathcal{F} \cap L^1$,*

$$||f||_2^{2+4/\theta} \le c_1(\mathcal{E}(f,f) + ||f||_2^2)||f||_1^{4/\theta}. \qquad (B.3.1)$$

(2) *For any $t > 0$, $T_t(L^1) \subset L^2$ and $T_t : L^1 \to L^2$ is a bounded operator. Moreover, there exists $c_2 > 0$ such that*

$$||T_t||_{1 \to 2} \le c_2 t^{-\theta/4} \qquad (B.3.2)$$

for any $t \in (0, 1]$.
(3) *For any $t > 0$, $T_t(L^2) \subset L^\infty$ and $T_t : L^2 \to L^\infty$ is a bounded operator. Moreover, there exists $c_3 > 0$ such that*

$$||T_t||_{2 \to \infty} \le c_3 t^{-\theta/4} \qquad (B.3.3)$$

for any $t \in (0, 1]$.
(4) *For any $t > 0$, $T_t(L^1) \subset L^\infty$ and $T_t : L^1 \to L^\infty$ is a bounded operator. Moreover, there exists $c_4 > 0$ such that*

$$||T_t||_{1 \to \infty} \le c_4 t^{-\theta/2} \qquad (B.3.4)$$

for any $t \in (0, 1]$.

Remark. As $||T_t||_p \le 1$ for $p = 2, \infty$, it follows that $||T_t||_{1 \to p} \le ||T_1||_{1 \to p}$ for any $t \ge 1$ and that $||T_t||_{2 \to \infty} \le ||T_1||_{2 \to \infty}$ for any $t \ge 1$. Hence (B.3.2), (B.3.3) and (B.3.4) are equivalent, respectively, to

$$||T_t||_{1 \to 2} \le c_2 \max\{1, t^{-\theta/4}\},$$

$$||T_t||_{2 \to \infty} \le c_3 \max\{1, t^{-\theta/4}\}$$

and

$$||T_t||_{1 \to \infty} \le c_4 \max\{1, t^{-\theta/2}\},$$

for any $t > 0$.

(B.3.1) is called the Nash inequality. In [144], Nash used essentially the same argument as (1) $\Rightarrow$ (4) to study the asymptotic behavior of the fundamental solution of a parabolic partial differential equation as $t \downarrow 0$. The present form of the theorem was that essentially obtained by Carlen, Kusuoka and Stroock [23]. See also Davies [25].

To prove the theorem, we need one lemma about a dual operator.

Lemma B.3.8. *Let $A : L^2 \to L^2$ be a bounded self-adjoint operator. Then the following two conditions are equivalent.*
(1) There exists a bounded linear operator B from L^1 to L^2 such that $A|_{L^1 \cap L^2} = B|_{L^1 \cap L^2}$.
(2) $A(L^2) \subset L^\infty$ and A is a bounded operator from L^2 to L^∞.
Moreover, if one of the above conditions holds, then $||A||_{2 \to \infty} = ||B||_{1 \to 2}$.

Proof. $(1) \Rightarrow (2)$: Since the dual spaces of L^1 and L^2 are L^∞ and L^2 respectively, the dual operator of B, B^*, is a bounded operator from L^2 to L^∞. If $u \in L^2$ and $v \in L^1 \cap L^2$, then

$$(v, Au) = (Av, u) = (Bv, u) = (v, B^*u).$$

Hence $Au = B^*u$ for any $u \in L^2$. Since $||B||_{1 \to 2} = ||B^*||_{2 \to \infty}$ (see, for example, [186, VII.1, Theorem 2']), we have $||B||_{1 \to 2} = ||A||_{2 \to \infty}$.
$(2) \Rightarrow (1)$ Let A^* be the dual operator of $A : L^2 \to L^\infty$. As $L^1 \subset (L^\infty)^*$, we see that $A^*u \in L^2$ for any $u \in L^1$. Set $B = A^*|_{L^1}$. If $u \in L^2$ and $v \in L^1 \cap L^2$ then $(u, Av) = (Au, v) = (u, Bv)$. Therefore, $A|_{L^1 \cap L^2} = B|_{L^1 \cap L^2}$. Hence we obtain (1). $\qquad\square$

Proof of Theorem B.3.7.
$(1) \Rightarrow (2)$: Let $f \in L^2 \cap L^1$ with $||f||_1 = 1$. Set $u(t) = (T_t f, T_t f)$. Then,
$$\frac{u(t+h) - u(t)}{h} = (T_{t+h} f + T_t f, (T_h - I)T_t f/h)$$
$$\to -2(T_t f, H T_t f) = -2\mathcal{E}(T_t f, T_t f)$$

as $h \to 0$. Hence $u'(t) = -2\mathcal{E}(T_t f, T_t f)$. Now, by the Nash inequality,

$$2u(t)^{1+2/\theta} \le c_1(-u'(t) + 2u(t))||T_t f||_1^{4/\theta} \le c_1(-u'(t) + 2u(t)),$$

where we used the fact that $||T_t f||_1 \le ||f||_1 = 1$. This implies

$$2(e^{-2t}u(t))^{(1+2/\theta)} \le 2e^{-2t}u(t)^{(1+2/\theta)} \le -c_1(e^{-2t}u(t))'.$$

Set $v(t) = (e^{-2t}u(t))^{-2/\theta}$. Then we obtain $v'(t) \ge 4/(c_1\theta)$. Since $v(t) \to u(0)^{-2/\theta} > 0$ as $t \downarrow 0$, it follows that $v(t) \ge 4t/(c_1\theta)$. Therefore

$$u(t) \le ce^{2t}t^{-\theta/2},$$

where $c = (c_1\theta/4)^{\theta/2}$. Hence

$$||T_t f||_2 \le ce^t t^{-\theta/4}||f||_1$$

for any $f \in L^2 \cap L^1$. This immediately implies (2).
$(2) \Leftrightarrow (3)$: Apply Lemma B.3.8.

$(2) \Rightarrow (4)$ As $(2) \Leftrightarrow (3)$, we see that $T_{t/2} : L^1 \to L^2$ and $T_{t/2} : L^2 \to L^\infty$ are bounded operators. Hence $T_t = T_{t/2} \circ T_{t/2} : L^1 \to L^\infty$ is a bounded operator and

$$||T_t||_{1 \to \infty} \leq ||T_{t/2}||_{1 \to 2} ||T_{t/2}||_{2 \to \infty} \leq c_2 c_3 t^{\theta/2}.$$

$(4) \Rightarrow (1)$: Assume that $T_t : L^1 \to L^\infty$ is a bounded operator and (B.3.4) holds for any $t \in (0, 1]$. By the remark immediately after Theorem B.3.7, it follows that $||T_t||_{1 \to \infty} \leq c_4 e^t t^{-\theta/2}$ for any $t > 0$. Let $f \in \mathcal{F} \cap L^1$. Then, if $t > \epsilon > 0$, we have

$$(e^{-t} T_t f, f) = (e^{-\epsilon} T_\epsilon f, f) + \int_\epsilon^t (\frac{\partial}{\partial s}(e^{-s} T_s f), f) ds$$
$$= (e^{-\epsilon} T_\epsilon f, f) - \int_\epsilon^t e^{-s}((I + H) T_s f, f) ds.$$

Now, using Lemma B.2.3 and Proposition B.2.4,

$$e^{-s}((I + H) T_s f, f) = e^{-s}(T_{s/2} f, T_{s/2} f) + e^{-s}(T_{s/2} H T_{s/2} f, f)$$
$$= e^{-s}(T_{s/2} f, T_{s/2} f) + e^{-s}\mathcal{E}(T_{s/2} f, T_{s/2} f)$$
$$\leq ||f||_2^2 + \mathcal{E}(f, f).$$

This, along with the fact that $(T_t f, f) \leq ||T_t||_{1 \to \infty} ||f||_1^2$, implies

$$c_4 ||f||_1^2 t^{-\theta/2} \geq (e^{-\epsilon} T_\epsilon f, f) - (t - \epsilon)(||f||_2^2 + \mathcal{E}(f, f)).$$

Letting $\epsilon \to 0$, we obtain

$$c_4 ||f||_1^2 t^{-\theta/2} + t(||f||_2^2 + \mathcal{E}(f, f)) \geq ||f||_2^2.$$

It is easy to calculate the value of $t_* > 0$ which gives the minimum value in the left-hand side of the above inequality. Then, substituting t_* for t in the above inequality, we immediately obtain (B.3.1). $\qquad \square$

From the arguments in the proof, in particular, for $(1) \Rightarrow (2)$, we see that the condition (1) can be replaced by the weaker condition
(1)′ There exits a positive constant c_1 such that (B.3.1) holds for any $f \in \text{Dom}(H) \cap L^1$ with $f \geq 0$.

In practice, however, this condition may be no easier to verify than the original one.

Corollary B.3.9. *Suppose that the Nash inequality* (B.3.1) *is satisfied. Let φ be an eigenfunction of H belonging to an eigenvalue $\lambda \geq 1$. Then*

$$||\varphi||_\infty \leq c \lambda^{\theta/4} ||\varphi||_2, \tag{B.3.5}$$

where $c > 0$ is a constant which is independent of φ and λ.

Remark. Under the assumptions of the above corollary, if $\varphi \in L^1$, we also obtain

$$\|\varphi\|_2 \leq c\lambda^{\theta/4}\|\varphi\|_1 \quad \text{and} \quad \|\varphi\|_\infty \leq c^2\lambda^{\theta/2}\|\varphi\|_1.$$

Proof. By Corollary B.2.7, $T_t\varphi = e^{-\lambda t}\varphi$. By Theorem B.3.7, $\|T_t\|_{2\to\infty} \leq c_2 t^{-\theta/4}$ for $t \in (0,1]$. Hence

$$e^{-\lambda t}\|\varphi\|_\infty = \|T_t\varphi\|_\infty \leq c_2 t^{-\theta/4}\|\varphi\|_2.$$

If $t = 1/\lambda$ and $c = c_2 e$, then we get (B.3.5). □

B.4 The renewal theorem

In this section, we will introduce the renewal theorem. One can find the renewal theorems (in various versions) in many places in the literature, for example, Feller [36], Rudin [159].

Definition B.4.1. A function $u : \mathbb{R} \to \mathbb{R}$ is directly Riemann integrable on $\mathbb{R}$ if and only if

$$\bar{\sigma}(h) = h \sum_{j=-\infty}^{\infty} \sup_{t\in[jh,(j+1)h]} u(t) \quad \text{and} \quad \underline{\sigma}(h) = h \sum_{j=-\infty}^{\infty} \inf_{t\in[jh,(j+1)h]} u(t)$$

are finite for any $h > 0$ and tend to the same limit as $h \to 0$.

The following is essentially the renewal theorem (alternative form) in section XI.1 p. 363 of [36]. We only present the case where the distribution is atomic.

Theorem B.4.2 (Feller's renewal theorem). *Let $t_* > 0$. Let f be a measurable function on $\mathbb{R}$ such that $f(t) = 0$ for $t < t_*$. Suppose f satisfies a renewal equation*

$$f(t) = \sum_{j=1}^{N} f(t - \alpha_j)p_j + u(t), \tag{B.4.1}$$

where $\alpha_1, \alpha_2, \ldots, \alpha_N$ are positive numbers, $\sum_{j=1}^{N} p_j = 1$ and $p_j > 0$ for all j and u is a non-negative directly Riemann integrable function on $\mathbb{R}$ with $u(t) = 0$ for $t < t_*$.
(1) *Arithmetic case (Lattice case): Suppose there exists $T > 0$ such that $\alpha_i = m_i T$ for all i, where $m_1, m_2, \ldots, m_N$ are positive integers whose*

greatest common divider is 1. Then $|f(t) - G(t)| \to 0$ as $t \to \infty$, where $G(t)$ is a T-periodic function given by

$$G(t) = (\sum_{j=1}^{N} m_j p_j)^{-1} \sum_{j=-\infty}^{\infty} u(t + jT). \qquad \text{(B.4.2)}$$

(2) Non-arithmetic case (Non-lattice case): *Suppose that $\sum_{i=1}^{N} \mathbb{Z}\alpha_i$ is a dense additive subgroup of $\mathbb{R}$. Then $f(t)$ is convergent as $t \to \infty$ and*

$$\lim_{t \to \infty} f(t) = (\sum_{i=1}^{N} \alpha_i p_i)^{-1} \int_{-\infty}^{\infty} u(t) dt.$$

Feller only stated the case where $t_* = 0$. Using a parallel translation, we can immediately obtain the present version. Also, in Feller's original version, the conclusion for the arithmetic case is that

$$\lim_{n \to \infty} f(t + nT) = G(t)$$

for any t. However, in equation (1.18), p. 362 of [36], Feller gives the following expression for $f(t)$:

$$f(t) = \sum_{k \geq 0} u(t - kT) v_k,$$

where $v_k \to (\sum_{j=1}^{N} m_j p_j)^{-1}$ as $k \to \infty$. Since u is directly Riemann integrable, a standard argument in calculus implies that $|f(t) - G(t)| \to 0$ as $t \to \infty$.

This version of the renewal theorem suffices to prove Theorem 4.1.5 on the asymptotic behavior of eigenvalue counting functions of Laplacians on p. c. f. self-similar sets. (The conditions of the renewal theorem in [93] were unfortunately incorrect. The author thanks D. Vassiliev for having pointed this out.)

Since Feller's version of the renewal theorem, much effort has been made to weaken the conditions, in particular, the condition that $f(t) = u(t) = 0$ for $t < t_*$. See, for example, M. Levitin & D. Vassiliev [115]. In [115], they proved the same conclusions as in the above theorem under the condition that $|u(t)|$ decays exponentially as $|t| \to \infty$, instead of assuming that $u(t) = 0$ for $t < t_*$ and that u is directly Riemann integrable.

The following renewal theorem for the arithmetic case contains information on the order of convergence of $|f(t) - G(t)|$ as $t \to 0$. This kind of error estimate allows us to deduce detailed information about the behavior of the eigenvalue counting function, (4.1.3).

Theorem B.4.3. *Let f be a measurable function on $\mathbb{R}$ with $f(t) \to 0$ as $t \to -\infty$. Suppose f satisfies a renewal equation (B.4.1), where there exists $T > 0$ such that $\alpha_i = m_i T$ for all i, where $m_1, m_2, \ldots, m_N$ are positive integers whose greatest common divisor is 1. Set $u_j(t) = u(t + jT)$ for $t \in [0, T]$. If $\sum_{j=-\infty}^{+\infty} |u_j(t)|$ converges uniformly on $[0, T]$, then $|f(t) - G(t)| \to 0$ as $t \to \infty$, where $G(t)$ is a T-periodic function given by (B.4.2). Moreover, set $Q(z) = (1 - \sum_{j=1}^{N} p_j z^{m_j})/(1 - z)$. Also define $\beta = \min\{|z| : Q(z) = 0\}$ and $m = \max\{$multiplicity of $Q(z) = 0$ at $w :$ $|w| = \beta, Q(w) = 0\}$. If there exist $C > 0$ and $\alpha > 1$ such that $|u(t)| \le C\alpha^{-t}$ for all t, then, as $t \to \infty$,*

$$|G(t) - f(t)| = \begin{cases} O(t^{m-1}\beta^{-t/T}) & \text{if } \alpha^T > \beta, \\ O(t^m \alpha^{-t}) & \text{if } \alpha^T = \beta, \\ O(\alpha^{-t}) & \text{if } \alpha^T < \beta. \end{cases}$$

Remark. If $Q(z) = 1$, then we set $\beta = +\infty$. Since $\sum_{j=1}^{N} p_j = 1$, Q(z) is a polynomial. Furthermore, $|\sum_{j=1}^{N} p_j z^{m_j}| < 1$ for $\{z : |z| \le 1, z \neq 1\}$. Hence we see that $\beta > 1$.

In the rest of this section, we will give a proof of Theorem B.4.3.

Lemma B.4.4. *Set $F(z) = \prod_{j=1}^{k}(1 - e^{i\theta_j}z)^{-1}$, where $0 \le \theta_j < 2\pi$ for $j = 1, 2, \ldots, k$. If $F(z) = \sum_{n=0}^{\infty} a_n z^n$, then $|a_n| = O(n^{m-1})$ as $n \to \infty$, where $m = \max\{\#\{j : \theta = \theta_j\} : 0 \le \theta < 2\pi\}$.*

Proof. We use induction on k. The conclusion is obvious when $k = 1$. Assume that the conclusion holds for k. For $F(z) = \prod_{j=1}^{k+1}(1 - e^{i\theta_j}z)^{-1} = \sum_{n=0}^{\infty} a_n z^n$, if $\theta_j = \theta$ for all j then it is easy to see that $|a_n| = O(n^k)$. If $\theta_p \neq \theta_q$ for some $p \neq q$, then $(1 - e^{i\theta_p}z)^{-1}(1 - e^{i\theta_q}z)^{-1} = a(1 - e^{i\theta_p}z)^{-1} + b(1 - e^{i\theta_q}z)^{-1}$ for some a, b. Hence the statement follows from the induction hypothesis. $\square$

Lemma B.4.5. *For $w = w_1 w_2 \ldots w_k \in W_k$, set $m(w) = \sum_{j=1}^{k} m_{w_j}$ and $p_w = p_{w_1} \cdots p_{w_k}$. Define $\tilde{M}(k) = \sum_{w \in W_* : m(w) = k} p_w$. Then*

$$|(\sum_{j=1}^{N} m_j p_j)^{-1} - \tilde{M}(n)| = O(n^{m-1}\beta^{-n})$$

as $n \to \infty$, where β and m are defined in the statement of Theorem B.4.3.

Remark. If $\{w \in W_* : m(w) = k\} = \emptyset$, then we set $\tilde{M}(k) = 0$. Hence $\tilde{M}(k) = 0$ for any $k < 0$. Also we set $\tilde{M}(0) = 1$.

Proof. Using the fact that $\tilde{M}(k) = \sum_{j=1}^{N} \tilde{M}(k-m_j)p_j$, we see that

$$\sum_{n=0}^{\infty} \tilde{M}(n)z^n = (1 - \sum_{j=1}^{N} p_j z^{m_j})^{-1}.$$

Note that $(\sum_{i=1}^{N} m_i p_i) = Q(1)$. Then we have that $\sum_{n=0}^{\infty} (\tilde{M} - \tilde{M}(n))z^n = R(z)/Q(z)$, where $\tilde{M} = Q(1)^{-1}$ and $R(z)$ is a polynomial defined by $R(z) = (Q(z)/Q(1) - 1)/(1 - z)$. If $Q(z) = a \prod_{j=1}^{deg(Q)}(z - z_j)$, it follows that

$$\sum_{n=0}^{\infty} (\tilde{M} - \tilde{M}(n))z^n = R(z)/Q(z) = (\sum_{j=0}^{\infty} c_n z^n) \prod_{j:|z_j|=\beta} (z - z_j)^{-1},$$

where the radius of convergence of $\sum_{n=0}^{\infty} c_n z^n$ is greater than β. Hence, applying Lemma B.4.4 to $\prod_{j:|z_j|=\beta}(z - z_j)^{-1}$, we obtain the required estimate of $|\tilde{M} - \tilde{M}(n)|$. □

Proof of Theorem B.4.3. By the renewal equation (B.4.1), we have

$$f(t) = \sum_{w \in W_n} f(t - m(w)T)p_w + \sum_{k=0}^{n} \sum_{w \in W_k} u(t - m(w)T)p_w.$$

Since $\lim_{t \to -\infty} f(t) = 0$ and $\sum_{n=0}^{\infty} u(t - nT)$ is absolutely convergent, we have

$$f(t) = \sum_{n=0}^{\infty} u(t - nT)\tilde{M}(n).$$

Hence we obtain

$$G(t) - f(t) = \tilde{M} \sum_{k>0} u(t + kT) + \sum_{k=0}^{\infty} u(t - kT)(\tilde{M} - \tilde{M}(k)). \quad (B.4.3)$$

As $\tilde{M} \leq 1$ and $\tilde{M}(k) \leq 1$,

$$|G(t) - f_n(t)| \leq 2 \sum_{k>n-m} |u_k(t)| + \sum_{k=m}^{\infty} |u_{n-k}(t)||\tilde{M}(k) - \tilde{M}|,$$

where $f_n(t) = f(t+nT)$ on $[0, T]$. For $\epsilon > 0$, choose m so that $|\tilde{M}(k) - \tilde{M}| < \epsilon$ for $k \geq m$. Then for sufficiently large n, we have $\sum_{k>n-m} |u_k(t)| \leq \epsilon$. Hence $|f_n(t) - G(t)| \leq (2+A)\epsilon$, where $A = \sup_{0 \leq t \leq T} \sum_{k=-\infty}^{\infty} |u_k(t)|$. Hence f_n is uniformly convergent to G as $n \to \infty$ on $[0, T]$. So $f(t) - G(t) \to 0$ as $t \to \infty$.

Now suppose $|u(t)| \leq C\alpha^{-t}$. Then, by (B.4.3) and Lemma B.4.5, it follows that

$$|G(t) - f(t)| \leq c_1\alpha^{-t} + c_2\alpha^{-t} \sum_{0 \leq k \leq t/T} k^{m-1}(\alpha^T/\beta)^k + c_3 \sum_{k > t/T} k^{m-1}\beta^{-k}$$
$$\leq c_1\alpha^{-t} + c_2\alpha^{-t} \sum_{0 \leq k \leq t/T} k^{m-1}(\alpha^T/\beta)^k + c_4(t/T)^{m-1}\beta^{-t/T},$$

where c_1, c_2, c_3 and c_4 are positive constants. From this inequality, it is easy to deduce the required estimate. In particular, if $\alpha^T > \beta$, we use the fact that there exists $c > 0$ such that

$$\alpha^{-nT} \sum_{0 \leq k \leq n} k^{m-1}(\alpha^T/\beta)^k \leq cn^{m-1}\beta^{-n}$$

for any $n \geq 1$. $\qquad\square$

Bibliography

[1] S. Alexander and R. Orbach, Density of states on fractals: fractons, *J. Physique Letters* **43** (1982), L623–L631.

[2] G. Ben Arous and T. Kumagai, Large deviations of Brownian motion on the Sierpinski gasket, *Stochastic Process. Appl.* **85** (2000), 225–235.

[3] M. T. Barlow, Random walks and diffusions on fractals, Proceedings of International Congress Math. Kyoto 1990 (Tokyo), Springer, 1991, pp. 1025–1035.

[4] ———, Harmonic analysis on fractal spaces, Séminaire Bourbaki Volume 1991/1992, Astérisque, vol. 206, 1992.

[5] ———, Random walks, electrical resistance, and nested fractals, Asymptotic problems in probability theory: stochastic models and diffusions on fractals (K. D. Elworthy and N. Ikeda, eds.), Pitman Research Notes in Math., vol. 283, Longman, 1993, pp. 131–157.

[6] ———, Diffusion on fractals, Lecture notes Math. vol. 1690, Springer, 1998.

[7] M. T. Barlow and R. F. Bass, The construction of Brownian motion on the Sierpinski carpet, *Ann. Inst. Henri Poincaré* **25** (1989), 225–257.

[8] ———, Local time for Brownian motion on the Sierpinski carpet, *Probab. Theory Related Fields* **85** (1990), 91–104.

[9] ———, On the resistance of the Sierpinski carpet, *Proc. R. Soc. London A* **431** (1990), 354–360.

[10] ———, Transition densities for Brownian motion on the Sierpinski carpet, *Probab. Theory Related Fields* **91** (1992), 307–330.

[11] ———, Coupling and Harnack inequalities for Sierpinski carpets, *Bull. Amer. Math. Soc. (N. S.)* **29** (1993), 208–212.

[12] ———, Brownian motion and harmonic analysis on Sierpinski carpets, *Canad. J. Math.* **51** (1999), 673–744.

[13] ———, Divergence form operators on fractal-like domains, *J. Functional Analysis* **175** (2000), 214–247.

[14] M. T. Barlow, R. F. Bass and K. Burdzy, Positivity of Brownian transition densities, *Elect. Comm. in Probab.* **2** (1997), 43–51.

[15] M. T. Barlow, R. F. Bass and J. D. Sherwood, Resistance and spectral dimension of Sierpinski carpets, *J. Phys. A: Math. Gen.* **23** (1990), L253–L258.

[16] M. T. Barlow and B. M. Hambly, Transition density estimates for Brownian motion on scale irregular Sierpinski gaskets, *Ann. Inst. H. Poincaré* **33** (1997), 531–557.

[17] M. T. Barlow, K. Hattori and T. Hattori, Weak homogenization of anisotropic diffusion on pre-Sierpinski carpets, *Comm. Math. Phys.* **188** (1997), 1–27.

[18] M. T. Barlow and J. Kigami, Localized eigenfunctions of the Laplacian on p.c.f. self-similar sets, *J. London Math. Soc. (2)* **56** (1997), 320–332.

[19] M. T. Barlow and T. Kumagai, Transition density asymptotics for some diffusion processes with multi-fractal structures, preprint.

[20] M. T. Barlow and E. A. Perkins, Brownian motion on the Sierpinski gasket, *Probab. Theory Related Fields* **79** (1988), 542–624.

[21] O. Ben-Bassat, R. S. Strichartz and A. Teplyaev, What is not in the domain of the Laplacian on a Sierpinski gasket type fractal, *J. Functional Analysis* **166** (1999), 197–217.

[22] M. Biroli and U. Mosco, Sobolev inequalities on homogeneous spaces, *Potential Anal.* **4** (1995), 311–324.

[23] E. Carlen, S. Kusuoka and D. Stroock, Upper bounds for symmetric Markov transition functions, *Ann. Inst. Henri Poincaré* **23** (1987), 245–287.

[24] K. Dalrymple, R. S. Strichartz and J. P. Vinson, Fractal differential equations on the Sierpinski gasket, *J. Fourier Anal. Appl.* **5** (1999), 203–284.

[25] E. B. Davies, *Heat Kernels and Spectral Theory*, Cambridge Tracts in Math. vol 92, Cambridge University Press, 1989.

[26] _____, *Spectral Theory and Differential Operators*, Cambridge Studies in Advanced Math. vol. 42, Cambridge University Press, 1995.

[27] M. Denker and H. Sato, Reflections on harmonic analysis of the Sierpinski gasket, preprint.

[28] _____, Sierpinski gasket as a Martin boundary I: Martin kernels, to appear in Potential Analysis.

[29] _____, Sierpinski gasket as a Martin boundary II: intrinsic metric, *Publ. Res. Inst. Math. Sci.* **35** (1999), 769–794.

[30] D. Dhar, Lattices of effectively nonintegral dimensionality, *J. Math. Phys.* **18** (1977), 577–583.

[31] P. G. Doyle and J. L. Snell, *Random Walks and Electrical Networks*, Math. Assoc. Amer., Washington, 1984.

[32] K. J. Falconer, *Geometry of Fractal Sets*, Cambridge University Press, 1985.

[33] _____, *Fractal Geometry*, Wiley, 1990.

[34] _____, *Techniques in Fractal Geometry*, Wiley, 1997.

[35] _____, Semilinear PDEs on self-similar fractals, *Comm. Math. Phys.* **206** (1999), 235–245.

[36] W. Feller, *An Introduction to Probabilistic Theory and its Application vol. II*, second ed., Wiley, New York, 1966.

[37] P. J. Fitzsimmons, B. M. Hambly and T. Kumagai, Transition density estimates for Brownian motion on affine nested fractals, *Comm. Math. Phys.* **165** (1994), 595–620.

[38] U. Freiberg and M. Zähle, Harmonic calculus on fractals – a measure theoretic approach I, to appear in Potential Analysis.

[39] T. Fujita, A fractional dimension, self-similarity and a generalized diffusion operator, Probabilistic Methods on Mathematical Physics, Proc. of Taniguchi International Symp. (Katata & Kyoto, 1985) (Tokyo) (K. Ito and N. Ikeda, eds.), Kinokuniya, 1987, pp. 83–90.

[40] _____, Some asymptotics estimates of transition probability densities for generalized diffusion processes with self-similar measures, *Publ. Res. Inst. Math. Sci.* **26** (1990), 819–840.

[41] M. Fukushima, Dirichlet forms, diffusion processes and spectral dimensions for nested fractals, Ideas and methods in mathematical analysis, stochastics and applications, Proc. Conf. in Memory of Hoegh-Krøhn (S. Albeverio *et al.*, eds.), vol. 1, Cambridge University Press, 1992, pp. 151–161.

[42] _____, On limit theorems for Brownian motions on unbounded fractal sets, Fractal Geometry and Stochastics II (C. Bandt *et al.*, eds.), Progress in Probability, vol. 46, Birkhäuser, 2000, pp. 227–237.

[43] M. Fukushima, Y. Oshima and M. Takeda, *Dirichlet Forms and Symmetric Markov Processes*, de Gruyter Studies in Math. vol. 19, de Gruyter, Berlin, 1994.

[44] M. Fukushima and T. Shima, On a spectral analysis for the Sierpinski gasket, *Potential Analysis* **1** (1992), 1–35.

[45] _____, On discontinuity and tail behavior of the integrated density of state for nested pre-fractals, *Comm. Math. Phys.* **163** (1994), 462–471.

[46] M. Fukushima, T. Shima and M. Takeda, Large deviations and related LILs for Brownian motion on nested fractals, *Osaka J. Math.* **36** (1999), 497–537.

[47] Y. Gefen, A. Aharony and B. Mandelbrot, Phase transitions on fractals I: Quasilinear lattices, *J. Phys. A* **16** (1983), 1267–1278.

[48] _____, Phase transitions on fractals III: Infinitely ramified lattices, *J. Phys. A* **17** (1984), 1277–1289.

[49] Y. Gefen, A. Aharony, Y. Shapir and B. Mandelbrot, Phase transitions on fractals II: Sierpinski gaskets, *J. Phys. A* **17** (1984), 435–444.

[50] M. Gibbons, A. Raj and R. S. Strichartz, The finite element method on the Sierpinski gasket, preprint.

[51] S. Goldstein, Random walks and diffusions on fractals, Percolation theory and ergodic theory of infinite particle systems (H. Kesten, ed.), IMA Math. Appl., vol. 8, Springer, 1987, pp. 121–129.

[52] P. Grabner, Functional iterations and stopping times for Brownian motion on the Sierpinski gasket, *Mathematika* **44** (1997), 374–400.

[53] P. Grabner and R. Tichty, Equidistribution and Brownian motion on the Sierpinski gasket, *Monatsh. Math.* **125** (1998), 147–164.

[54] P. Grabner and W. Woess, Functional iterations and periodic oscillations for simple random walk on the Sierpinski graph, *Stochastic Process. Appl.* **69** (1997), 127–138.

[55] B. M. Hambly, Brownian motion on a homogeneous random fractal, *Probab. Theory Related Fields* **94** (1992), 1–38.

[56] _____, Brownian motion on a random recursive Sierpinski gasket, *Ann Probab.* **25** (1997), 1059–1102.

[57] _____, Heat kernels and spectral asymptotics for some random Sierpinski gaskets, Fractal Geometry and Stochastics II (C. Bandt *et al.*, eds.), Progress in Probability, vol. 46, Birkhäuser, 2000, pp. 239–267.

[58] B. M. Hambly, J. Kigami and T. Kumagai, Multifractal formalisms for the local spectral and walk dimensions, to appear in Math. Proc. Cambridge Phil. Soc.

[59] B. M. Hambly and T. Kumagai, Heat kernel estimates and homogenization for asymptotically lower dimensional processes on some nested fractals, *Potential Analysis* **8** (1998), 359–458.

[60] _____, Transition density estimates for diffusion processes on post critically finite self-similar fractals, *Proc. London Math. Soc. (3)* **78** (1999), 431–458.

[61] B. M. Hambly, T. Kumagai, S. Kusuoka and X. Y. Zhou, Transition density estimates for diffusion processes on homogeneous random Sierpinski carpets, *J. Math. Soc. Japan* **52** (2000), 373–408.

[62] B. M. Hambly and T. J. Lyons, Stochastic area for Brownian motion on the Sierpinski gasket, *Ann. Probab.* **26** (1998), 132–148.

[63] B. M. Hambly and V. Metz, The homogenization problem for the Vicsek set, *Stochastic Process. Appl.* **76** (1998), 167–190.

[64] M. Hata, On the structure of self-similar sets, *Japan J. Appl. Math.* **3** (1985), 381–414.

[65] K. Hattori, Self-avoiding processes on the Sierpinski gasket, Asymptotic problems in probability theory: stochastic models and diffusions on fractals (K. D. Elworthy and N. Ikeda, eds.), Pitman Research Notes in Math., vol. 283, Longman, 1993, pp. 183–200.

[66] ———, Fractal geometry of self-avoiding processes, *J. Math. Sci. Univ. Tokyo* **3** (1996), 379–397.

[67] K. Hattori and T. Hattori, Self-avoiding process on the Sierpinski gasket, *Probab. Theory Related Fields* **88** (1991), 405–428.

[68] K. Hattori, T. Hattori and S. Kusuoka, Self-avoiding paths on the pre-Sierpinski gaskets, *Probab. Theory Related Fields* **84** (1990), 1–26.

[69] ———, Self-avoiding paths on the three-dimensional Sierpinski gasket, *Publ. Res. Inst. Math. Sci.* **29** (1993), 455–509.

[70] K. Hattori, T. Hattori and T. Watanabe, Gaussian field theories on general networks and the spectral dimensions, *Progr. Theoret. Phys. Suppl.* **92** (1987), 108–143.

[71] ———, Asymptotically one-dimensional diffusions on the Sierpinski gasket and the *abc*-gaskets, *Probab. Theory Related Fields* **100** (1994), 85–116.

[72] T. Hattori, Asymptotically one-dimensional diffusions on scale-irregular gaskets, *J. Math. Sci. Univ. Tokyo* **4** (1997), 229–278.

[73] T. Hattori and S. Kusuoka, The exponent for the mean square displacement of self-avoiding random walks on the Sierpinski gasket, *Probab. Theory Related Fields* **93** (1992), 273–284.

[74] T. Hattori and H. Watanabe, Anisotropic random walks and asymptotically one-dimensional diffusion on *abc*-gaskets, *J. Statist. Phys.* **88** (1997), 105–128.

[75] S. Havlin and D. Ben-Avraham, Diffusion in disordered media, *Adv. Phys.* **36** (1987), 695–798.

[76] J. E. Hutchinson, Fractals and self similarity, *Indiana Univ. Math. J.* **30** (1981), 713–747.

[77] F. John, *Partial Differential Equations*, fourth ed., Applied Math. Science vol. 1, Springer, 1982.

[78] A. Jonsson, Brownian motion on fractals and function spaces, *Math. Z.* **222** (1996), 495–504.

[79] A. Jonsson and H. Wallin, Boundary value problems and Brownian motion on fractals, *Chaos Solitons & Fractals* **8** (1997), 191–205.

[80] A. Kameyama, Self-similar sets from the topological point of view, *Japan J. Indust. Appl. Math.* **10** (1993), 85–95.

[81] T. Kato, *Perturbation Theory for Linear Operators*, Classics in Math., Springer, 1995, originally published as Grundlehren der mathematischen Wissenschaften band 132.

[82] J. Kigami, A harmonic calculus on the Sierpinski spaces, *Japan J. Appl. Math.* **6** (1989), 259–290.

[83] ———, Harmonic calculus on p.c.f. self-similar sets, *Trans. Amer. Math. Soc.* **335** (1993), 721–755.

[84] ———, Harmonic metric and Dirichlet form on the Sierpinski gasket, Asymptotic problems in probability theory: stochastic models and diffusions on fractals (K. D. Elworthy and N. Ikeda, eds.), Pitman

Research Notes in Math., vol. 283, Longman, 1993, pp. 201–218.

[85] _____, Effective resistances for harmonic structures on p.c.f. self-similar sets, *Math. Proc. Cambridge Phil. Soc.* **115** (1994), 291–303.

[86] _____, Laplacians on self-similar sets (analysis on fractals), *Amer. Math. Soc. Transl. Ser. 2* **161** (1994), 75–93.

[87] _____, Harmonic calculus on limits of networks and its application to dendrites, *J. Functional Analysis* **128** (1995), 48–86.

[88] _____, Hausdorff dimensions of self-similar sets and shortest path metrics, *J. Math. Soc. Japan* **47** (1995), 381–404.

[89] _____, Laplacians on self-similar sets and their spectral distributions, Fractal Geometry and Stochastics (C. Bandt *et al.*, eds.), Progress in Probability, vol. 37, Birkhäuser, 1995, pp. 221–238.

[90] _____, Distributions of localized eigenvalues of Laplacians on p.c.f. self-similar sets, *J. Functional Analysis* **156** (1998), 170–198.

[91] _____, Markov property of Kusuoka-Zhou's Dirichlet forms on self-similar sets, *J. Math. Sci. Univ. Tokyo* **7** (2000), 27–33.

[92] J. Kigami and M. L. Lapidus, Self-similarity of volume measures for Laplacians on p. c. f. self-similar fractals, to appear in Comm. Math. Phys.

[93] _____, Weyl's problem for the spectral distribution of Laplacians on p.c.f. self-similar fractals, *Comm. Math. Phys.* **158** (1993), 93–125.

[94] J. Kigami, D. Sheldon and R. S. Strichartz, Green's function on fractals, to appear in Fractals, 1999.

[95] S. M. Kozlov, Harmonization and homogenization on fractals, *Comm. Math. Phys.* **153** (1993), 339–357.

[96] W. B. Krebs, A diffusion defined on a fractal state space, *Stoch. Proc. Appl.* **37** (1991), 199–212.

[97] _____, Hitting time bounds for Brownian motion on a fractal, *Proc. Amer. Math. Soc.* **118** (1993), 223–232.

[98] T. Kumagai, Construction and some properties of a class of non-symmetric diffusion process on the Sierpinski gasket, Asymptotic problems in probability theory: stochastic models and diffusions on fractals (K. D. Elworthy and N. Ikeda, eds.), Pitman Research Notes in Math., vol. 283, Longman, 1993, pp. 219–247.

[99] _____, Estimates of the transition densities for Brownian motion on nested fractals, *Probab. Theory Related Fields* **96** (1993), 205–224.

[100] _____, Regularity, closedness and spectral dimensions of the Dirichlet forms on p.c.f. self-similar sets, *J. Math. Kyoto Univ.* **33** (1993), 765–786.

[101] _____, Rotation invariance and characterization of a class of self-similar diffusion processes on the Sierpinski gasket, Algorithms, fractals, and dynamics (Okayama/Kyoto, 1992) (New York), Plenum, 1995, pp. 131–142.

[102] _____, Short time asymptotic behavior and large deviation for Brownian motion on some affine nested fractals, *Publ. Res. Inst. Math. Sci.* **33** (1997), 223–240.

[103] _____, Brownian motion penetrating fractals: an application of the trace theorem of Besov spaces, *J. Functional Analysis* **170** (2000), 69–92.

[104] _____, Stochastic processes on fractals and related topics, *Sugaku Expositions* **13** (2000), 55–71.

[105] T. Kumagai and S. Kusuoka, Homogenization on nested fractals, *Probab. Theory Related Fields* **104** (1996), 375–398.

[106] S. Kusuoka, A diffusion process on a fractal, Probabilistic Methods on Mathematical Physics, Proc. of Taniguchi International Symp. (Katata &

Kyoto, 1985) (Tokyo) (K. Ito and N. Ikeda, eds.), Kinokuniya, 1987, pp. 251–274.

[107] _____, Dirichlet forms on fractals and products of random matrices, *Publ. Res. Inst. Math. Sci.* **25** (1989), 659–680.

[108] _____, Lecture on diffusion process on nested fractals, Lecture Notes in Math., vol. 1567, pp. 39–98, Springer, 1993.

[109] S. Kusuoka and X. Y. Zhou, Dirichlet forms on fractals: Poincaré constant and resistance, *Probab. Theory Related Fields* **93** (1992), 169–196.

[110] _____, Waves on fractal-like manifolds and effective energy propagation, *Probab. Theory Related Fields* **110** (1998), 473–495.

[111] M. L. Lapidus, Fractal drum, inverse spectral problems for elliptic operators and a partial resolution of the Weyl-Berry conjecture, *Trans. Amer. Math. Soc.* **325** (1991), 465–529.

[112] _____, Analysis on fractals, Laplacians on self-similar sets, noncommutative geometry and spectral dimensions, *Topological Methods in Nonlinear Analysis* **4** (1994), 137–195.

[113] _____, Towards a noncommutative fractal geometry?—Laplacians and volume measures on fractals—, *Contemporary Mathematics* **208** (1997), 211–252.

[114] M. L. Lapidus and M. van Frankenhuysen, *Fractal Geometry and Number Theory*, Birkhäuser, 2000.

[115] M. Levitin and D. Vassiliev, Spectral asymptotics, renewal theorem, and the Berry conjecture for a class of fractals, *Proc. London Math. Soc. (3)* **72** (1996), 188–214.

[116] T. Lindstrøm, Brownian motion on nested fractals, *Mem. Amer. Math. Soc.* **420** (1990).

[117] _____, Brownian motion penetrating the Sierpinski gasket, Asymptotic problems in probability theory: stochastic models and diffusions on fractals (K. D. Elworthy and N. Ikeda, eds.), Pitman Research Notes in Math., vol. 283, Longman, 1993, pp. 248–278.

[118] S. H. Liu, Fractals and their applications in condensed matter physics, *Solid State Physics* **39** (1986), 207–273.

[119] L. Malozemov, The integrated density of states for the difference Laplacian on the modified Koch graph, *Comm. Math. Phys.* **156** (1993), 387–397.

[120] _____, Spectral theory of the differential Laplacian on the modified Koch curve, Geometry of the spectrum (Seattle, WA, 1993), Contemp. Math., vol. 173, Amer. Math. Soc., 1994, pp. 193–224.

[121] L. Malozemov and A. Teplyaev, Pure point spectrum of the Laplacian on fractal graphs, *J. Functional Analysis* **129** (1995), 390–405.

[122] B. B. Mandelbrot, *Fractals: Form, Chance and Dimension*, Freeman, 1977.

[123] _____, *The Fractal Geometry of Nature*, Freeman, 1982.

[124] P. Mattila, *Geometry of Sets and Measures in Euclidean Spaces*, Cambridge Studies in Advanced Math. vol. 44, Cambridge University Press, 1995.

[125] V. Metz, Potentialtheorie auf dem Sierpinski gasket, *Math. Ann.* **289** (1991), 207–237.

[126] _____, How many diffusions exist on the Vicsek snowflake, *Acta Appl. Math.* **32** (1993), 227–241.

[127] _____, Hilbert projective metric on cones of Dirichlet forms, *J. Functional Analysis* **127** (1995), 438–455.

[128] _____, Renormalisation of finitely ramified fractals, *Proc. Roy. Soc. Edinburgh Sect. A* **125** (1996), 1085–1104.

[129] _____, Renormalization contracts on nested fractals, *J. Reine Angew. Math.* **480** (1996), 161–175.

[130] _____, Renormalization contracts on nested fractals, *C. R. Acad. Sci. Paris Sér. I Math.* **322** (1996), 1037–1042.

[131] V. Metz and K. Sturm, Gaussian and non-Gaussian estimates for heat kernels on the Sierpinski gasket, Dirichlet forms and stochastic processes (Beijing, 1993) (Berlin), de Gruyter, 1995, pp. 283–289.

[132] R. Meyers, R. S. Strichartz and A. Teplyaev, Dirichlet forms on the Sierpinski gasket, preprint.

[133] J. Milnor and W. Thurston, On iterated maps of the interval, Dynamical Systems, Lecture Note in Mathematics, vol. 1342, Springer, 1988, pp. 465–563.

[134] P. A. P. Moran, Additive functions of intervals and Hausdorff measure, *Proc. Cambridge Phil. Soc.* **42** (1946), 15–23.

[135] U. Mosco, Metric properties of degenerate and fractals media, Homogenization and applications to material sciences (Nice, 1995), GAKUTO Internat. Ser. Math. Sci. Appl., vol. 9, Gakutoshoseki, Tokyo, 1995, pp. 291–308.

[136] _____, Variational metrics on self-similar fractals, *C. R. Acad. Sci. Paris Sér. I Math.* **321** (1995), 715–720.

[137] _____, Variational fractals. Dedicated to Ennio De Giorgi, *Ann. Scoula Norm. Sup. Pisa CI. Sci. (4)* **25** (1997), 683–712.

[138] _____, Dirichlet forms and self-similarity, New directions in Dirichlet forms, AMS/IP Stud. Adv. Math., vol. 8, Amer. Math. Soc., 1998, pp. 117–155.

[139] _____, Lagrangian metrics on fractals, Recent advances in partial differential equations, Venice 1996, Proc. Sympos. Appl. Math., vol. 54, Amer. Math. Soc., 1998, pp. 301–323.

[140] _____, Self-similarity and the calculus of variations, *Atti. Sem. Mat. Fis. Univ. Modena, suppl.* **46** (1998), 295–313.

[141] U. Mosco and L. Notarantonio, Homogeneous fractal spaces, Variational methods for discontinuous structures (Como, 1994), Progr. Nonlinear Differential Equations Appl., vol. 25, Birkhäuser, 1996, pp. 155–160.

[142] J. Murai, Diffusion processes on mandala, *Osaka J. Math.* **32** (1995), 887–917.

[143] _____, Percolation in high-dimensional Menger sponges, *Kobe J. Math.* **14** (1997), 49–61.

[144] J. Nash, Continuity of solutions of parabolic and elliptic equations, *Amer. J. Math.* **80** (1958), 931–954.

[145] S. O. Nyberg, Brownian motion on simple fractal spaces, *Stochastics and Stochastics Reports* **55** (1995), 21–45.

[146] M. Okada, T. Sekiguchi and Y. Shiota, Heat kernels on infinite graph networks and deformed Sierpinski gaskets, *Japan J. Appl. Math.* **7** (1990), 527–554.

[147] H. Osada, Isoperimetric dimension and estimates on the heat kernels of pre-Sierpinski carpets, *Probab. Theory Related Fields* **86** (1990), 469–490.

[148] _____, Self-similar diffusions on a class of infinitely ramified fractals, *J. Math. Soc. Japan* **47** (1995), 591–616.

[149] _____, A family of diffusion processed on Sierpinski carpets, preprint, 2000.

[150] K. Pietruska-Paluba, The Lifschitz singularity for the density of states on

the Sierpinski gasket, *Probab. Theory Related Fields* **89** (1991), 1–34.

[151] _____ , The reflected Brownian motion on the Sierpinski gasket, *Probab. Math. Statist.* **16** (1996), 99–112.

[152] M. Protter and H. Weinberger, *Maximum Principles in Differential Equations*, Springer, 1984, originally published by Prentice-Hall in 1967.

[153] R. Rammal, Spectrum of harmonic excitations on fractals, *J. Physique* **45** (1984), 191–206.

[154] R. Rammal and G. Toulouse, Random walks on fractal structure and percolation cluster, *J. Physique Letters* **44** (1983), L13–L22.

[155] M. Reed and B. Simon, *Methods of Modern Mathematical Physics II: Fourier Analysis, Self-Adjointness*, Academic Press, New York, 1975.

[156] _____ , *Methods of Modern Mathematical Physics I: Functional Analysis*, revised and enlarged ed., Academic Press, New York, 1980.

[157] C. A. Rogers, *Hausdorff Measures*, Cambridge Math. Library, Cambridge University Press, 1998, First published in 1970, Reissued with a foreword by K. Falconer in 1998.

[158] W. Rudin, *Real and Complex Analysis*, third ed., McGraw-Hill, 1987.

[159] _____ , *Functional Analysis*, second ed., McGraw-Hill, 1991.

[160] C. Sabot, Existence and uniqueness of diffusions on finitely ramified self-similar fractals, *Ann. Sci. École Norm. Sup. (4)* **30** (1997), 605–673.

[161] _____ , Density of states of diffusions on self-similar sets and holomorphic dynamics in $\mathbb{P}^k$: the example of the interval $[0, 1]$, *C. R. Acad. Sci. Paris Sér. I Math.* **327** (1998), 359–364.

[162] _____ , Density of states of diffusions on self-similar sets and holomorphic dynamics in higher dimension : the example of the interval $[0, 1]$, preprint, 1998.

[163] _____ , Espaces de Dirichlet reliès par des points et application aux diffusions sur les fractals finiment remifiès, *Potential Analysis* **11** (1999), 183–211.

[164] _____ , Integrated density of states of self-similar Sturm-Liouville operators and holomorphic dynamics in higher dimension, preprint, 1999.

[165] _____ , Pure point spectrum for the Laplacian on unbounded nested fractals, *J. Functional Analysis* **173** (2000), 497–524.

[166] T. Shima, On eigenvalue problems for the random walks on the Sierpinski pre-gaskets, *Japan J. Indust. Appl. Math.* **8** (1991), 124–141.

[167] _____ , Lifschitz tails for random Schrödinger operators on nested fractals, *Osaka J. Math.* **29** (1992), 749–770.

[168] _____ , The eigenvalue problem for the Laplacian on the Sierpinski gasket, Asymptotic problems in probability theory: stochastic models and diffusions on fractals (K. D. Elworthy and N. Ikeda, eds.), Pitman Research Notes in Math., vol. 283, Longman, 1993, pp. 279–288.

[169] _____ , On eigenvalue problems for Laplacians on p.c.f. self-similar sets, *Japan J. Indust. Appl. Math.* **13** (1996), 1–23.

[170] M. Shinoda, Percolation on the pre-Sierpinski gasket, *Osaka J. Math.* **33** (1996), 533–554.

[171] J. Stanley, R. S. Strichartz and A. Teplyaev, Energy partition on fractals, preprint.

[172] R. S. Strichartz, Harmonic mappings of fractals I: mapping the Sierpinski gasket to the circle, preprint.

[173] _____ , The Laplacian on the Sierpinski gasket via the method of averages, to appear in Pacific J. Math.

[174] _____, Analysis on fractals, *Notices of Amer. Math. Soc.* **46** (1999), no. 10, 1199–1208.

[175] _____, Some properties of Laplacians on fractals, *J. Functional Analysis* **164** (1999), 181–208.

[176] _____, Taylor approximations on Sierpinski gasket type fractals, *J. Functional Analysis* **174** (2000), 76–127.

[177] R. S. Strichartz and M. Usher, Splines on fractals, *Math. Proc. Cambridge Phil. Soc.* **129** (2000), 331–360.

[178] A. Teplyaev, Spectral analysis on infinite Sierpinski gaskets, *J. Functional Analysis* **159** (1998), 537–567.

[179] _____, Gradients on fractals, *J. Functional Analysis* **174** (2000), 128–154.

[180] H. Triebel, *Fractals and Spectra*, Monographs in Math., Birkhäuser, 1991.

[181] S. Wainger and G. Weiss (eds.), *Harmonic Analysis in Euclidean Spaces*, Proc. Symposia in Pure Math., vol. 35, Amer. Math. Soc., 1979.

[182] H. Weyl, Das asymptotisce Verteilungsgesetz der Eigenwerte linearer partieller Differentialgleichungen, *Math. Ann.* **71** (1912), 441–479.

[183] _____, Über die Abhängigkeit der Eigenschwingungen einer Membran von deren Bergrenzung, *J. Angew. Math.* **141** (1912), 1–11.

[184] M. Yamaguti, M. Hata and J. Kigami, *Mathematics of Fractals*, Translations of Math. Monographs, vol. 167, Amer. Math. Soc., 1997.

[185] M. Yamaguti and J. Kigami, Some remarks on Dirichlet problem of Poisson equation, Analyse Mathématique et Application (Paris), Gautier-Villars, 1988, pp. 465–471.

[186] K. Yosida, *Functional Analysis, sixth ed.*, Classics in Math., Springer, 1995, originally published in 1980 as Grundlehren der mathematischen Wissenschaften band 123.

[187] M. Zähle, Harmonic calculus on fractals – a measure theoretic approach II, preprint.

Index of Notation

Index